The Freshwater Fishes of Europe
Vol. 5/III

The Freshwater Fishes of Europe

Vol. 1/I
Petromyzontiformes
Edited by Juraj Holcík

Vol. 1/II
General Introduction to Fishes
Acipenseriformes
Edited by Juraj Holcík

Vol. 2
Clupeidae, Anguillidae
Edited by Henri Hoestlandt

Vol. 3
Salmonidae, Coregonidae, Thymallidae
Osmeridae, Umbridae, Esocidae

Vol. 4/I, 4/II, 5/I and 5/II
Cyprinidae 1/I, 1/II, 2/I and 2/II
Edited by Petru Bănărescu

Vol. 5/III
Cyprinidae 2/III and Gasterosteidae
Edited by Petru Bănărescu and Hans-Joachim Paepke

Vol. 6
Cobitidae, Siluridae, Ictaluridae
Edited by K.D. Vasilieva and V.P. Vasiliev

Vol. 7
Gadidae, Syngnathidae, Cyprinodontidae,
Poecilidae, Atherinidae, Cottidae, Centrarchidae, Percidae

Vol. 8
Gobiidae, Blenniidae, Pleuronectidae
Edited by Peter Miller

Vol. 9
Threatened Fishes of Europe
Written by Anton Lelek
Edited by the European Committee for the Conservation of Nature
and Natural Resources – Council of Europe

The Freshwater Fishes of Europe

Vol. 5/III

Cyprinidae 2

Part III: *Carassius* to *Cyprinus*

Gasterosteidae

Edited by Petru M. Bănărescu
and Hans-Joachim Paepke

With contributions by

Petru M. Bănărescu, Vlastimil Baruš, Nina G. Bogutskaya,
Klaus Kohlmann, Alexander M. Naseka, Hans-Joachim Paepke,
Milan Peňáz, Vladimir G. Poznjak, Jan A. Szczerbowski,
Andrzej J. Szczerbowski

With 58 Figures
and 41 Tables

AULA-Verlag

Dr. Petru Bănărescu
Institutel de Stiinte Biologice
Laboratorul Biologie Evolution
Spl. Independentei 296
R-79651 Bucuresti, Romania

Dr. Hans-Joachim Paepke
Lennéstraße 14b
D-14469 Potsdam

Die Deutsche Bibliothek - CIP-Einheitsaufnahme

The freshwater fishes of Europe. - Wiebelsheim : Aula-Verl.

Vol. 5. Cyprinidae. - 2
Pt. 3. Carassius to Cyprinus : Gasterosteidae ; 41 tables /
ed.: Petru Banarescu ; Hans-Joachim Paepke. - 2001
ISBN 3-89104-658-8

Linguistic revision and translations from Russian: Charles Heckman
Cover design: Klaus Neumann
Layout: Claudia Huber
Printed and bound in Germany
Complete production by AALEX Druck GmbH, Großburgwedel

ISBN 3-89103-658-8

Publisher's Foreword

After publication of Volume 5/I, which presents the first parts of the sections on four subfamilies, Acheilognathinae, Gobioninae, Tincinae, and Cyprininae, it was originally planned to publish only one more volume on these subfamilies. This was to be Volume 5/II on the genera *Barbus, Carassius*, and *Cyprinus*, as well as on a few additional species. After most of the manuscripts had been received, however, it became evident that the genus *Barbus* alone would fill an entire volume because of the large number of species it encompasses. Furthermore, there are so many points to clarify in the sections on the individual species that a rapid preparation of the manuscripts for publication was not possible. For these reasons, the publisher decided to include the contributions on *Carassius, Cyprinus*, and the introduced genus *Catla*, which have already been completed, together with addenda to Volume 5/I, in a separate volume – this part 5/III. An addendum is the section on a newly described species in the subfamily Gobioninae: *Romanogobio pentatrichus*. This section clearly illustrates the systematic and taxonomic difficulties with which the publisher has been confronted. Contrary to the opinion of the editor, the authors of the section placed several species that have been considered members of the genus *Gobio*, under which they appear in Volume 5/I, in a new genus, *Romanogobio* (formerly ascribed by BĂNĂRESCU only as subgenus). The authors also placed the newly described species in this genus. This classification has not been accepted by the editor. A decision on which of these opinions is correct must await future investigations.

Because there were not enough sections on species left that do not belong to the genus *Barbus* – to fill an entire volume – the publisher has decided to incorporate the family of the sticklebacks, Gasterosteidae, in the newly prepared Volume 5/III, even though it does not belong to the Cypriniformes. The sections on this family, which were originally intended for Volume 7, had already been submitted to the publisher in their final form. Since no other manuscripts for Volume 7 had been submitted to that time, it seemed justified to deviate from the original plan and make the completed text for this family available to the users of the handbook immediately.

Concerning the correct spelling for the authors' names and geographical locations, the Foreword to Volume 5/I should be consulted, in which we have provided a detailed discussion of the relevant problems. Also discussed are the different methods of determining the standard length, which has been designated by two different abbreviations, *Sl* and *Sl**.

The Publishers Foreword and the editor's Introduction to Volume 5/I encompass the genera treated in this volume, as well. Its contents are therefore also applicable to this Volume 5/III.

Wiebelsheim, June 2001
AULA-Verlag GmbH

Contents

List of symbols and abbreviations

ad–C	distance between adipose fin base and caudal fin base
an–A	distance between anus and base of anal fin
A	anal fin
A–C	distance between base of anal fin and base of caudal fin
Ab	branched (soft) rays of anal fin
Au	unbranched rays [spinous rays, spines] of anal fin
b–mc	distance between base of barbels and cartilaginous arch of mouth
BR	branchiostegal rays
C	caudal fin
C_1, C_2, C_3	upper lobe, median part and lower lobe of caudal fin, respectively
Cb	branched rays of caudal fin
Cu	unbranched rays of caudal fin
D (D_1, D_2)	dorsal fin [first D, second D]
D–ad	distance between end of D base and beginning of adipose fin base
DB	branched rays of dorsal fin
D.ph.	pharyngeal teeth
Du	unbranched rays [spinous rays, spines] of dorsal fin
Fl	fork length [Smitt's length]
Fu	fulcrae
h	minimum body depth [least depth of caudal peduncle]
hA	depth of anal fin
hD (hD_1, hD_2)	depth of dorsal fin [depth of D_1, depth of D_2]
hDf	depth of posterior part of dorsal fin
hc	head depth [at nape]
hco	head depth [at centre of eye]
hmx	depth of upper jaw
hpc	depth of caudal peduncle [maximum depth of caudal peduncle]
H	body depth [maximum body depth]
ina	distance between nostrils
io	interorbital distance
K	Fulton's coefficient of condition
l	body length (see also *Sl*)
lab	width of snout at base of barbels
lac	head width
laco	body width
lad	length of adipose fin base
lam	width of mouth
lapc	width of caudal peduncle

las	width of snout
lb	length of barbel
lbr	length of gill arch
lc	length of head
liu	length of throut
l.l.	lateral line [pored] scales
l.l.–ad	scales between lateral line and adipose fin base
lmd	lower jaw length
lmx	upper jaw length
lpc	length of caudal peduncle
lA	length of anal fin base
lC (lC_1, lC_2, lC_3)	length of caudal fin [length of upper lobe, middle part and lower lobe of C, respectively]
lD (lD_1, $l D_2$)	length of dorsal fin [length of D_1, length of D_2]
lP	length of pectoral fin
Lpbs	length of pectoral fin base
lSp. br.	length of gill raker
lV	length of ventral [pelvic] fin
lVbs	length of ventral fin base
L.	lake
Oh	horizontal diameter of eye
Ov	vertical diameter of eye
Or	diameter of orbit
poD	postdorsal distance
prO	preorbital distance [snout length]
pA	preanal distance
pD	predorsal distance
pP	prepectoral distance
pV	preventral distance
P	pectoral fin
Pb	branched [soft] rays of pectoral fin
Pu	unbranched rays [spinous rays, spines] of pectoral fin
PC	pyloric caeca
P–A	distance between pectoral fin base and anal fin base
P–V	distance between pectoral fin base and ventral fin base
R.	river
Res.	reservoir [man-made lake, impoundment]
Ripo	pouch rings
Risd	subdorsal rings
Rita	tail rings
Rito	total rings
Ritr	trunk rings
s–b	distance between tip of snout and barbels
s–m	distance between tip of snout and mouth
s–mc	distance between tip of snout and cartilaginous arch of mouth

*Sl, Sl**	standard length (the length, from which the different body proportions are calculated). In Cyprinids in the rule this length is measured as distance between the anteriormost part of the head and the outward visible base of the caudal fin (= body length, also defined as *l*). Some authors used as standard length the distance between the anteriormost part of the head and the end of the hypural plate, which forms the true base of the caudal fin. This measurement is here defined as *Sl**.
Sp. br.	branchial spines [gill rakers]
Squ.	transverse rows above lateral line
Squ.sup.	scale rows above lateral line
Squ.inf.	scale rows below lateral line
SA	abdominal scutes
SL	lateral scutes
SV	ventral scutes
Tl	total length
TU	temperature units [degrees/days]
V	ventral [pelvic] fin
Vb	branched [soft] rays of ventral fin
Vert.	vertebrae
Vu	unbranched rays [spinous rays, spines] of ventral fin
V–A	distance between central fin base and anal fin base
w	individual weight

Cyprinidae 2
Part III: *Carassius* to *Cyprinus*

Edited by Petru Bănărescu

This first part of the volume is dedicated to the memory of Grigore Antipa (1867–1944) in appreciation of his studies on the systematics and biology of the carp from the Lower Danube.

Carassius Jarocki, 1822

E = carassius, crucian carp; F = carassin; G = Karausche, R = karaœ

JAN A. SZCZERBOWSKI

Carassius Jarocki, 1822: 54, 71 (type species, by tautonomy: *Cyprinus carassius* Linnaeus
Carassius Nilsson, 1832: 32 (type species, by tautonomy: *Cyprinus carassius* Linnaeus).
Cyprinopsis Fitzinger, 1832: 334 (type species, by monotypy: *Cyprinus auratus* Linnaeus).
Neocarassius Castelnau 1872: 236 (type species, by monotypy: *Neocarassius ventricosus* Castelnau = *Cyprinus auratus* Linnaeus).

Etymology: the generic name *Carassius* is derived from the Latin name of the type species, *Cyprinus carassius.*

Typical cyprinid fishes of medium or small size, with deep or oblong bodies, compressed laterally, without a ventral or dorsal keel. The mouth is terminal; there are no barbels. The dorsal fin is long; it has 3 or 4 unbranched rays and 14 to 20 branched ones. The anal fin is short, with two or three unbranched and five, six, or rarely seven branched rays. The last unbranched rays of the dorsal and anal fins are strongly ossified and serrated. The insertion the pelvic (ventral) fin lies below the vertical line through the anterior edge of the dorsal fin; that of the anal fin, below the vertical line through the posterior margin of the dorsal. The lateral line is straight; it comprises 21 to 38 scales. The scales are firmly rooted in the skin. The four pharyngeal teeth on each side form a single row; they are narrow and laterally compressed (Fig. 1; BERG 1949, BĂNĂRESCU 1964, BRYLIŃSKA *et al.* 1991, SZCZERBOWSKI and SZCZERBOWSKI 1996). The brain is characterized by a strong development of the facial and vagal lobes (Fig. 2). It is similar to the brain of *Tinca* (KOTRSCHAL and JUNGER 1988).

It is difficult to establish the native range of the genus and those of its two species because they have been introduced to many areas. The range of the genus encompasses the whole Palearctic region, including Europe, the Arctic watershed of Siberia, and the temperate and warm parts of East Asia. The species *C. carassius* is native only to Europe and the Arctic Ocean watershed of Siberia, where it is found east to the Lena River. It is absent from China. *C. auratus* is almost surely native only to continental East Asia, from the Amur Basin to southern China, and possibly also to North Vietnam. The problem of the origin and historical age of *C. auratus gibelio* in Europe is controversial. Fossils of *Carassius* and of the so-called *Paleocarassius* dating from the Miocene and later have been found frequently, especially in the European part of the former Soviet Union (LEBEDEV 1960). Bones and scales of *Carassius* have been found in Quaternary beds in the catchment areas of

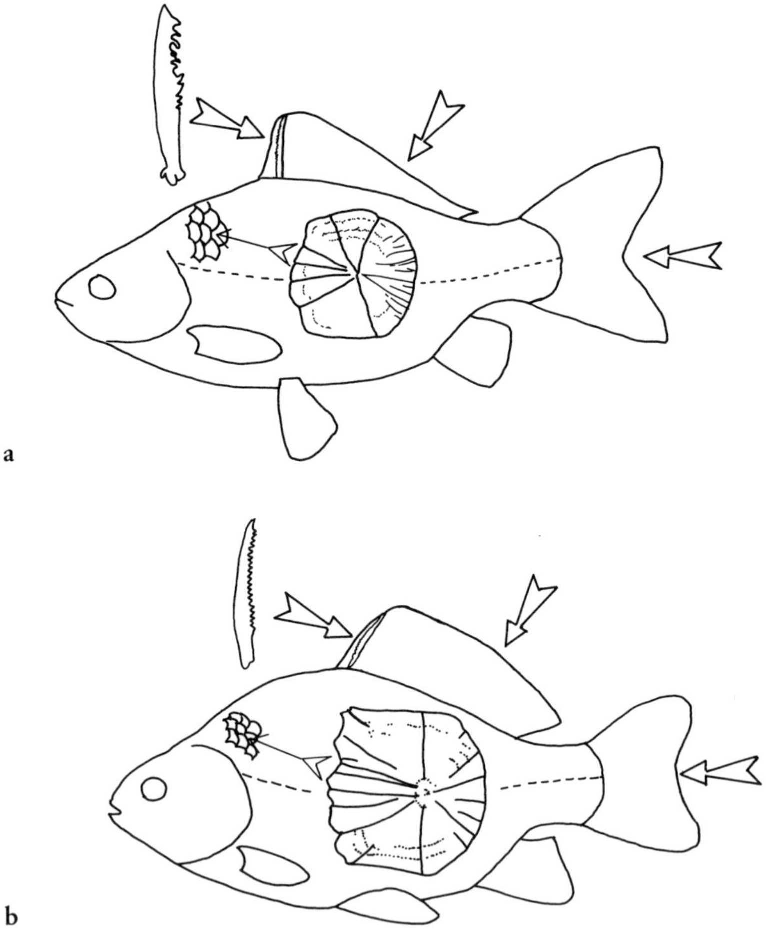

Fig. 1: Schematic illustration of the main morphological features by which *Carassius auratus* (**a**) and *Carassius carassius* (**b**) are distinguished.

the Baltic, Black, and Caspian Seas. The scales belonged to specimens four to eleven years old. The lengths of the skeletons usually range between 25.6 and 29 cm. The largest one, from Riazan, measured 35 cm.

The generic name *Carassius* was first used by JAROCKI in 1822. The genus was fully described by NILSSON (1832), who has usually been considered to be the author who first described the genus. The changing of the name to *Cyprinopsis* by FITZINGER (1832) and BLANCHARD (1866) is unjustified.

Both species of *Carassius* are present in Europe.

Key to the species of *Carassius*

1 Upper edge of the dorsal fin slightly convex; caudal fins slightly emarginate; 21 to 38 scales along the lateral line; 6 branched anal rays usually present; 23 to 33 gill rakers on the first branchial arch; 31 to 34 vertebrae, usually 32 or 33; 28 or 29 fine denticles along the posterior edge of the spiny ray of the dorsal fin (Fig. 3). Peritoneum light; a blackish spot at the base of the caudal fin, more evident in juveniles but sometimes also present on adults. Body coppery gold . . . . *C. carassius*

2 Upper edge of the dorsal fin straight or slightly concave; caudal fin deeply emarginate; 21 to 36 scales along the lateral line; the anal fin always has five branched rays; 37 to 53 gill rakers on the first branchial arch; 28 to 31 vertebrae, usually 29 or 30; 10 to 11 denticles on the posterior edge of the spiny last unbranched dorsal fin ray; peritoneum blackish; no blackish spot at the base of the caudal fin. Body silvery, pinkish gold, gold, or red. . . . . *C. auratus*

Literature

BĂNĂRESCU, P., 1964: Pisces – Osteichthyes. – Fauna Republicii Populare Romine 13: 1–961.

BERG, L. S., 1949: (Freshwater fishes of the USSR and Adjacent Countries.). Vol. **2**, 1, Izd. Akad. Nauk SSSR. Moscow – Leningrad. pp. 496–925. (in Russian).

BLANCHARD, C. E., 1866: Les poissons des eaux douces de la France. Paris. **8**, 656 pp.

BRYLIŃSKA, M., E. BRYLIŃSKI, J. TERLECKI, W. BIAŁOKOZ, J. RADZIEJ, B. MŁYNIEC, R. BARTEL, W. KOPIEJEWSKA, M. TADAJEWSKA, T. KRZYWOSZ, and K. GORYCZKO, 1991: (Freshwater Fishes of Poland.). PWN, Warsaw. pp. 289–298. (in Polish).

CASTELNAU, F. L., 1872: Contributions to the Ichthyology of Australia. – Proc. Zool. Acclim. Soc. Victoria **1**: 29–247.

FITZINGER, L. J., 1832: Prodrom. Faune Austriae. Vienna. 334 pp.

JAROCKI, F. P., 1822: (Zoology or notes on animals arranged according to the latest system.). Warsaw. 331 pp. (in Polish).

KOTRSCHAL, K., and H. JUNGER, 1988: Patterns of brain morphology in Mid-European Cyprinidae (Pisces, Teleostei): A quantitative histological study. – J. Hirnforsch. **29**(3): 341–352.

LEBEDEV, V. D., 1960: (Quaternary freshwater ichthyofauna in the European part of the USSR.). – Izdat. Mosk. Univ. **1**: 401. (in Russian).

NILSSON, S., 1832: Prodromus Ichthyologiae Scandinaviae. Berlingska Tryckeriet, Lund. 32 pp.

SZCZERBOWSKI, J. A., and A. J. SZCZERBOWSKI, 1996: (*Carassius.*). IRS, Olsztyn. pp. 1–175. (in Polish).

Carassius auratus (Linnaeus, 1758)

E = goldfish, F = poisson rouge, G = Goldfisch, R = zolotaia rybka

JAN A. SZCZERBOWSKI

Synonyms

Cyprinus auratus Linnaeus, 1758: 322 (China)
Carassius auratus Blecker, 1871: 1 (China)

The species has been divided into five subspecies. It is represented by two subspecies in Europe (BĂNĂRESCU 1964, STERBA 1977, BRYLIŃSKA *et al.* 1991, SZCZERBOWSKI and SZCZERBOWSKI 1996).

To distinguish them they used the following criteria:

1. Along the lateral line 21 to 36 scales. Pink or gold colour. Occurs in natural waters and in a variety of artificial water bodies into which it had been introduced as an ornamental fish. *C. auratus auratus*
2. Along the lateral line 27–35 scales. Yellowish silver colour. Occurs in natural waters and in various small artificial water bodies, in which it is cultured. *C. auratus gibelio*

1. *Carassius auratus auratus* (Linnaeus, 1758)

E = goldfish, F = poisson rouge, G = Goldfisch, R = zolotaia rybka

Synonyms
Cyprinus auratus Linnaeus, 1758 (China)
Carassius gibelioides Sauvage and Dabry, 1874: 8 (Europe)
Cyprinopsis auratus, Lunel, 1874: 128 (France)

2. *Carassius auratus gibelio* (Bloch, 1783)

E = Prussian carp, F = –, G = Giebel, R = serebrianyi karas'

Synonyms

Cyprinus gibelio Bloch, 1783: 90 (Central Europe)
Carassius gibelio, Lorek, 1837: 13 (Central Europe)
Carassius var. *gibelio,* Tobias, 1865: 95 (Austria)
Cyprinopsis oblongus Taczanowski, 1877: 121 (Poland)
Carassius vulgaris var. *Kolenty* Dybovski, 1877: 27 (Amur River watershed)
Carassius vulgaris, (non Nilsson) Kessler, 1877: 346 (Ukraine)
Carassius vulgaris var. *oblongus,* Varpakhovski, 1886: 54 (Kazakhstan)
Carassius carassius, (non Linnaeus) Berg, 1905: 49 (Amur River watershed)
Carassius vulgaris gibelio, Roule, 1914: 485 (France)
Carassius carassius morpha *gibelio* Berg, 1916: 33 (France)
Carassius carassius morpha *humilis,* (non Heckel) Lindberg, 1927: 527 (China)
Carassius auratus gibelio Berg, 1932: 531 (Northern Asia)
Carassius carassius gibelio, Lukash, 1933: 72 (Viatka River watershed)
Carassius carassius gibelio morpha *vovki* Ioganzen, 1945: 10 (Siberia)
Carassius carassius jacuticus Kirillov, 1972: 254 (Yakutia)

Holotype

The holotype of *Carassius auratus* is not preserved. Its *terra typica* is said to be China, specifically the catchments of rivers flowing into the Pacific Ocean.

Etymology

The generic name originates from the Latin word *carassius* (GRATZIANOV 1907), while the specific name, *auratus,* refers to the golden colouration of the fishes that were first brought to Europe (MULERTT 1902).

Description

1. *Carassius auratus auratus* – goldfish

Du 3–4, *Db* 14–19, *Au* 3–4, *Ab* 5–6, scales along the *l.l.* 21–36, *Sq. sup.* 5–7, *Sq. inf.* 5–8, *Vert.* 25–31, *Sp. br.* 37–47 (Table 1).

Table 1: Meristic features of *Carassius auratus*

Locality	Author	Branchiostegal rays				*Vert.*	*l.l.*	*Sp.br.*
		Du	*Db*	*Au*	*Ab*			
Goldfish (*C. a. auratus*)								
Ukrainian Ponds	KISEĽOV 1962	III-IV	13-19	-	-	29-31	$26\ \frac{5-7}{5-8}\ 31$	38-47 (42.6)
China	RADZIMOWSKI 1973	III	16-18	III	5-6	26-31 (28.42)	$28\ \frac{6-7}{6}\ 30$	37-46
Ukrainian Ponds	RADZIMOWSKI 1973	III	15-19	III	5-6	25-31 (28.8)	$26\ \frac{5-7}{6-8}\ 31$	38-45 (40.88)
Prussian carp (*C. a. gibelio*)								
Amur R. System	BERG 1949	III-IV	15-19 (14)	II-III	5-6	-	$28\ \frac{5-7}{5-7}\ 33$	39-50
	NIKOLSKI 1956b	III-IV	15-19	III-IV	5-6	-	30-33	45.1
L. Chervonnoe	SAVINA 1958	III-IV	16-18	III	5-6	29-31	$28\ \frac{7-8}{6-7}\ 33$	42-54 (50.74)
Polish Ponds	GĄSOWSKA 1962	III-IV	15-19	II-III	5-6	-	$29\ \frac{5-7}{5-7}\ 33$	39-52
Volma Pond	DOMBROVSKI 1964	III	13-19	III	5-6	-	$29\ \frac{6-7}{6-7}\ 32$	37-50 (44.2)
Belorussia	ZHUKOV 1965	-	16-19	-	5-6	28-31	29-32	40-48
Polish Ponds	SKÓRA 1971	III	15-18	III	5-6	28-31	$29\ \frac{6-7}{5-7}\ 31$	38-49
Ukrainian Ponds	RADZIMOVSKI 1973	III-IV	16-19	III	5-6	29-31	$29\ \frac{5-7}{5-7}\ 33$	38-45
Yakutia	SILIN 1983	II-IV	12-18	II-IV	5-6	25-34	27-35	35-54
Volga R.	GUDKOV 1985	III	15-17	III	6	-	30-32 (33)	41-53

2. *Carassius auratus gibelio* – Prussian carp

Du 3–4, *Db* (13) 15–19, *Au* 2–3, *Ab* 5–6 most frequently 5, *l.l.* 27–35, *Sq. sup.* 5–7, *Sq. inf.* 5–7, *Vert.* (25) 28–31 (34) most frequently 30 and 29, *Sp. br.* 37–54 (Table 1).

The goldfish is a subspecies that reaches 35 cm (*Tl*) and 1 to 2 kg (Mulertt 1902). It is highly variable, especially when cultured. Fishes 15 to 20 cm long and weighing 100 to 300 g are most abundant in the catches.

The Prussian carp reached no more than 45 cm in length (Nikolski 1956). Bănărescu (1964) reported a length of 36.5 cm and a weight of 900 g for this fish in Europe. Throughout its entire range, specimens 22 cm in length and 330 g in weight are most frequently caught (Volskis 1989). Smaller fishes occur at higher densities. This led to the designation of a form *humilis,* as in the case of *C. carassius.* At low densities, the fish may develop a greater body height, producing a form that has been called *vovki* (Ioganzen 1945).

Morphology: According to Bănărescu (1964), the maximal body height (*H*) of fishes belonging to the subspecies *C. a. auratus* (Fig. 2) is equal to 34.5 % to 45.5 % of their standard length (*Sl*). The predorsal distance (*pD*) is 50 % to 52.5 %; the minimal body height (*h*) is 12.6 % to 16.4 %, and the eye diameter is 6.1 % to 9.8 % of *Sl.* The snout is 25.6 % to 37.0 % of the head length. Similar data were

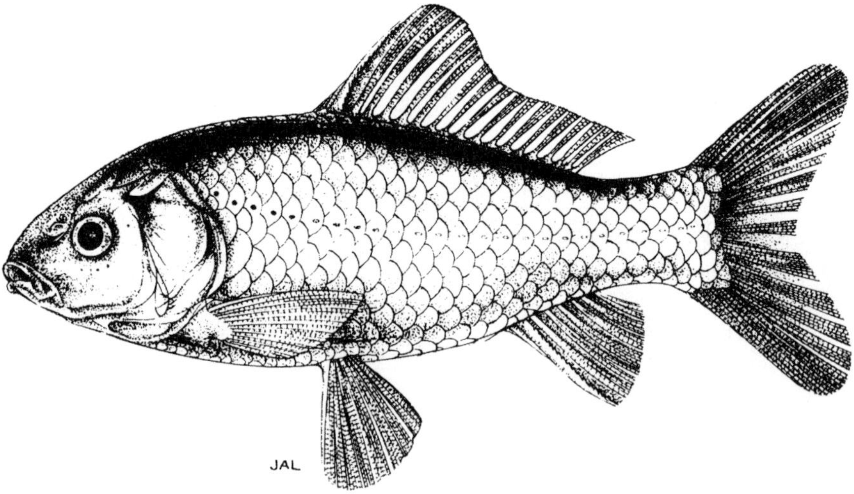

Fig. 2: Goldfish – *Carassius auratus auratus* (from Maitland 1972).

reported by Radzimovski (1973) for fishes living in the Ukraine. Under natural conditions, the wild form of *C. auratus auratus* is similar in colour to *C. auratus gibelio*, usually being greenish brown, grayish green, or gray. However, there are populations in which the colour of the body is deep orange dorsally, becoming

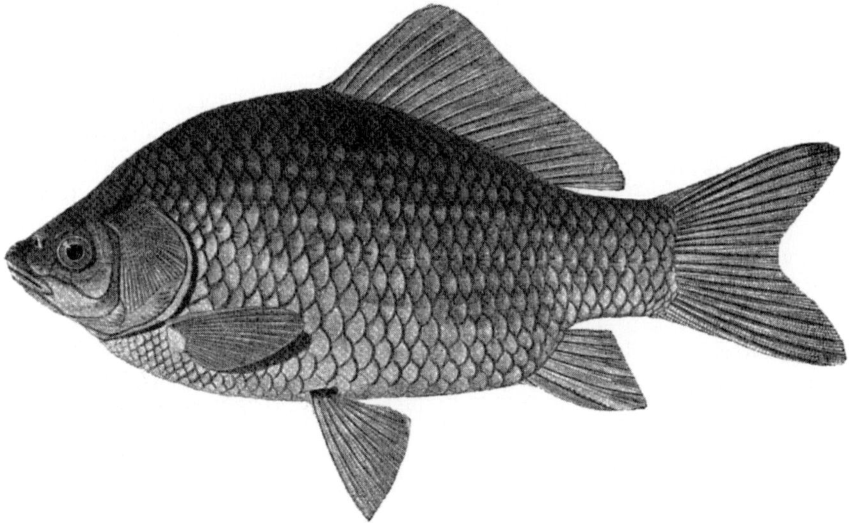

Fig. 3: Prussian carp – *Carassius auratus gibelio* (Berg 1949).

Fig. 4: *Carassius auratus gibelio* morpha *humilis* (Berg 1949).

Table 2: Morphometric features of *Carassius auratus*.
P – pond, R – river, L – lake.

	Goldfish "Pushcha Voditsa" - P - Ukraine		**Prussian carp** Amur R. system Bolon'	Tir	Bal'zinskoe	L. Chervonnoe mature	juvenile	Volma P.	L. Revuchee	Volga R.
	RADZIMOVSKI 1973		NIKOLSKI 1956			SAVINA 1958		DOMBROVSKI 1964	ZHUKOV 1965	GUDKOV 1985
% of *Sl*										
H	38.8	39.5	42.7	48.7	41.7	39.1	44.6	41.5	40.9	49.6
h	15.0	15.8	16.4	17.0	16.1	14.1	15.5	15.4	14.1	17.4
pD	52.3	48.0	51.3	52.9	52.9	45.4	33.7	52.1	50.4	-
poD	19.6	21.1	-	-	-	19.9	15.7	22.2	22.8	-
lpc	16.7	16.8	17.9	17.3	17.2	15.2	15.8	17.4	17.5	-
lD	35.5	38.1	38.2	39.4	34.1	35.9	39.4	36.6	35.9	-
hD	22.7	18.5	19.8	19.1	17.8	17.8	23.7	18.3	16.8	18.1
lA	12.7	12.5	11.2	12.0	10.5	10.8	13.8	10.9	11.1	-
hA	19.6	16.2	17.6	16.8	15.4	14.7	20.1	16.6	17.2	-
lP	21.6	19.1	19.2	18.9	20.6	17.5	18.8	18.5	20.1	-
lV	23.2	20.9	21.1	21.5	20.7	19.5	20.9	20.6	21.7	-
P-V	21.1	22.1	20.5	21.0	21.5	22.0	22.0	20.7	21.3	22.3
V-A	23.2	31.4	-	-	-	30.3	31.3	29.6	30.6	30.9
lc	33.0	29.3	26.8	26.4	30.0	22.8	26.6	28.6	27.3	26.0
Oh	-	-	5.5	5.3	5.5	-	5.5	5.5	-	-
% of *lc*										
Oh	24.4	23.2	20.6	20.1	18.4	16.8	25.0	20.1	22.3	-
PrO	26.7	27.5	28.5	29.0	31.4	-	26.0	31.3	27.9	-
Sl in cm	5.67	11.8	17-28	17-28	11-14	24-33	8.6-12.5	10.7	8.3	-
n	31	42	9	27	10	50	25	69	75	25

lighter ventrally and with a bright golden gloss laterally. In Europe, goldfishes cultured in ponds are usually pinkish gold, but there are a number of varieties of other colouration, including black and variegated.

The body height (*H*) of the Prussian carp (Fig. 3) is 37 % to 48 % of *Sl.* Smaller fishes were classified as morpha *humilis* (Fig. 4). The body height is quite variable, but it is usually 40 % to 50 % of the body length (Table 2). In the Novo-Marievski area, it constitutes only 38.9 % of *Sl* (ASTANIN and PODGORNYI 1963), but at the mouth of the Volga River, it is 49.6 %. According to BĂNĂRESCU (1964), the body height varies between 34.1 % and 48.8 % of the body length, and it is most frequently between 37 % and 44 %. The body thickness is 40 % to 62 % of the length. The head length is 24.7 % to 30.5 % of the body length. The snout length ranges from 7.5 % to 9.1 %, and the eye diameter, from 5.8 % to 9.1 % of *Sl.* The caudal fin length is 16.2 % to 23 %; the predorsal distance (*pD*), 45.6 % to 53.4 %; the base of the dorsal fin, 30.5 % to 39 %; and the postdorsal distance (*poD*), 17 % to 23.3 % of *Sl.*

Many data indicate that there are considerable variations in these dimensions (BALON 1962, ZHUKOV 1965, MIŠÍK and HOLČÍK 1962, BAUCH 1966). The lateral line is straight or only tightly arched upward anteriorly. In the form *humilis,* it does not reach the caudal region, or it is incomplete. The Prussian carp is a dark steel colour with a dark blue or greenish gloss dorsally. Laterally, it is silver-gray, and ventrally, it is white. Young fishes are lighter in colour. The colouration differs according to environmental conditions. The fins are more rigid than those of crucian carp. The dorsal and caudal fins are dark gray. The paired fins and the anal fin are light pinkish.

Karyotype: The chromosome numbers of the goldfish and Prussian carp were estimated by a several authors. MAKINO (1939) stated that this number was n = 47 and 2n = 94 in the goldfish. He also mentioned possible individual variations. He found that the chromosomes were in the form of "equal bars." LIEDER (1955, 1957) also found V and J-shaped chromosomes in female Prussian carp, which are similar to those observed in other cyprinids. CHERFAS (1966) showed that the number of chromosomes in females from bisexual populations numbered 2n = 94, whereas in unisexual, gynogenetic populations, it was 3n = 141. Females from bisexual and unisexual, gynogenetic populations did not differ from one another. According to CHERFAS (1969), females from bisexual populations were for the most part already mature toward the end of their second season of growth, whereas most of those from gynogenetic populations were still immature at that time.

VASIL'EV (1985) reviewed publications dealing with morphological characteristics of the karyotypes of 12 varieties of *Carassius auratus,* and citing the report of OJIMA *et al.*, he reported the number of chromosomes to be 2n = 100 with the number of branches (NF) being 148, according to the following formula:

12 m (metacentric)
36 sm (submetacentric)
52 sta (subteloacrocentric),

Citing Kobayasi *et al.*, Vasil'ev (1985) reported that *Carassius auratus gibelio* also has 2n = 100 chromosomes with 160 branches (NF): 20 m + 40 sm + 40 a. Chiarelli *et al.* (1969) found that the goldfish has 2n = 104 chromosomes: 62 m and 42 a.

Triploid forms that reproduce gynogenetically have been reported for all species in the genus *Carassius* as well as in the subspecies of *Carassius auratus*, except *Carassius auratus auratus*. Their karyotypes are more variable. One clone of *C. auratus langsdorfi* from Lake Biwa in Japan and *C. auratus buergeri* have 3n = 156 chromosomes, of which there are 34 m, 62 sm and st, and 60 a. In the former Czechoslo-

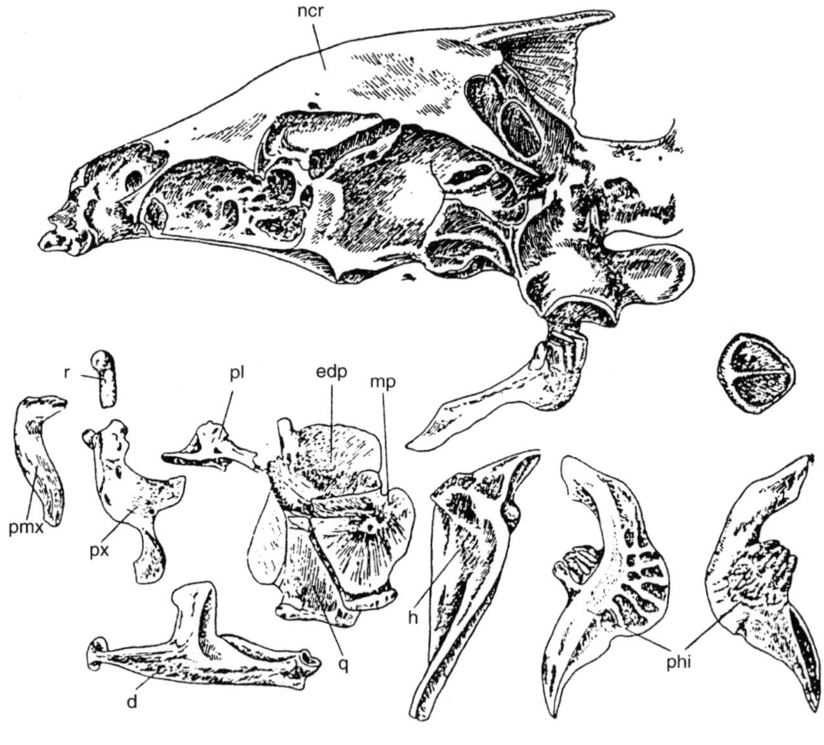

Fig. 5: Most important bones of the Prussian carp head.
ncr – neurocranium, **r** – rostrum, **pmx** – praemaxillare, **px** – maxillare, **d** – dentale, **pl** – palatinum, **edp** – endopterygoideum, **mp** – mesopterygoid, **q** – quadratum, **h** – hyomandibulare, **phi** – os pharyngeum inferior

vakia, these numbers were 160: 42, 82, and 32, respectively (PEŇÁZ *et al.* 1979). Triploid chromosomes were also found during hybridization in the hybrid *Carassius auratus* x *Cyprinus carpio*. In *C. a. langsdorfi* and *C. a. grandoculis*, tetraploid individuals were also found, in which 4n = 206 with 44 m + 82 sm + 80 a. These seem to have been produced as a result of hybridization between a triploid and a diploid form. VASIL'EV (1985) stated that the goldfish was the only representative of the Cyprinidae which possesses morphologically differentiated sex chromosomes. Females of 13 goldfish varieties possess two large submetacentric chromosomes with short arms containing blocks of C-positive heterochromatin. Males had only one such chromosome containing heterochromatin.

Nine groups in the Cyprinidae have been distinguished on the basis of their distinctive karyotypes. VASIL'EV (1985) placed the goldfish in Group 8, in which the diploid chromosome number ranges from 98 to 100 or 150 to 162. The largest chromosomes are subcentric and metacentric. This author thinks that they all originated from a tetraploid ancestor.

Protein specifity: According to VALENTA *et al.* (1977), separation of the serum transferrins has shown that there are two variants among the transferrins in the goldfish. The mobilities of these transferrin variants differed from those of the transferrins from crucian carp and other species in the family Cyprinidae.

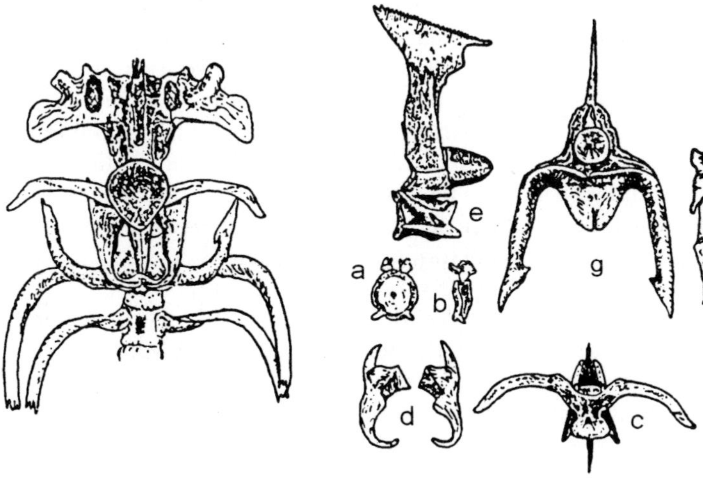

Fig. 6: The first four vertebrae of Prussian carp.
a – Frontal vertebra, **b** – Lateral view of the frontal vertebra, **c** – 2nd and 3rd vertebrae fused, **d** – Processi of the 3rd vertebra, **e** – Lateral view of the 2nd and 3rd vertebrae, **f** – 4th vertebra, back view, **g** – Lateral view of the 4th vertebra.

Osteological characteristics: The osteology of the goldfish has been described in only a few publications, which compare it with that of the crucian carp. Information is usually provided on the number of vertebrae and pharyngeal teeth and sometimes also of the gill rakers. The characteristics of the skull bones (Fig. 5) were discussed by DMITRIEVA (1957), and the first four vertebrae (Fig. 6) were described by MIASKOWSKI (1959b).

Sexual dimorphism: Sexual dimorphism of *Carassius auratus* is not very pronounced (DOMBROVSKI 1964). It can only be pointed out that the male is characterized by a slightly higher body and a higher dorsal fin. Males of *Carassius auratus* in the Amur River Basin are slightly smaller than females (SYSOEVA 1956), have fewer gill rakers, and, contradicting DOMBROVSKI's (1964) findings, are characterized by a lower dorsal fin and shorter anal and pectoral fins (NIKOLSKI 1956).

Variations: In addition to being highly variable in its morphometric features, the goldfish also varies greatly meristically and in colouration, probably more so than any other subspecies (HERVEY and HEMS 1948). The domesticated ornamental form occurs in many varieties. The earliest reports on this fish in China date back to 265 to 419 A.D., whereas the oldest literature providing data on these fishes was published during the first years of the SUNG dynasty, between 968 and 975 A.D. By that time, the goldfish and crucian carp had already been distinguished. By the first half of the 11th century, *C. auratus* was known to be lightly pinkish to deep red. Since 1163, this fish has been cultured in China, and the selection of varieties began at that time (LANDOWSKI 1968). SUVOROV (1954) mentioned that 16 basic varieties of this fish were known in Japan. Most of them are also found in Europe. They are usually called Prussian carp (STERBA 1977, WOLIŃSKI 1967) or, wrongly, crucian carp (NIKOLSKI 1956, REMBISZEWSKI 1971). Among the common cultured forms, the "fringetail" is also mentioned. Its standard body proportions are shown in Figure 7. The individual fins of this variety are very long and frequently fringed. Other varieties include the comet, egg-fish, shubunkin, celestial, telescope–veiltail, and lionhead. In the last of these forms, bony projections develop on the head in the form of a cap or a mane. Some varieties have no dorsal fins. In the telescope variety, large eyes become noticeable by the second month of life, and after 4 or 5 months, they have attained their characteristic size. These features, which were produced by selection in breeding, are inheritable.

The colouration of the goldfish is also highly variable. The classic form is reddish golden dorsally, golden laterally, and reddish yellow ventrally. The fins are reddish yellow or yellow. However, a wide variety of white, red, and golden or black, red, and golden fishes exists, and there are also light pinkish or silverish which used to be called "silver". Some have no scales.

The Prussian carp is not so variable. Variations were reported for individuals found in different places (NIKOLSKI 1956, SAVINA 1958, DOMBROVSKI 1964, ZHUKOV

1965) and also for fishes taken from the same water bodies (Radzimovski 1973). The shape of the body may vary slightly, and the number of gill rakers ranges from 37 to 53 (Table 1).

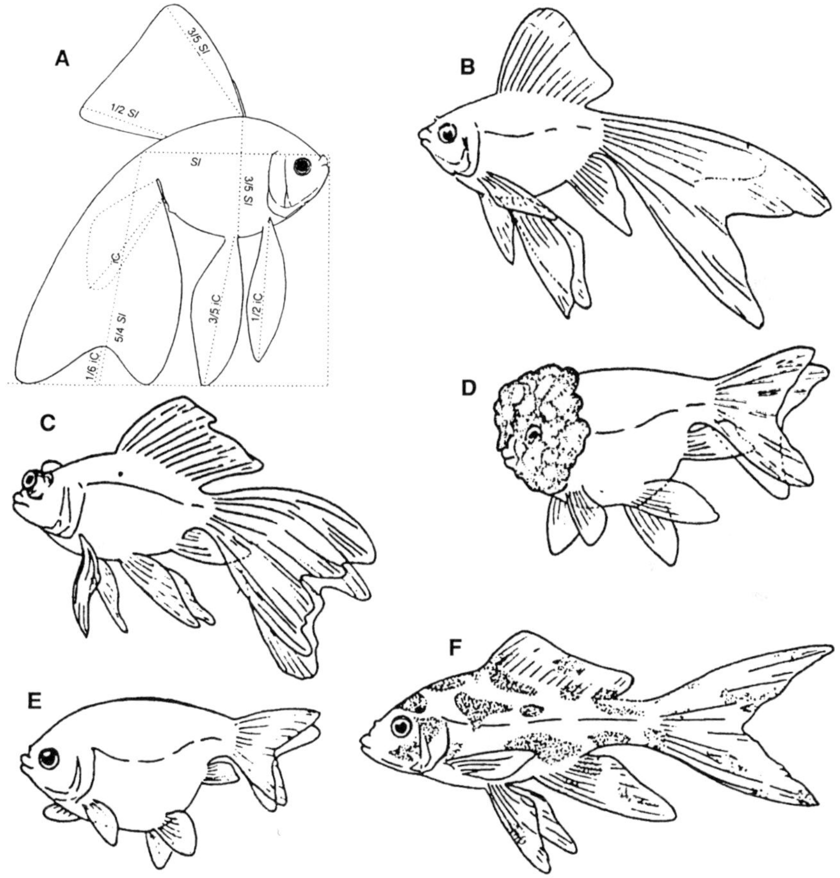

Fig. 7: Standard measurements for the Veiltail **(A)** and the main domesticated varieties of goldfish: **(B)** comet; **(C)** telescope-veiltail; **(D)** lionhead; **(E)** egg-fish; **(F)** shubunkin (Sterba 1977).

Age and size variability: The meristic features change with growth in both the goldfish and the Prussian carp. The increase in the number of gill rakers is especially interesting. On fish farms in Belorussia, the average number of gill rakers

Table 3: The number of gill rakers in Prussian carp of various length classes (Dombrovski 1964).

Fish farm - BELOE										
Sl., cm	8.4-9.4	9.4-10.4	10.4-11.4	11.4-12.4	12.4-13.4	-	-	15.4-16.4	16.4-17.4	17.4-18.4
Gill rakers	42.7	42.7	44.1	44.3	44.7	-	-	48.2	48.6	49.7
n of fish	27	53	28	8	3	-	-	5	9	4
Fish farm - VOLMA										
Sl., cm	8.9-9.9	9.9-10.9	10.9-11.9	11.9-12.9	12.9-13.9	13.9-14.9	14.9-15.9	15.9-16.9	16.9-17.9	17.9-18.9
Gill rakers	43.0	44.6	45.0	45.0	45.3	47.0	47.2	48.4	48.2	49.0
n of fish	20	45	24	3	3	7	21	14	4	1

(Table 3) increases from 42.7 to 49.7 as the fishes grow from 8.4 cm to 18.4 cm (DOMBROVSKI 1964). The relative eye diameter and height of the dorsal and anal fins (Table 2) decrease with age in both the goldfish and the Prussian carp. The head length of juvenile fishes equals 33.0 % of the body length, but it decreases to 29.3 % in adult fishes (RADZIMOVSKI 1973). The distance between the ventral and anal fins (*V–A*) increases, and so does the postocular head length. The length of the digestive tract is 77 % to 100 % of the body length in fry 1.3 to 1.4 cm, but 250 % to 275 % in fishes 3 cm long, and 350 % to 410 % in older fishes. This was found in an experimental fish culture facility (DOMBROVSKI 1964) as well as under natural conditions (SAVINA 1958). Changes in colour are also very noticeable. Both the goldfish and the Prussian carp are usually greyish green when young and attain their characteristic colouration only after 12 to 18 months (STERBA 1977).

Hybrids

The goldfish frequently produces hybrids. This has been known for a long time. The most popular are hybrids with the common carp, *Cyprinus carpio*, referred to as: *Cyprinus carpio carpio* x *Carassius auratus auratus* [BĂNĂRESCU (1964) citing COSTEA and POJOGA]. See also for hybrids page 118 and Fig. 29, p. 119.

This hybrid is characterized by the following features: *D* III 21, *A* III 6, *V* II 8 (9), scales along the *l.l.* $38 \frac{5}{6} 39$, *Sp. br.* 24, *Vert.* 35, *Ph. F.* 1.1.3–3.1.1. The body shape is similar to that of the carp. The body height is 26.3 % to 30.3 %, and the head length, 27 % to 32.3 % of *Sl.* The fins are of the same colour as those of the carp, while the body colouration resembles that of the goldfish. These hybrids are abundant within the distribution range of *C. auratus.* They are also cultured in ponds for food. It was demonstrated that diploid hybrids but not tetraploid ones (GOMEL'SKI *et al.* 1985a, b) grow more rapidly than the Prussian carp and are more resistant to low temperatures and oxygen deficiencies than the common carp, just as hybrids between the crucian carp and the common carp are (NIKOLIUKIN 1952).

C. auratus can also produce hybrids with the crucian carp, *C. carassius.* Therefore, MULERTT (1902) and later SCHÄPERCLAUS (1953) hypothesized that the goldfish developed as a hybrid between the common and crucian carps. Attempts were also made to hybridize *C. auratus* with other species (NIKOLIUKIN 1952), but the hybrids were characterized by high mortality rates.

Distribution

Carassius auratus is native to China, the Amur River Basin in Manchuria, Korea, Japan, and Taiwan. The general native range is therefore said to be East and Central Asia. In Europe, it is represented by two introduced subspecies (BĂNĂRESCU

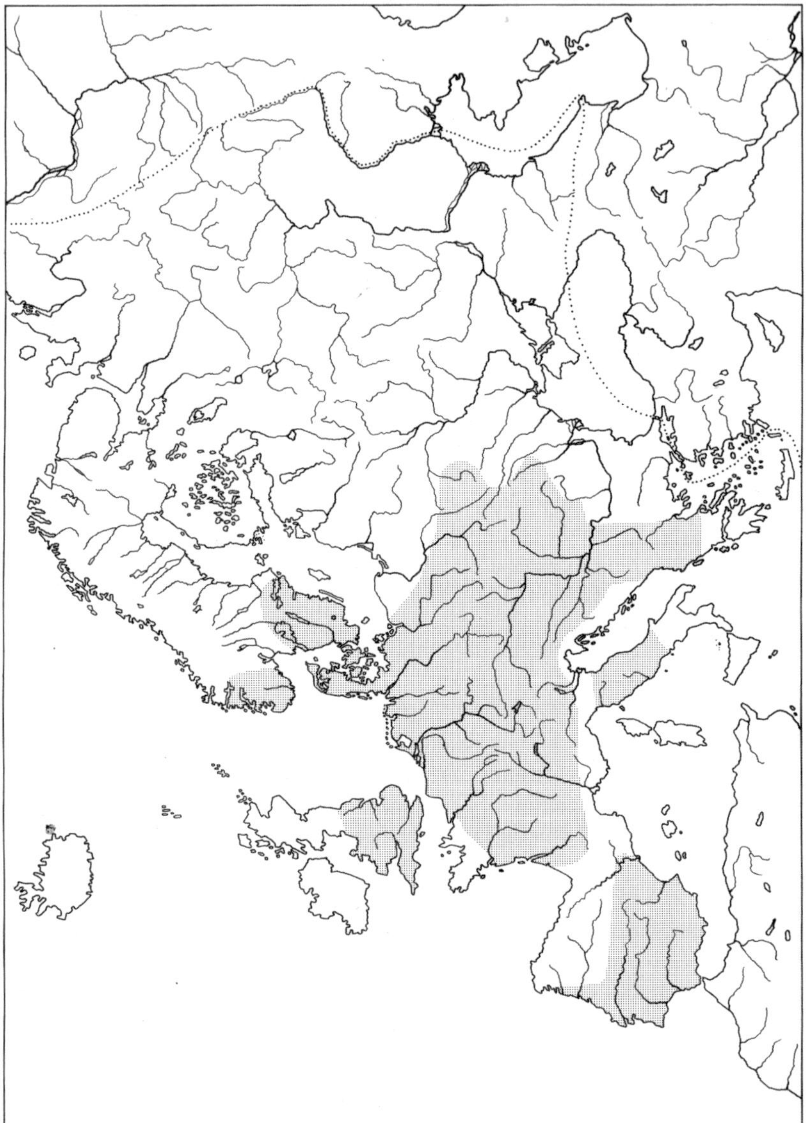

Fig. 8: Geographical distribution of *Carassius auratus auratus* in Europe.

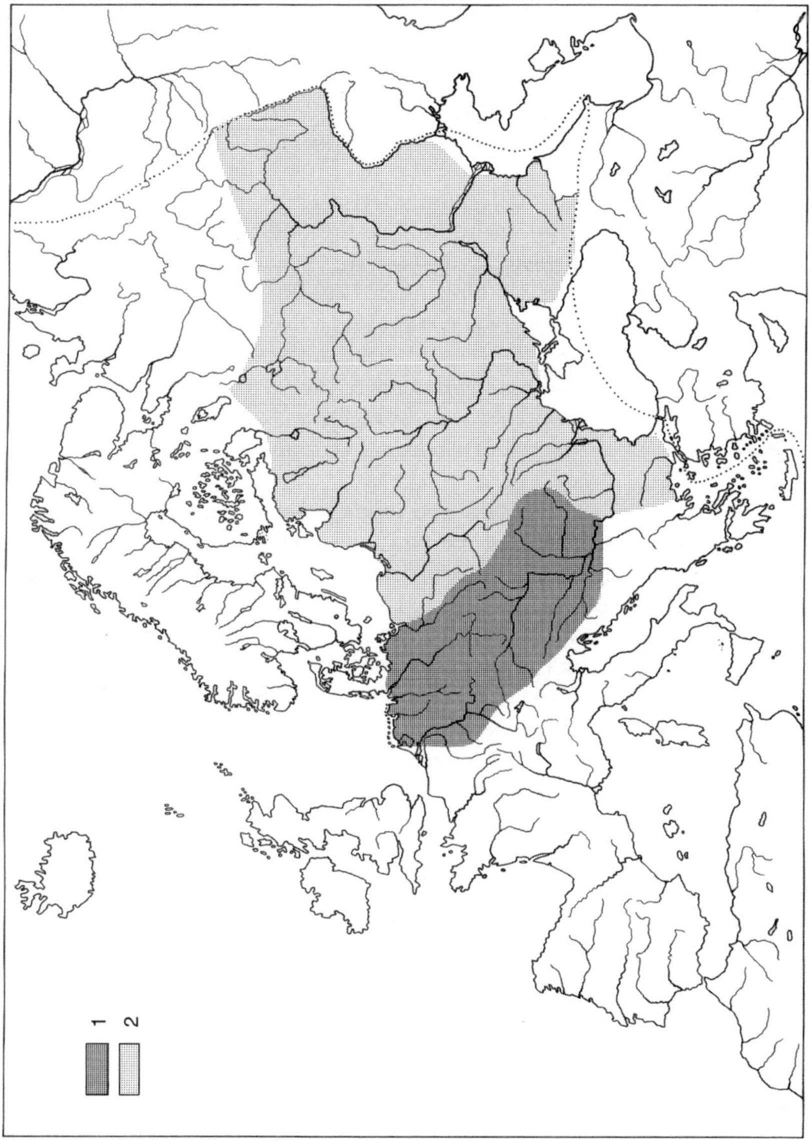

Fig. 9: Geographical distribution of *Carassius auratus gibelio*.
1 – Successful introduction in the last century; 2 – Area of distribution till the end of the 19th century.

1960, Holčík *et al.* 1989), which have spread rapidly. Holčík (1980) stated that in the Danube Basin, they have become very abundant, and Gudkov (1985) reported that they were caught in large numbers in the estuarine section of the Volga River.

The first introduction of the goldfish into Europe has been reported to have occurred in various years, including 1611, 1691, and 1728 A.D. (Mulertt 1902). They were first seen in Portugal. In England, it has been known since 1691, and in France, since 1750 (Woliński 1967). Soon afterward, the goldfish became quite common, and it was found to thrive in the waters of Southern Europe, especially in Portugal, where it was considered a delicacy for the table. The first successful breeding of the goldfish occurred in Holland in 1728 (Sterba 1977). According to the range information reported by Blanc *et al.* (1971), these fishes are not found in Northeast Europe (Fig. 8), in northern Spain, Italy, or the former Yugoslavia. They also do not occur in most southern regions bordering on the Mediterranean Sea. According to Welcomme (1988), they were brought to Hungary in 1891. In many cases, wild populations of the goldfish become very numerous in natural water bodies (Sterba 1977).

The range of *Carassius auratus gibelio* in Europe is not so great as that of *C. a. auratus*. This subspecies is distributed more in the southeastern regions of Europe than *C. a. auratus* (Fig. 9). It does not occur in Scandinavia, the British Islands, France, Spain, Italy, southern Yugoslavia, or Greece. Holčík *et al.* (1989) mentioned that it first appeared in Romania as late as 1920. Borcea (1935), Ziemiankowski (1947), Bănărescu (1964), and Holčík (1980) stated that this fish was known only from the lower Danube, from the Iron Gate downstream to the estuaries, including the lower reaches of its Romanian tributaries, the Jiu, Olt, Vedea, Arges, Ialomita, and the watersheds of the Siret and Prut until approximately the end of the first half of this century. Since the begining of the 1960s, this fish has regularly been found in the upper parts of the Danube Basin. In Poland, it is regarded as a fish of the Far East, and it first appeared only at the begining of this century (Gąsowska 1934, 1936).

Introductions: The goldfish has acquired a pan-global distribution through widespread introduction, generally for ornamental purposes. Escapes and deliberate releases have resulted in natural populations developing in more than 20 European countries, and its actual distribution is probably far wider than known (Welcomme 1988). Outside of Europe, the goldfish has been introduced to South Africa, and to Madagascar and Mauritius, and it is widespread in Asia. It was introduced in 1874 to North America (Landowski 1968), to Oceania, and from 1856 to 1930, to various countries in South and Central America and the Caribbean, including Argentina, Bolivia, Brazil, Chile, Colombia, Mexico, Peru, Puerto Rico, and Uruguay. Escaping from ornamental pools in parks and gardens, the goldfish has become naturalized in many ponds and streams in the eastern United States and in the Potomac River. In some localities, it occurs in sufficient abundance to be marketed as a food fish.

The Prussian carp has been less widely introduced, although it has expanded its range westward into much of Europe. This subspecies is among the most widely introduced fishes within the territory of the former USSR (BURMAKIN 1963, BERDICHEVSKI *et al.* 1968, KARPEVICH 1975). The results of its introductions show that its success has been poor. During the period from 1930 to 1957, it was released into 156 natural lakes, 5 man-made lakes, and 4 rivers, but only in 3 of these water bodies is it known to have reproduced and formed self-sustaining populations. However, KARPEVICH (1975) wrote that the biological adaptation, naturalization, and production of populations capable of supporting a commercial catch of *Carassius auratus gibelio* occurred in almost all of the 250 water bodies into which this subspecies has been released. According to ZHUKOV (1965), the introduction of the Prussian carp into natural water bodies of Belorussia was successful only in one lake, in spite of the fact that from 1948 to 1961, almost 93 million one-year-olds and fingerlings and 182,000 adults had been released on a regular basis.

Ecology

Biotope: *C. auratus auratus* and *C. auratus gibelio* inhabit ponds, rivers, and lakes. They are most numerous in water bodies similar to carp ponds and natural ponds and in shallow lakes of floodplains of large rivers, as the middle and the lower Danube, while they are rare in rivers and lakes. Observations on their feeding habits in ponds have revealed that these fish were more shy than carp. They also feed less intensively. On warm, dry, sunny days, they gather in shallow water overgrown with plants. On windy, cold days, they move into deeper waters.

There are some reports in the literature that *C. auratus gibelio* may bury itself in the mud when in danger. DOMBROVSKI (1964) cited some authors who stated that when the water froze in autumn, they found these fishes buried in the mud, even as deep as 1 to 2 m. He performed some experiments by scaring two-year-old fishes weighing 120 g each, and these buried themselves 7 to 15 cm deep. However, this is not a common phenomenon, and the fishes remained buried for only 2 to 10 minutes. This behaviour may protect the fishes when tow nets are used.

Migrations: The goldfish and Prussian carp are considered to be non-migratory fishes which do not change their habitats. The Prussian carp tends to be more active than the goldfish. It frequently moves from one place to other in search of food. Such migrations were observed most often in rivers, and they noticeably enlarged the distribution range of this fish (HOLČÍK 1980). It seems likely that some populations or strains of this species have a tendency to migrate, which is not shown by most members of the species (ROVNIN *et al.* 1977). In spring and summer, the Prussian carp inhabits more shallow, warmer water than the goldfish does. At such sites, it feeds and reproduces, moving only within the immediate area, but in autumn and winter, it moves into deeper water and frequently congregates in deep holes. It has returned to water near the shore by the end of April, and in May, where it begins intensive feeding. During spawning, the fishes move along the shore but return to deeper waters between the releases of consecutive egg batches.

Hardiness: The subspecies *Carassius a. auratus* is among the most resistant fishes (BERG 1949, CHERFAS 1965, HOLČÍK 1980). Like the carp, it tolerates high water turbidity. The introduction of suspended matter in the form of montmorylomite clay at a concentration of 100,000 mg/l did not cause any mortalities for one week (WALLEN 1951), and some fishes tolerated a concentration of 225,000 mg/l for one to three weeks.

Eggs, fry, and adults tolerate extreme fluctuations in temperature, thermal shock, and low oxygen levels (SANTA and MARINESCU 1970, SPOTILA *et al.* 1979, DRJAGIN 1949, JÁSZFALUSI 1959, GOR'UNOVA 1960, BURMAKIN 1963, ZHUKOV 1965, BUGAI and KOVAL 1976). These subspecies are frequently the only inhabitants of shallow lakes with total oxygen depletion during some years (GOR'UNOVA 1960). The threshold oxygen concentrations for the Prussian carp and goldfish are as follows:

10 °C	–	0.5 mg/l O_2
20 °C	–	0.6 mg/l O_2
30 °C	–	0.7 mg/l O_2

The goldfish tolerates pH variations within the range from 4.5 to 10.5, but it avoids water with the pH between 5.5 and 7.0. The optimal pH level is 7.2 to 8.4, that is, about 1 unit more than that of *Carassius carassius*. It can survive in water at pH 4.5 for about two weeks. These fishes are highly resistant to such toxic agents as crude oil, phenols, pesticides, heavy metals, and some salts (JOHNSON 1968, WEISS 1961, ADELMAN *et al.* 1976, VOSILENE 1977). *C. auratus auratus* survives in an acetic acid solution of 100 mg/l at a pH of 6.8 for 2 to 4 days and in sulfuric acid at 134 mg/l and a pH of 4.0 for 4.5 days (ŁUKJANIENKO 1974).

A hydrogen sulfide concentration of 4.3 to 25 mg/l causes death of the goldfish after only 24 hours. In hard water, the fishes may live longer at an H_2S concentration of 25 mg/l. An ammonia concentration of 2.0 to 2.5 mg/l is tolerated by goldfish for one to four days. At a chloride concentration of 7341 mg/l, the LC_{50} for the goldfish is 96 hours. A similar rate of survival was observed at 12 °C at a chlorine concentration of 1180 mg/l. Phenol is toxic and lethal to the goldfish depending on its concentration, as the following survival times show:

10 mg/l	–	more than 72 hours
51 mg/l	–	90–140 minutes
333 mg/l	–	less than 15 minutes

The resistance of the goldfish to heavy metals, as indicated by the LC_{50}, was higher than that of other fishes (HELLAWELL 1989):

Cd	–	48 h	–	2.76 mg/l in soft water
Cd	–	96 h	–	2.13 mg/l in soft water
Cr^{3+}	–	96 h	–	4.1 mg/l in soft water
Cu	–	96 h	–	0.46 mg/l in hard water
Pb	–	96 h	–	31.5 mg/l in soft water
Ni	–	96 h	–	4.5 mg/l in soft water
Se	–	7 d	–	12.0 mg/l
Hg	–	6 d	–	0.37-0.74 mg/l

The resistance of the goldfish to organochlorine insecticides can be inferred from the following LC_{50} values for 96 hours:

Chlordane	–	50–82 mg/l
DDT	–	27 mg/l
HCH or BHC	–	230 mg/l
Lindane	–	152 mg/l
Heptachlor	–	230 mg/l
Methoxychlor	–	56 mg/l
Toxaphene	–	5.6 mg/l
Aldrin, Dieldrin, Endrin	–	1.9–3.7 mg/l
Guthion	–	2400 mg/l

Feeding habits: Goldfish and Prussian carp feed on planktonic crustaceans, insect larvae, molluscs, plants, and the eggs and fry of other fishes, depending upon the kind of food available and the food supply (DRJAGIN 1949, CSÁKÁNY 1958, JÁSZFALUSI 1959, KUKURADZE and MARIIASH 1975, ROVNIN *et al.* 1977). Thanks to their dense gill rakers, they can even feed on small planktonic organisms. They feed mostly in shallow waters overgrown with vegetation. Larger fishes may feed in deeper waters, and sometimes they are facultative cannibals (RADZIMOVSKI 1970).

SYSOEVA (1956) showed that stomachs of Prussian carp 5.5 to 7.0 mm in length contained rotifers, copepods, and unicellular algae. Larvae 7 to 10 mm long feed mostly on nauplii, but those 10 to 12 mm long have already consumed small planktonic crustaceans and chironomid larvae. Fry 15 to 29 mm long also consume plants and detritus. After reaching 29 to 38 mm, the fishes again feed intensively on plankton. Prussian carps over 40 mm in length show the same omnivorous feeding behaviour as adults. It has not been possible to determine any seasonal preferences. The fishes feed on all available foods, and detritus may sometimes constitute 85 % to 90 % of the food weight. Typically, these fishes do not feed from December until March. Most intensive feeding is observed from June through August. The index of fullness reaches 70 at that time (NIKOLSKI 1956). Feeding is slightly less intensive during reproduction.

Longevity: The goldfish lives under natural conditions for 10 to 16 years (MULERTT 1902) or may even reach 18 to 20 years (LANDOWSKI 1968). In ponds, it can sometimes reach an age of 25 years. CELLER (1976) mentioned that in China, 40 year old individuals were found. However, so old goldfishes are very rare. Most specialists agree that the goldfish can live for 8 to 9 years. Among Prussian carp populations in natural European water bodies, the oldest fishes caught were 10 to 11 years of age (PUSHKIN 1976).

Growth: The growth rates of the goldfish and Prussian carp are highly variable, and they usually depend on the population density and food supply (SUKHOVERKHOV 1951a, DRJAGIN 1953, SCHÄPERCLAUS 1961). Wild goldfishes reach 15 to 30 cm. In Africa, individuals 35 cm long are sometimes found, and in the USA, goldfishes 60 cm long have been encountered (CELLER 1976). Data collected by NIKOLSKI (1956) within the natural distribution range of the Prussian carp suggest that commercial catches of this fish include specimens 15 to 33 cm in length, and in some regions, 39 cm is reached. The average length of the Prussian carp was 27.9 cm and its average weight, 865 g in rivers. In lakes, the corresponding averages were 26 cm and 750 g.

In the European part of the former USSR, the Prussian carp is characterized by a similar growth rate to that in the basins of the Amur and Ussuri Rivers (Table 4). In Lake Chervonnoe, four-year-old fishes attained 521 g, and six-year-olds, 971 g (SAVINA 1958). The average lengths of the Prussian carp in 28 water bodies scattered over its whole range in Europe varied from 19.4 to 24.5 cm, and its average weight ranged from 300 to 350 g (VOLSKIS 1989). The growth rate in ponds at high stocking densities of 700 to 1200 individuals per hectare together with 1400 to 2100 common carp per hectare was noticeably lower. The fishes reached from 111 to 133 g by autumn of the second year, from 135 to 140 g during the third year, and from 74 to 135 g during the fourth year (Table 5). The growth rate was much lower than in Savvinskom Pond, in which the fishes attained 150 to 170 g during the second year of life (SUKHOVERKHOV 1951b). It was also lower than in many natural water bodies.

The coefficient of condition of FOULTON for the goldfish in the Danube River varied from 3.19 to 4.65 and averaged 4.25. The corresponding values for fishes from Nemana were 3.50 to 4.60 and 4.00. In the Drissa River, the coefficient ranged from 4.20 to 5.49 and averaged 4.68 (ZHUKOV 1965). KUKURADZE and MARIIASH (1975) observed that there were Prussian carp populations on the lower Danube which shared the same habitats but had different growth rates and attained different sizes. The differences amounted to as much as 10 cm and 600 g by the fifth year of life. During the first year of life, the length differed by 1.7 to 10.4 cm

Table 4: The length of *Carassius auratus gibelio* (*Sl* in cm) in various age classes. L – lake, R – river.

Author	Location	Age 1	2	3	4	5	6	7	8	9	10	11
DRJAGIN 1949	L. Krugloe	-	11.4	14.2	15.0	20.8	-	-	-	-	-	-
	L. Bolon'	-	-	21.2	24.4	29.0	33.0	35.0	-	-	-	-
BERG 1949	L. Golodovka	3.7	7.8	10.2	12.9	14.9	18.1	22.8	-	-	-	-
	L. Hanka	4.2	10.1	14.9	19.2	22.5	25.0	-	-	-	-	-
	L. Sudotse	6.6	11.4	15.2	18.3	20.4	-	-	-	-	-	-
NIKOLSKI 1956	**L. Bolon'**											
	Nakki R. 1945	5.9	10.2	15.0	19.4	22.8	24.1	26.4	-	-	-	-
	Nakki R. 1946	5.1	11.4	15.8	19.0	22.0	23.6	24.6	-	-	-	-
	Nakki R. 1947	5.3	12.0	17.8	21.2	22.5	23.3	-	-	-	-	-
	Nakki R. 1948	5.6	12.0	16.5	-	-	-	-	-	-	-	-
SYSOEVA 1956	**Amur R. - upper**											
	L. Bal'zinskoe	4.0	6.7	8.9	12.5	-	-	-	-	-	-	-
	L. Kenon	8.8	16.1	21.1	25.7	27.7	28.5	29.3	-	-	-	-
	Various lakes	4.8	9.8	14.6	18.8	21.8	23.2	-	-	-	-	-
	Amur R. - lower											
	Elabuga region	5.7	11.2	15.5	18.4	21.1	22.5	-	-	-	-	-
	L. Gassi	6.1	11.7	15.7	18.9	20.7	22.4	23.7	26.8	28.1	-	-
	L. Dzalunskoe	5.7	11.2	15.5	18.7	21.3	22.9	24.7	26.7	27.4	-	-
	L. Bolon'	5.9	13.1	18.7	21.9	24.7	25.9	27.9	31.4	-	-	-
	L. Zworon	5.5	10.5	14.3	17.5	19.9	21.8	22.9	-	-	-	-
	Tyr R.	6.3	13.0	19.0	22.0	24.3	26.6	27.9	28.9	33.7	-	-
	Ussuri R. region											
	Czernaia R.	6.4	13.0	17.6	21.3	23.8	-	-	-	-	-	-
	Czernaia R. - lower	4.4	9.8	15.1	18.4	20.7	22.3	26.8	-	-	-	-
	Iman R. - lower	5.4	11.5	16.7	18.1	20.4	22.3	-	-	-	-	-
	L. Hanka	5.5	11.6	15.8	19.3	23.1	24.9	25.5	-	-	-	-
DMITRIEVA 1957	L. Bolshoi Aksuat	2.7	5.8	9.2	12.7	15.6	17.9	20.2	22.5	24.8	26.5	28.3
	L. Kurkovo	1.7	3.8	7.5	9.9	-	-	-	-	-	-	-
GAĽCOVA 1957	L. Kitsino	-	13.2	17.8	22.4	27.0	-	-	-	-	-	-
SAVINA 1958	L. Chervonnoe	7.7	11.3	21.8	25.8	28.1	31.2	-	-	-	-	-
ZHUKOV 1965	Belorussia	7.7	11.3	21.8	25.8	28.1	31.2	-	-	-	-	-
KUKURADZE and MARIIASH 1975	Danube R. - lower	10.4	16.2	23.8	26.0	29.1	32.4	34.8	-	-	-	-
GUDKOV 1985	Volga R. - mouth	-	18.2	19.8	22.8	25.0	28.8	31.0	33.0	-	-	-

Table 5: The weight of *Carassius auratus gibelio* (g) in various age classes.

Author	Area	Age 1	2	3	4	5	6	7	8
DRJAGIN 1949	L. Krugloe	-	70	102	139	313	-	-	-
	L. Bolon'	-	-	266	329	500	700	900	-
SAVINA 1958	L. Chervonnoe	13	48	325	521	694	971	-	-
DOMBROVSKI 1964	Volma Pond	111	135	135	-	-	-	-	-
	Beloe Pond	120	136	248	-	-	-	-	-
	Tremlja Pond	133	140	74	-	-	-	-	-
KUKURADZE. MARIIASH 1975	Danube R. - mouth	120	285	487	700	838	917	985	-
GUDKOV 1985	Volga R. - mouth	-	100	270	443	653	913	1217	1468

Table 6: The length (*Sl* in cm) of male (♂) and female (♀) *Carassius auratus gibelio* in various age classes in the Amur River system (SYSOEVA 1956).

Area	Sex	Age 1	2	3	4	5	6	7
L. Shilka	♂	4.8	9.0	13.5	17.6	-	-	-
	♀	5.3	10.5	14.9	18.7	21.8	-	-
L. Oshalunskoe	♂	4.8	9.4	12.9	16.4	19.3	21.2	-
	♀	5.7	11.3	15.7	18.8	21.4	23.4	25.2
L. Bolon'	♂	5.7	13.1	19.2	21.9	25.3	28.8	30.9
	♀	7.0	14.4	19.6	22.4	25.7	29.6	32.1
Canals of L. Bolon'	♂	5.7	11.7	17.2	20.1	22.8	25.7	-
	♀	5.5	12.8	18.2	20.1	22.9	24.7	-
Ussuri R.	♂	5.5	11.5	16.8	19.6	22.0	-	-
	♀	5.9	12.8	18.6	20.3	22.2	23.2	-

Table 7: The percentages of Prussian carp populations in different age classes (KUKURADZE and MARIIASH 1975).

Age	Rapid growth: Danube, mouth	Rapid growth: Kagul Liman	Rapid growth: Lake Kartal	Slow growth: Danube, mouth	Slow growth: Kagul Liman
1	14.4	2.6	0.4	-	-
2	10.5	24.7	45.7	30.5	1.6
3	42.0	49.1	23.5	33.4	25.4
4	17.5	20.7	18.3	25.0	44.2
5	10.5	2.6	10.3	8.3	23.7
6	2.0	0.3	1.8	1.4	5.1
7	3.1	-	-	1.4	-
n	353	381	1500	72	59

Table 8: The size distribution of *Carassius* species caught using different mesh sizes (Pushkin 1976).

Mesh sizes	*Tl*, cm											
mm	8-10	10-12	12-14	14-16	16-18	18-20	20-22	22-24	24-26	26-28	n	$\bar{x}$
24	2	34	18	4	9	3	3	6	1	-	80	13.8
30	-	5	34	22	3	8	21	24	6	2	125	17.8
36	-	-	22	27	40	24	42	51	19	5	230	19.6
50	-	-	-	-	14	73	282	345	127	30	871	22.5
65	-	-	-	-	-	-	7	35	36	30	108	24.6
n	2	39	74	53	66	108	355	461	189	67	1414	21.2

(Table 4). The differences are more pronounced in sexually mature fishes and reach 17.9 to 33.0 cm among six year old fishes in natural waters. Males grow somewhat slower than females, the differences reaching as much as 2.4 cm, depending on the region (Table 6).

Population dynamics: Changes in the numbers and biomass of *C. auratus auratus* and *C. auratus gibelio* populations are very large due to a number of factors. During intensive fishing, the greatest percentages that can be caught are accounted for by fishes aged three or four years (Table 7), which constitute about 60 % to 70 % of the total catch (KUKURADZE and MARIIASH 1975). Systematic, intensive fishing results in a decrease in the average length during subsequent years (NIKOLSKI 1956). Analyses of the size distribution of Prussian carps caught with nets of different mesh sizes (Table 8) revealed that 22 to 24 cm specimens were most numerous (PUSHKIN 1976). Beginning with the fourth year of life, the mortality rate becomes very high, exceeding 60 % to 70 % annually.

Reproduction biology

Maturity: Within its natural distribution range, *C. auratus auratus* attains sexual maturity at an age of one to two years and reproduces each year for about 6 to 7 years. *C. auratus gibelio* attains maturity at the age of 3+, 4+, or even 5+ at a length (*Sl*) of 18 to 23 cm (SYSOEVA 1956). In Romania, sexually mature fishes were 2+. Fishes introduced to waters located farther south, which are warmer, matured earlier and commenced spawning at an age of 1+ or 2+ and an average length (*Sl*) of 9.4 cm and weight of 30.2 g (DOMBROVSKI 1964). In Belorussian ponds, females spawned for the first time at a length of 7.9 cm and a weight of 16 g. In the Belorussian Lake Chervonnoe, spawning occurred at an age of 2+ and length of 18 cm (SAVINA 1958). It was observed that in the same region, in the Danube Delta, a rapidly growing form of *C. auratus gibelio* matured at an age of two to three years and at an average length of 20.5 cm. In contrast, the slowly growing form matured at an age of one to two years and an average length of 10.5 to 15.0 cm (KUKURADZE and MARIIASH 1975).

Gonads: The fecundity of the goldfish and the Prussian carp changes with age and fish size. The total number of eggs ranges from 30,000 to almost 400,000 in females 45 to 290 mm of standard length (DRJAGIN 1949, CSÁKÁNY 1958, KUKURADZE and MARIIASH 1975, ART'USHCHIK 1975, BUGAI and KOVAL 1976). Statistically, the most significant correlation was found between fecundity and fish weight by ASTANIN and PODGORNYI (1968). The fecundity of fishes weighing 100 to 130 g was about 21,400 eggs, while that of fishes 250 to 280 g was about 173,300 eggs (Fig. 10). BĂNĂRESCU (1964) stated that absolute fecundity during the third year of life varied

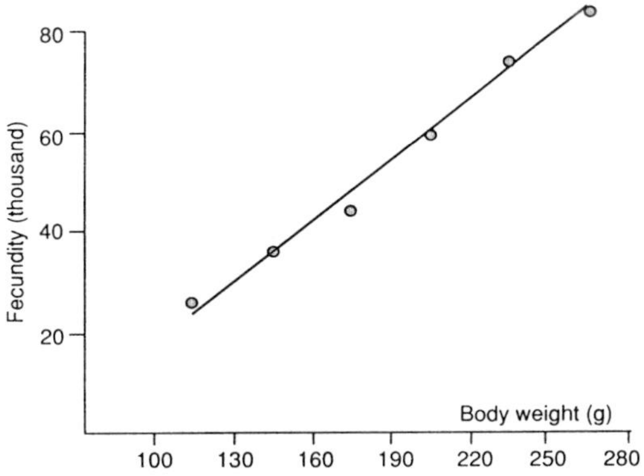

Fig. 10: Relationship between body weight and fecundity of *Carassius auratus gibelio* (ASTANIN and PODGORNYI 1968).

from 100,000 to 860,000 eggs. The individual fecundity of *C. auratus gibelio* in the region of the Amur River ranged from 44,900 to 222,500 eggs (NIKOLSKI 1956). It amounted to 100,000 eggs on the average in fishes 15 to 25 cm. According to SAVINA (1958), in the Belorussian Lake Chervonnoe, the fecundity of 29 cm fishes was 685,700 eggs, and according to ZHUKOV (1965), the maximum value was 641,700, the eggs accounting for 28.7 % of the body weight (Table 9). In ponds, the fecundity varied from 3,000 to about 205,000 (DOMBROVSKI 1964). The ovaries contain three or sometimes four batches of eggs of different sizes (Table 10). These develop

Table 9: The fecundity of *Carassius auratus gibelio* in Lake Chervonnoe (ZHUKOV 1965).

Sl in cm	Weight of fish in g	Weight of eggs in g	Mass of eggs in g			Fecundity		Weight of eggs as % of fish weight
			First portion	Second Portion	Third portion	Absolute	Relative	
21.5	314	32.0	1285	232	2834	139,232	443	10.2
22.0	315	20.0	1321	342	2972	92,700	294	6.3
22.0	360	44.0	1321	213	2456	175,560	488	12.2
22.0	302	45.0	1374	281	2583	190,710	631	14.8
22.0	339	57.5	1251	135	2928	248,055	732	16.9
22.5	335	40.0	1125	313	1981	136,760	408	11.9
22.5	352	50.0	1194	311	2355	193,500	550	14.2
23.5	375	40.9	1045	225	2848	167,603	447	13.3
24.0	372	51.0	1215	288	2114	184,467	496	13.6
24.0	394	53.0	1392	129	3000	283,823	723	13.4
24.0	435	98.0	1326	354	1872	348,096	800	22.5
24.5	435	84.0	1436	258	1931	304,500	700	19.4
27.0	549	79.0	1543	267	1727	279,423	529	14.4
28.0	597	101.0	1010	205	1733	297,748	499	16.8
29.0	754	217.0	1144	185	1628	641,669	908	28.7

Table 10: Fecundity data for female *Carassius auratus gibelio* from the Belorussian SSR (DOMBROVSKI 1964).

Age class	Size class (mm)	Body weight (g)	Egg weight (g)	Diameter of large eggs (mm)	Diameter of medium-sized eggs (mm)	Diameter of small eggs (mm)	Absolute fecundity (n)	Relative fecundity, eggs per g of body weight
2+	79-100	16-38	0.93-3.5	0.7-1.38	0.45-0.97	0.33-0.56	3 362-14 531	210-432
3+	128-160	71-157	2.7-10.5	0.67-1.04	0.52-0.76	0.39-0.53	17 752-68 956	250-554
4+	160-181	161-240	25.2-54.0	0.98-1.28	0.62-0.90	0.43-0.58	78 000-205 200	447-866

consecutively and are released in portions. The diameter of the large eggs is 0.67 to 1.38 mm. The diameter of average ones to 0.45 to 0.97 mm, and that of small ones is 0.33 to 0.58 mm (DOMBROVSKI 1964).

The eggs of the Prussian carp are a little bigger than those of the crucian carp. The egg diameter prior to spawning is 1.0 mm in March, 0.96 to 0.97 in April, 0.95 mm in May, and 0.97 to 1.00 mm in June. On the average, the egg size in members of a spawning Prussian carp population in Novo-Marievski Reservoir was 0.973 mm ± 0.003 (ASTANIN and PODGORNYI 1968). They were smaller than those of Prussian carps in West Siberia, which were 1.25 to 1.60 mm (KRIVOSHCHEKOV 1953). According to SAVINA (1958), the index of gonad maturity of the Prussian carp was 20.7 % to 42.8 % in March, 62.2 % to 100 % by the end of May and during June, and 10.0 % to 18.6 % in September.

Spawn: The goldfish and Prussian carp are characterized by batch spawning, and two or three egg batches are released during the reproductive season. KUKURADZE and MARIIASH (1975) stated that the first egg batch was the largest, accounting for 58 % of the total. The second usually accounted for 22 %, and the third, for 20 %. SAVINA (1958) and ZHUKOV (1965) stated that the first egg batch constituted 25.4 % to 43.6 % of the total, while the third one was the largest and reached 69.2 % in fishes 23.5 cm long (Table 9). Three egg batches are released by rapidly growing fishes. Individuals characterized by slower growth release only two egg batches. DOMBROVSKI (1964) suggested that older females (3+) released all eggs in only one batch, whereas young ones, which were spawning for the first time, did not lay small eggs at all. Batch spawning results in differences in the size of fry in autumn. Two groups of fishes are usually easily distinguishable (NIKOLSKI 1956, DOMBROVSKI 1964), their size differing by about 3 cm on the average. The eggs are released on plants.

Spawning period: Reproduction begins by the end of May or during June, at the same time as that of the crucian carp, *Carassius carassius*, and common carp, *Cyprinus carpio*, at a water temperature of 18 to 19 °C (BUSNITA and CRISTIAN 1955, DOMBROVSKI 1964). It lasts until the end of July. It should be emphasized that younger fishes commence spawning later in the season at higher temperatures. The time interval between the first and last spawning of fish may reach up to one month. In water bodies located in the northern part of the range, spawning of the Prussian carp begins later and at lower temperatures. Spawning is most intensive at 19 to 25 °C. The fishes spawn only during the daytime and usually in the early morning. The most intensive spawning activity is observed between 9:00 and 10:00 a.m., but a few fishes spawn throughout the day.

Spawning sites: The goldfish and Prussian carp spawn in shallow, warm places that are overgrown with aquatic vegetation. Eggs are laid on plants, to which they adhere for the whole period of embryonic development.

Mating habits: In summer, most fishes caught are females, while in winter, males are more numerous in the catch. This suggest that the two sexes do not always occur in the same places. Males attain maturity a little earlier than females. They move to spawning grounds earlier than females and remain there for shorter periods. Hence, more females than males are observed on the spawning grounds (LIEDER 1957). According to SYSOEVA (1956), in the Amur Basin near Bolonia, the spawning population was composed on the average of 32.5 % males and 67.5 % females. This was confirmed by observations in many other places. Usually, the ratio between male and female Prussian carps is 1:3 or 1:2 (MIKULICH 1939, KUKURADZE and MARIIASH 1975). However, in many places, the same numbers of male and female Prussian carps are present during reproduction. The sex ratio of the progeny is the same as that of the spawning population. BERG (1949), GOLOVINSKAIA and ROMASHOV (1947), DRJAGIN (1949), NIKOLSKI (1956), ROMASHOV and GOLOVINSKAIA (1960), HOLČÍK (1980), and HENSEL (1971) reported that the number of males in Prussian carp populations decreases continuously toward the borders of its range. Therefore, many European populations of this species consist almost exclusively of females. GOLOVINSKAIA and ROMASHOV (1947) and NIKOLIUKIN (1952) proved experimentally that Prussian carp eggs can be activated by the sperm of other cyprinid species. These authors, as well as DRJAGIN (1949), BUSNITA and CRISTIAN (1955), GOLOVINSKAIA *et al.* (1965), BUGAI and KOVAL (1976), and PEŇÁZ *et al.* (1979), found, partly through experiments, that the males of *Cyprinus carpio*, *Carassius carassius*, *Tinca tinca*, *Rutilus rutilus*, *Scardinius erythrophthalmus*, *Abramis brama*, *Leuciscus cephalus*, *Barbus barbus*, and *Misgurnus fossilis* may participate in the reproduction of the *C. auratus gibelio* subspecies. Nuclei of the reproductive cells do not fuse in these cases. Spermatozoa only adhere to the egg membrane, and the egg develops without fertilization. The progeny are all females, and these do not possess any genetic features of the sperm donors. The same phenomenon was observed in Belorussian ponds (GOLOVINSKAIA *et al.* 1965) and in southern Poland (SKÓRA 1971). The process is called gynogenesis. Unisexual, gynogenetic populations attain sexual maturity somewhat later but grow more rapidly than bisexual populations (CHERFAS 1965, GOLOWINSKAIA *et al.* 1965).

Another remarkable phenomenon related to gynogenesis is hermaphroditism, according to HOLČÍK (1980). NIKOLIUKIN (1952), GOR'UNOVA (1960), and ART'USHCHIK (1975) found that hermaphrodites constitute some percentage of the unisexual populations. GOR'UNOVA's (1960) experiments and observations showed that the male gonad of the hermaphrodites functions normally. During spawning, the number of hermaphrodites increases. They display the breeding habits typical of males and function as normal males. Hermaphrodites originate through a gradual

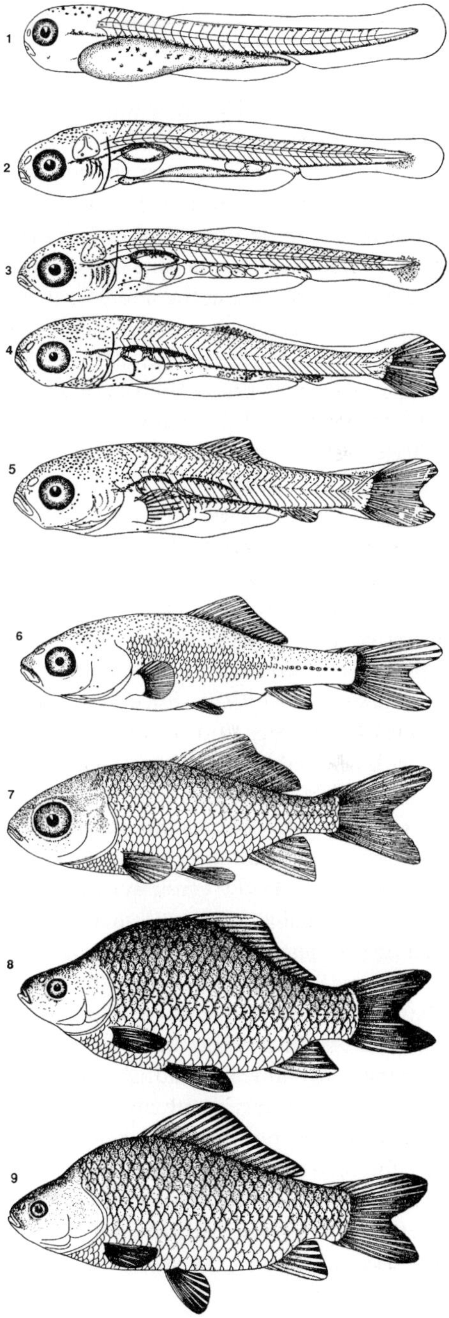

Fig. 11: Development of Prussian carp (DMITRIEVA 1957).
Sl in mm: **1** – A (5.2), **2** – B (5.5), **3** – C_1 (6.0), **4** – C_2 (9.0), **5** – D_2 (10.6), **6** – E (12.0), **7** – F (21.0), **8** – H (165.0), **9** – K (250.0).

masculinization of females and appear just after the winter kill or after water bodies inhabited by populations of these subspecies have completely desiccated.

Spawning habits: The spawning season of the goldfish and the Prussian carp occurs in spring or summer, depending on the temperature. As the season approaches, the colours become brighter; the abdomen of the female enlarges due to enlargement of the ovaries, and the males often develop temporary, minute excrescences, each about the size of a pinhead, on the gill covers and sometimes also on the dorsum and the pectoral fins. Before spawning, the fishes become extremely active, the males chasing the females, at first aimlessly but gradually more purposefully. Two, three, or more males follow one female, chasing her into shallow waters where aquatic plants are abundant. The fishes engage in lively splashing, twisting the posterior portions of their bodies energetically and shooting through the water near the surface while making short, tremulous movements of the fins. In some places, they gather together to form a compact mass, one tumbling over the other. This is the moment when the female releases her eggs, which are immediately fertilized by the males.

This process is repeated throughout the summer with periods of rest taken when the weather is hottest.

Early ontogeny: Eggs spawned by the goldfish and Prussian carp are greenish yellow. They adhere strongly to plants. Embryonic development is first similar to that described by OZIM for carp (LIEDER 1959). Spermatids penetrate the membranes of the eggs and move through the egg plasma toward their nuclei. Two polar bodies are formed in the egg. These move to the perivitelline space formed after reductive division of the nucleus, which contains half of the chromatin and half of the chromosomes (n). This mature egg nucleus comes into contact with the sperm nucleus, and they fuse to form one body containing the 2n number of chromosomes. In the Prussian carp, this process requires about 40 minutes. After this, polylecital striation commences. The first division takes place after about 60 minutes, depending on the temperature (LIEDER 1959), or 100 minutes (KIRILLOV 1972) after fertilization. Further divisions occur at 15, 20, and 40 minute intervals. Blastomeres are thus formed, and gastrulation begins, which is completed after 9 to 30 hours. Embryonic buds develop into an embryo. The embryos of both the Prussian carp and the goldfish hatch from the egg membrane after 5 to 8 days, when their length is about 5.8 mm. It continues to develop for the next three days while adhering tightly to water plants close to the water surface. During this time, the embryos resorb the yolk sac content. Active life begins 8 to 10 days after hatching, when the larvae are about 7 mm long. The digestive tract opens, and the caudal fin develops. Individual fishes begin to feed and gather into small groups. The anetrior part of the swim bladder fills with air after 25 days, and the ventral fins develop (Fig. 11).

Important parasites and diseases

Like the crucian carp, the goldfish and Prussian carp are attacked by many parasites (GUSEV 1985). In both subspecies of *Carassius auratus*, BYKHOVSKAIA-PAVLOVSKAIA *et al.* (1962) found 9 species in the genus *Myxobolus* and 11 species in the genus *Dactylogyrus*. LOM *et al.* (1989) distinguished 17 protozoan species, including such pathogenic ones as *Ichthyophthirius multifiliis*, *Myxobolus dispar*, *Trichodina* sp. and *Trypanosoma carassi*. It was shown that the number of *Dactylogyrus* sp. parasites on Prussian carps changed throughout the year. The least infestation was observed in June and the greatest in July and August (ZHARIKOVA *et al.* 1980). Initially, as the number of parasites increased, the immunological activity of the blood serum also increased. Later, antibacterial properties decreased, and lipid levels of the blood dropped. The greatest numbers of the parasites were found on the middle section of the gills, and the least on the fourth gill arch (IZIUMOVA and ZHARIKOVA 1982). Among the parasites of the Prussian carp in Romania, BĂNĂRESCU (1964) included *Dactylogyrus formosus*, *D. intermedius*, *D. vastator*, *D. wegeneri*, and among the Protozoa, he mentioned *Myxobolus bramae* and *M. carassi*.

Economic importance

The goldfish and Prussian carp are important for food mainly in Southeast Asia. In Europe and on other continents, they are important mainly as ornamental fishes. The goldfish is kept in pools, ponds in parks, and aquaria especially frequently. Concerning the use of these fishes in Europe, as many as 19 farms bred goldfishes between 1960 and 1978 in Italy alone. On an area of 156.5 ha, these farms produced about 22 million fishes annually, mainly for export (MELOTTI 1982, 1986). The goldfish is produced as an ornamental fish in categories including "commons", "fans", "shubunkins", "calicoes", and "black-moors". Its production is a well developed industry in southern France (STERBA 1977).

These fishes are also important as prey for predators and are consumed by pike, perch, and catfish. Therefore, these fishes are also cultured as live bait. Members of both subspecies are eaten by water fowl. In the Amur River Basin, catches of Prussian carp reach 7000 tons annually. This was one of the most abundantly caught fishes in the USSR from 1941 to 1950, when its catch ranged from 1,681.5 to 4,818 tons (NIKOLSKI 1956). In 1977, the Prussian carp catch in the area around the Black Sea was 2,218 tons (ZAMBRIBORSHCH and NABIL KHADŽ KHAMMUD 1981). Total world catches are estimated at 30,000 tons annually. BĂNĂRESCU (1964) estimated the catch in Romania before 1964 to be from 309 to 883 tons. The fish is caught in towed seine nets as well as in trap nets set from April through July on the migration routes from the winter quarters to the feeding grounds, as well as on the spawning grounds.

In polyculture with the carp, these fishes increase natural productivity, probably through better utilization of the planktonic crustacean resources than in monoculture. Except for Romania and the southern part of the former USSR, these fishes have no great economic importance in Europe, mainly due to their low population densities. They are caught sporadically by anglers. There is little interest in these fishes due to their slow growth, early maturation, and the taste of their bony flesh. The fat content of the flesh varies from 1.6 % to 9.5 %. However, in some regions of southeastern Europe, the Prussian carp is highly valued as a food fish (SUKHOVERKHOV 1951a, b). According to this author, one kg of its flesh has a caloric value of 1,189 kcal, which is higher than that of crucian carp (810 kcal) and common carp (1,103 kcal) flesh. In relation to its body weight, the caloric value is 511 kcal per kg.

Literature

ADELMAN, I. R., Jr., L. L. SMITH, and G. D. SIESENNOP, 1976: Acute toxicity of sodium chloride, pentachlorophenol, guthion and hexavalent chromium to fathead minnow (*Pimephales promelas*) and goldfish (*Carassius auratus*). – J. Fish. Res. Board Canada **33** (2): 203–208.

ART'USHCHIK, S. T., 1975: (Biological traits of Dnieper Prussian carp.). – Ryb. Choz. Mosk. **6:** 18–19. (in Russian).

ASTANIN, L. P., and M. I. PODGORNYI, 1968: [Characteristics of *Carassius carassius* (L.) and *C. auratus gibelio* (Bloch) fecundity.]. – Vopr. Ikhtiol. **8**(2): 266–273. (in Russian).

BALON, E., 1962: Ökologische Bemerkungen über die Standarten der Donaufische mit einer Beschreibung des Funds des *Carassius auratus gibelio* (Bloch, 1783) und *Alburnoides bipunctatus* (Bloch, 1782). – Vest. Česk. Sp. Zool. **26**(4): 333–351.

BĂNĂRESCU, P., 1960: Einige Fragen zur Herkunft und Verbreitung der Süßwasserfischfauna der europäisch-mediterranen Unterregion. – Arch. Hydrobiol. **57**(1/2): 16–134.

BĂNĂRESCU, P., 1964: Pisces – Osteichthyes. – Fauna Republicii Populare Romine 13: 1–961.

BAUCH, G., 1966: Die einheimischen Süßwasserfische. Neumann, Leipzig. 199 pp.

BERDICHEVSKI, L., A. F. KARPEVICH, and I. I. LOKSHINA, 1968: (Results and effectiveness of fish and invertebrate acclimatization in USSR waters during 15 years.). In: Akklimatizaciia ryb i bezpozvonochnykh v vodoiemakh SSSR. Nauka, Moscow. pp. 5–28. (in Russian).

BERG, L.S., 1905: (Fishes of Turkestan.). S. Petersburg. 260 pp. (in Russian).

BERG, L.S., 1916: (Freshwater fishes of Russia.). Depart. Zemledelia, Moscow. 563 pp. (in Russian).

BERG, L.S., 1932: Über *Carassius carassius* und *C. gibelio*. – Zool. Anzeiger **98:** 531.

BERG, L.S., 1949: (Freshwater Fishes of the USSR and Adjacent Countries.). Vol. **2**. Izd. Akad. Nauk SSSR., Moscow – Leningrad. pp. 469–925 (in Russian).

BLANC, M., P. BĂNĂRESCU, J. L. GAUDET, and J. C. HUREAU, 1971: European Inland Water Fish. A multilingual catalogue. – Fishing News. England. 24 pp.

BLOCH, M. E., 1783: Naturgeschichte der Ausländischen Fische. **3**, Berlin. 90 pp.

BORCEA, I., 1935: Etude sur la gibéle: *Carassius gibelio* Bloch. – Ann. Sci. l'Univ. Yassy **21:** 526–536.

Brylińska, M., E. Bryliński, J. Terlecki, W. Białokoz, J. Radziej, B. Młyniec, R. Bartel, W. Kopiejewska, M. Tadajewska, T. Krzywosz, and K. Goryczko, 1991: (Freshwater fishes of Poland.). PWN, Warsaw. pp. 289–298. (in Polish).

Bugai, K. S. and N. V. Koval', 1976: [Biology of Prussian carp (*Carassius auratus gibelio* Bloch) reproduction in waters of Dneper – Bug district.] – Gidrobiol. Zh. **12:** 53–58. (in Russian).

Burmakin, I. V., 1963: (Acclimatization of freshwater fish in the USSR.). – Izv. GosNIORKH **53:** 3–317. (in Russian).

Busnita, T., and A. Cristian, 1955: Contributiuni la cunoasterea reproducerii la carasul argintiu (*Carassius auratus gibelio* Bloch). – Bull. Inst. Cerc. Piscicole **14**(4): 15–24.

Bykhovskaia-Pavlovskaia, I. E., A. W. Gusev, M. W. Dubinina, N. A. Iziumova, T. S. Smirnova, I. L. Sokolovskaia, G. A. Shtein, S. S. Shul'man, and M. E. Epshtein, 1962: (Key to the parasitic fauna of freshwater fish in the USSR.). Izd. Akad. Nauk SSSR., Moscow – Leningrad. 776 pp. (in Russian).

Celler, M., 1976: (It is possible to reestablish goldfish culture.). – Gosp. Ryb. **5:** 22–23. (in Polish).

Cherfas, N. B., 1965: (Development of female gonads in uni- and bisexual Prussian carp.). – Tr. Uses. m.-i. in. prud. ryb. khoz. **13:** 105–112. (in Russian).

Cherfas, N. B., 1966: [Natural triploidy of unisex Prussian carp (*Carassius auratus gibelio* Bloch) females.] – Genetika **1**(5): 16–24. (in Russian).

Cherfas, N. B., 1969: (Basic results of cytogenetic analysis of uni- and bisexual Prussian carp.). – Sbornik: Genetika, Sekecia i gibridizacia ryb (Moscow) **4:** 85–98. (in Russian).

Chiarelli, B., O. Ferantelli, and C. Cucchi, 1969: The karyotype of some teleostean fish obtained by tissue culture in vitro. – Experientai vol. **25;** 426 pp.

Csákány, I., 1958: Különös uj halunk az ezüstkarász. – Halászat **5:** 238.

Dmitrieva, E. N., 1957: (Morpho-ecological analysis of two species of *Carassius*.). –Tr. Inst. Morf. Zhivot. **16:** 102–170. (in Russian).

Dombrovski, V.K., 1964: [Morphological traits of Prussian carp, *Carassius auratus gibelio* (Bloch), in the Belorussian Soviet Republic.]. – Tr. Belorussk. n.-i. In-ta rybn. Khoz-va **5:** 62–82. (in Russian).

Drjagin, P. A., 1949: (Crucian carp. Commercial fishes of the USSR.). Pishchepromizdat. Moscow: 443–448 (in Russian).

Drjagin, P. A., 1953: (Fish acclimatization in USSR inland waters.). – Izv. VNIORKH, T. **32:** 10–98. (in Russian).

Dybowski, B. I., 1869: Vorläufige Mittheilungen über die Fischfauna des Onon–Flusses und die Ingoda in Transbaikalien. – Verh. Zool. Bot. Ges. Wien **19:** 945–948.

Dybovski, B., 1877: (Fishes of the Amur River system.). – Izv. Sib. otd. IRGO **8**(1–2): 1–127. (in Russian).

Gal'cova, M. Z., 1957: (Materials on growth and feeding of Prussian carp.). Ucheniie zapiski Vitebskogo pediinstituta im. S.M.Kirova; 18 pp. (in Russian).

Gąsowska, M., 1934: [Prussian carp – *Carassius gibelio* (Bloch), a new component of the ichthyofauna in Polish waters.]. – Prz. Ryb., Warsaw **7:** 261–266. (in Polish).

Gąsowska, M., 1936: Der Giebel – eine ostasiatische Silberkarausche (*Carassius auratus gibelio* Bloch). – Z. f. Fischerei **34:** 719–725.

Gąsowska, M., 1962: (Keys to Polish vertebrates.). Cz. I. PWN, Warsaw – Kraków. 240 pp. (in Polish).

GOLOVINSKAIA, K. A., and D. D. ROMASHOV, 1947: (Studies on gynogenesis in the Prussian carp.). – Tr. VNIORKH **4:** 73-113. (in Russian).

GOLOVINSKAIA, K. A., D. D. ROMASHOV, and N. B. CHERFAS, 1965: [Uni- and bisexual forms of the Prussian carp (*Carassius auratus gibelio* Bloch).]. –Vopr. Ikhtiol. **5** (4): 614–628. (in Russian).

GOMEL'SKI , B.I., O. V. EMEL'IANOVA, and A. V. REKUBRATSKI, 1985a: (Production and some biological properties of amphiploidal hybrids of crucian carp and common carp.). – Dokl. Akad. Nauk SSSR. **301**(5): 1210–1213. (in Russian).

GOMEL'SKI , B. I., N. B. CHERFAS, and O. V. EMEL'IANOVA, 1985b: (On the possibility of producing bisexual spermatozoid hybrids of carp and crucian carp.). – Dokl. AN SSSR **282**(5): 1255–1258. (in Russian).

GOR'UNOVA, A. I., 1960: (Reproduction of Prussian carp.). – Vopr. Ikhtiol. **15:** 106–110. (in Russian).

GRATZIANOV, V. I., 1907: (A review of the fish in the Russian empire.). Moskow; 567 pp. (in Russian).

GUDKOV, P. K., 1985: [Data on the Prussian carp, *Carassius auratus gibelio* (Bloch) (Cyprinidae) biology of the mouth of the Volga River.] – Vopr. Ikhtiol. **25** (3): 517–520. (in Russian).

GUSEV, A. V., 1985: (Key to freshwater fish parasites in the USSR.). Izd. Nauka, Leningrad. 424 pp. (in Russian).

HELLAWELL, J. M., 1989: Biological indicators of freshwater pollution and environmental management. Elsevier Applied Science, London and New York. 546 pp.

HENSEL, K., 1971: Some notes on the systematic status of *Carassius auratus gibelio* (Bloch, 1782) with further record of the fish from the Danube River in Czechoslovakia. – Vést. Čs. Spol. Zool. **35** (3): 186–198.

HERVEY, G. F., and J. HEMS, 1948: The goldfish. The Batchworth Press, London. 323 pp.

HEUSCHMANN, O., 1957: Die Weißfische (Cyprinidae). In: Handb. d. Binnenfischerei Mitteleuropas, Bd. III. Stuttgart. Schweitzerbart'sche Verlagsbuchhandlung.

HOLČÍK , J., 1980: *Carassius auratus* (Pisces) in the Danube River. – Acta Sc. Nat. Brno **14**(11): 1-43.

HOLČÍK , J., P. BĂNĂRESCU, and D. EVANS, 1989: General Introduction to fishes. In: The Freshwater Fishes of Europe. Vol. **1/II**. AULA-Verlag, Wiesbaden. pp. 18–147.

IOGANZEN, B. G., 1945: (New fish forms in West Siberia.). – V. sb. Zametki po faune i flore Sibiri. Tomsk, **5:** 1–16. (in Russian).

IZIUMOVA, N. A., and T. I. ZHARIKOVA, 1982: (Cases of *Dactylogyrus anchoratus* and *D. chranilow* occurrance on crucian carp gills.). – Tr. Inst. Biol. Vnutr. Vod. **46:** 89–100. (in Russian).

JÁSZFALUSI, L., 1959: Holtágaink jövö hala az ezüstkárász. – Halaszát **6**. 207 pp.

JOHNSON, D. W., 1968: Pesticides and fishes – a review of selected literature. –Transact. Am. Fish. Soc. **97** (4): 389–424.

KARPEVICH, A. F., 1975: (Theory and practice of the acclimatization of aquatic organisms.). Pischevaia promyshlennost', Moscow. 432 pp. (in Russian).

KESSLER, K., 1877: (Fishes living and occuring in the Aral-pontic ichthyological region). Tr. Aralo-Casp. Exp. St. Peterburg. 346 pp. (in Russian).

KIRILLOV, F. N., 1972: (Fishes of Yakutia.). M. Nauka. 360 pp. (in Russian).

KISEL'OV, I. V., 1962: (Prussian carp hybrids.). Praci inst. gidrob. – AN USSR **38:** 72–83. (in Russian).

KRIVOSHCHEKOV, G. M., 1953: (Crucian carp in West Siberia.). - Tr. Barabinsk. otdel. VNIORKH **6**(2): 71–124. (in Russian).

KUKURADZE, A. M., and L. F. MARIIASH, 1975: [Data on Prussian carp, *Carassius auratus gibelio* (Bloch), ecology in the lower Danube.]. - Vopr. Ikhtiol. **15**(3): 456–462. (in Russian).

LANDOWSKI, J., 1968: (The goldfish in its native land.). - Gosp. Ryb. **2:** 22–23. (in Polish).

LIEDER, U., 1955: Die Karpfkarausche. - Deutsche Fisch. H. **3:** 80–84.

LIEDER, U., 1957: Die Ergebnisse der im Jahre 1956 durchgeführten Karpfen - Karauschenkreuzungen. - Zeitschr. f. Fischerei **6**(1–7): 283–299.

LIEDER, U., 1959: Über die Eientwicklung bei männchenlosen Stämmen der Silberkarausche *Carassius auratus gibelio* (Bloch). - Biolog. Zentral. B. **78**(2): 284–291.

LINNAEUS, C., 1758: Systema Naturae per Regna tria Naturae secundum classes, ordines, genera, species, cum characteribus, differentiis, synonymis, locis. X. ed. Holmiae. pp. 1–316.

LOM, J., and I. DYKOVA, 1989: (Protozoological parasites of commercially exploited fish.). Česky rybařsky Svaz. Prague. 312 pp. (in Czech).

LOREK, C. C., 1837: Zur Fauna Prussica. Königsberg. 13 pp.

LUKASH, B. S., 1933: (Fish of the lower stretch of the Viatka River.). - Tr. Viatsk. n.-i. inst. kraeved. **6:** 5–110. (in Russian).

LUNEL, G., 1874: Histoire naturelle des poissons du bassin du Léman. Geneva. 128 pp.

LUKJANIENKO, W. I., 1974: (Fish Toxicology.). PWRiL Warsaw. 315 pp. (in Polish).

MAKINO, S., 1939: The chromosomes of the carp, *Cyprinus carpio*, including those of some related species of Cyprinidae for comparison. - Cytologia **9:** 430-440.

MAKINO, S., 1941: A karyological study of goldfish of Japan. - Cytologia **12**.

MATSUI, Y., 1934: Genetical studies on goldfish of Japan. - Journ. Imp. Fish. Institute **30**(1).

MELOTTI, P., 1982: Allevamento e commercializzazione del carassio dorato (*Carassius auratus* L.) in Italia. - Riv. Ital. Piscicolt. Ittiopat. A. **17**(3): 91–102.

MELOTTI, P., 1986: Goldfish (*Carassius auratus* L.) farming in Italy. In: R. BILLARD and J. MARCEL (Eds.), Aquaculture of Cyprinids. INRA Paris. pp. 71–80.

MIKULICH, L. V., 1939: [Data on the crucian carp, *Carassius auratus gibelio* (Bloch), and Amur barbel, *Hemibarbus labeo* in sp. *maculatus* (Bleck.), in Lake Hanka.]. - Ucz. zap. Permsk. gos. ped. In-ta **4:** 73–112. (in Russian).

MIŠÍK, V., and J. HOLČÍK, 1962: A Note on a *Carassius auratus* in Cechoslovakian Silesia. - Vést. Čs. Spol. Zool. **26**(4): 329–332.

MULERTT, H., 1902: The goldfish and its systematic culture. Brooklyn, New York; 143 pp.

NIKOLIUKIN, N. I., 1952: (Interspecies fish hybridization.). Saratov. Otdel. Kasp. Fil. VNIRO: 1–312. (in Russian).

NIKOLSKI, G. W., 1956: (Special ichthyology.). PWN Warsaw. pp. 1–624. (in Polish).

NIKOLSKI, G. W., 1956: (Fish of the Amur basin.). Izd-vo AN SSSR. 551 pp. (in Russian).

OHNO, S., 1974: Animal Cytogenetics. Vol. 4. Chordata 1, Gebrüder Borntraeger, Berlin, Stuttgart.

PEŇÁZ, M., P. RAB, and M. PROKEŠ, 1979: Cytological analysis, gynogenesis and early development of *Carassius auratus gibelio*. - Acta Sc. Nat. Brno **13**(7): 3–33.

PUSHKIN, Iu. A., 1976: (Characteristics of a commercial stock of crucian carp and the dynamics of its exploitation in Botkin.). - Rybokhoz. Izuch. Vnutr. Vod. **17:** 47–56. (in Russian).

RADZIMOVSKI, V. D., 1970: (Cannibalism amongst goldfish juveniles.). - Ryb. Khoz. Kiev **11:** 123–126. (in Russian).

RADZIMOVSKI, V. D., 1973: (Morphological characteristics of goldfish from ponds in "Pushcha-vodica" fish farm.). – Ryb. Khoz. Kiev **17:** 35–39. (in Russian).

REMBISZEWSKI, J. M., 1971: (Goldfish.). – Gosp. Ryb. **12:** 20–21. (in Polish).

ROMASHOV, D. D., and K. A. GOLOVINSKAIA, 1960: (Gynogenesis and hybridization of distant fish forms.). In: "Otdalennaia gibridicaciia rastenii i zhivotnykh". Izd. AN SSSR, Moskow. pp. 496–510. (in Russian).

ROULE, L., 1914: Traité de la Pisciculture et des Pêches. Paris: 734 pp.

ROVNIN, A. A., A. M. KUKURADZE, and N. I. STAKHORSKAIA, 1977: (Some problems of *Carassius auratus gibelio* biology and its role in the ecosystem of the Danube Basin.). – Ryb. Khoz. **2:** 9–11. (in Russian).

SANTA, N., and A. G. MARINESCU, 1970: Der Einfluß einiger hypothermischer Variationen auf die Resistenzadaptation und den Sauerstoffverbrauch des Giebels (*Carassius auratus gibelio* Bloch 1783). – Rev. Roum. Biologie–Zoologie **15:** 415–425.

SAVINA, N.O., 1958: (Prussian carp – *Carassius auratus gibelio* (Bloch) – in new conditions of its occurrence.). – Tr. Belorussk. n.-i. in-ta ryb. khoz. **2:** 60–94. (in Russian).

SCHÄPERCLAUS, W., 1953: Die Züchtung von Karauschen mit höchster Leistungsfähigheit. – Z. f. Fisch. **1/2:** 19–69.

SCHÄPERCLAUS, W., 1961: Lehrbuch der Teichwirtschaft. P. Parey, Berlin, Hamburg. 582 pp.

SILIN, B.V., 1983: [Details of the species status of Crucian carp (Genus *Carassius*, Cyprinidae) in Yakutian water bodies.]. – Vopr. Ikhtiol. **23**(2): 186–192. (in Russian).

SKÓRA, S., 1971: The German carp (*Carassius auratus gibelio* Bloch) from the River Ilownica stocked in a carp pond. – Acta Hydrobiol. Kraków **13**(2): 217–232.

SPOTILA, J. R., K. M. TERPIN, R. R. KOONS, and R. L. BONATI, 1979: Temperature requirements of fishes from eastern Lake Erie and upper Niagara River. – Env. Biol. Fish. **4** (3): 281–307.

STERBA, G., 1977: Süßwasserfische aus aller Welt. Verlag J. Neumann-Naudamm. Melsungen., Berlin, Basel, Wien. 558 pp.

SUKHOVERKHOV, F.M., 1951a: (New race of Prussian carp for pond farming.). – Sovetskaia zootekhnika **3**. (in Russian).

SUKHOVERKHOV, F. M., 1951b: (Economic importance of Prussian carp in pond fish culture.). – Ryb. Khoz. **4:** 36–37. (in Russian).

SUVOROV, E., 1954: Principles of ichthyology. PWN Warsaw; 968 pp. (in Polish).

SUZUKI, R., 1953: Studies on the interfamiliar crossing between loach (*Misgurnus anguilicaudatus*) and goldfish (*Carassius auratus*). – Japan. J. Ichthyol. **5**.

SUZUKI, R., 1955: Note on the interfamiliar hybrid larvae between mud loach (*Misgurnus anguilicaudatus*) and crucian carp (*Carassius carassius*) or goldfish (*Carassius auratus*). – Japan. J. Ichthyol. **4**(1–3).

SYSOEVA, T. K., 1956: [Data on age composition and growth rate of Prussian carp (*Carassius auratus gibelio* Bloch) in the Amur basin.]. –Tr. Amurskoi ikht. ekspedicii 1945–1949: **4**. Moscow (in Russian).

SZCZERBOWSKI J. A., and A. J. SZCZERBOWSKI, 1996: (The carassius.). IRS, Olsztyn. 175 pp. (in Polish).

TACZANOWSKI, L., 1877: Liste des vertébrés de Pologne. – Bull. Soc. Zool. France, Paris **2:** 121–174.

TOBIAS, R., 1865: Die Wirbeltiere der Oberlausitz. – Abh. Naturf. Ges., Görlitz **12:** 57–96.

VALENTA, M., V. ŠLECHTA, V. ŠLECHTOVA, and L. KÁLAL, 1977: Genetic polymorphism and isoenzyne patterns of lactate dehydrogenase in tench (*Tinca tinca*), crucian carp (*Carassius carassius*) and carp (*Cyprinus carpio*). – Anim. Blood Grps biochem. Genet. **8:** 217–230.

VARPAKHOVSKI, N., 1886: (A review of the ichthyofauna of Kazan District.). – Zap. Akad. Nauk, L. **2**(3): 1–70. (in Russian).

VASIL'EV , V. P., 1985: (Evolutionary caryology of fish.). Izd. Nauka, Moscow. 300 pp. (in Russian).

VOLSKIS, R. S., 1989: Procedings of the 24th Meeting of the International Working Group on the Project "Species and its Productivity in the Distribution Area" for the UNESCO Programme "Man and Biosphere." Lithuanian Acad. Sci. 68 pp.

VOSILENE, M. E., 1977: (The effect of oil on noradrenaline content in brain tissue of the Prussian carp and mottled sea bass.). – Tr. AN Lit. SSR, **B (1):** 93–97. (in Russian).

WALLEN, I. E., 1951: The direct effect of turbidity on fishes. – Bull. Okla. Agric. Mech. Coll. **2:** 1-48.

WEISS, C. M., 1961: Physiological effect of organic phosphorus insecticides on several species of fish. – Transact. Am. Fish. Soc. **90**(2): 143–152.

WELCOMME, R. L., 1988: International introductions of inland aquatic species. – FAO Fish. Tech. Pap. (294). 318 pp.

WOLIŃSKI, Z., 1967: (Truth about goldfish.). – Gosp. Ryb. **6:** 20–21. (in Polish).

ZAMBRIBORSHCH, V. S., and NABIL'KHADŽ KHAMMUD, 1981: [Prussian carp *Carassius auratus gibelio* (Bloch) from lower sections of the rivers of the North-West part of the Black Sea.]. – Vopr. Ikhtiol. **21**(1): 160–164. (in Russian).

ZHARIKOVA, T. I., N. I. SILKINA, and M. A. STEPANOVA, 1980: [Dependence between *Dactylogyrus* spp. (Dactylogyridae, Monogenea) numbers and immuno-physiological state of *Carassius auratus* (L.).]. – Dokl. Akad. Nauk SSSR **2:** 510–512. (in Russian).

ZHUKOV, P. I., 1965: (Fishes of Belorussia.). Nauka i Tekhnika, Mińsk. 372 pp. (in Russian).

ZIEMIANKOWSKI, V., 1947: Fauna pestilor din Bucovina. – Analel Institului de Crcetari Piscicole al. Romaniei **3:** 115–220.

Carassius carassius (Linnaeus, 1758)

E = crucian carp, F = carassin commun, G = Karausche, R = zolotoi karas'

Jan A. Szczerbowski and Andrzej J. Szczerbowski

Synonyms

Cyprinus carassius Linnaeus, 1758: 321 (Europe)
Cyprinus gibelio (non Bloch), Reisinger, 1830: 59 (Hungary)
Cyprinopsis carassius, Fitzinger, 1832: 334 (Austria)
Carassius vulgaris Nilsson, 1832: 178 (Northern Europe)
Carassius charax Lesniewski, 1837: 53 (Poland)
Carassius humilis Heckel, 1837: 156 (Palermo, Sicily)
Carassius moles Nordmann, 1840: 480 (Rivers of the Black Sea watershed)
Carassius gibelio (non Bloch) Bielz, 1853: 178 (Transylvania)
Carassius gibelio var. *minutus* Kessler, 1856: 38 (Ukraine)
Carassius moles (Agassiz), Heckel and Kner, 1858: 72 (Danube River watershed, Poland)
Carassius oblongus Heckel and Kner, 1858: 73 (Danube River watershed, Poland)
Carassius carassius var. *kolenty* Dybovski, 1877: 11 (Amur River watershed)
Carassius vulgaris var. *oblongus,* Varpakhovski, 1886: 54 (Kazakhstan)
Carassius vulgaris crassior Walecki, 1889: 16 (Poland, Ukraine)
Carassius carassius, Berg, 1909: 70 (Amur River watershed)
Carassius vulgaris lacustris Roule, 1914: 484 (France)
Carassius carassius morpha *humilis,* Menshikov and Revnivykh, 1937: 151 (Om River watershed)
Carassius carassius typ. Maslovski, 1940: 149 (Don River watershed)

Holotype

The holotype of the crucian carp has not been preserved. However, it can be assumed that the fishes in Central Europe are typical.

Etymology

According to GRATZIANOV (1907), the name *Carassius* originates from Latin. However, it might also be suggested that it is a latinization of a French or German vernacular name.

Description

Du 3–4, *Db* 14–25 most frequently 15–19, *Au* 2–3, *Ab* 5–8 most frequently 6, Scales along the *l.l.* 21–38, *Sq. sup.* 6–8, *Sq. inf.* 5–7, *Sp. br.* 23–33 most frequently 26–31, *Vert.* 31–34 ($\bar{x}$ – 32.5; Table 11).

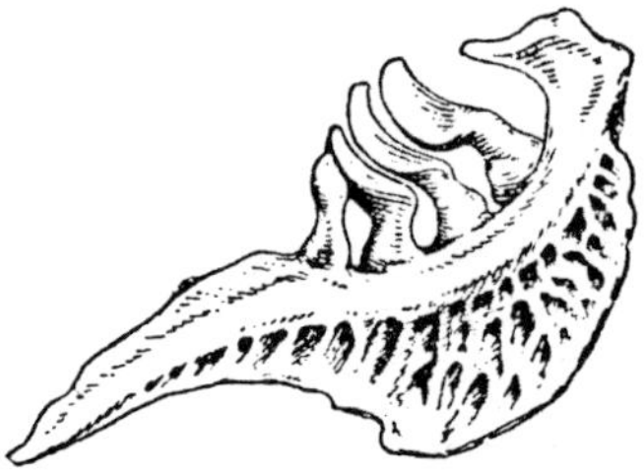

Fig. 12: Pharyngeal teeth of *Carassius* sp.

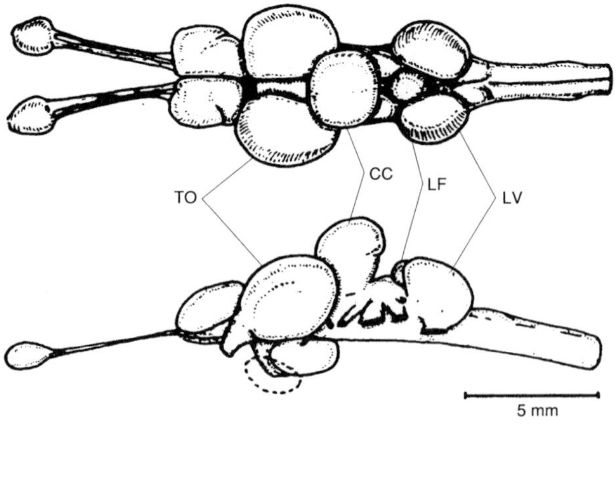

Fig. 13: Dorsal (top) and lateral (below) views of the brain of *Carassius carassius:* **TO** – Tectum opticum, **CC** – Cerebellum, **LF** – Lobus facialis, **LV** – Lobus vagus (KORTSCHAL, JUNGER 1988).

The maximal length (*Tl*) is 500 mm, and the weight reaches 5 kg (BERG 1949). Fishes 15 to 20 cm in *Tl* and with body weights of 100 to 200 g usually have deep bodies. Smaller ones with elongated bodies are also frequently caught. Small fishes of this species are very well known and are often called *Carassius carassius* morpha *humilis* (MENSHIKOV and REVNIVYKH 1937), which is the substitute for the names *Carassius gibelio* var. *minutus* (KESSLER 1856) and *Carassius oblongus* (HECKEL and KNER 1858). This fish rarely grows larger than 15 cm in *Tl* and 100 g in weight (SZCZERBOWSKI and SZCZERBOWSKI 1996).

Morphology: Body laterally compressed and highly peaked dorsally (Fig. 15). The body width (*laco*) is usually 31.5 % to 44.5 % of its depth (*H*). The body depth (*H*) of normally nourished fishes is 1.8 to 2.2 times *H*, and according to BĂNĂRESCU (1964), it equals 43 % to 56 % of the length. Specimens with lower dorsal peaks are also encountered (Table 12), and in the form referred to as *humilis*, the ratio between the length (*Sl*) and depth (*H*) is from 2.4 : 1 to 3.0 : 1. In Germany, the lowest reported body height of an individual was 24.5 % of the length, and the highest was 55.9 %. The average values for distinct populations ranged from 36.2 to 51.4 % (LIBOSVARSKY 1964). BĂNĂRESCU (1964) stated that in the *humilis* form, this ratio is 30.3 % to 36.0 %. The lateral line is almost straight and frequently does not reach the end of the body, or it is interrupted, most frequently in *humilis* form (Fig. 16). The head is not very large. Its length equals 24.3 % to 31.0 % of the body length (*Sl*), and its depth is about 30 % of its length (SZCZERBOWSKI 1995, SZCZERBOWSKI *et al.* 1997a). The lower jaw is directed upward. The mouth is centrally located. The snout length is 24.0 % to 32.6 % of the head length. The eye diameter is 18.4 % to 26.0 % of the head length.

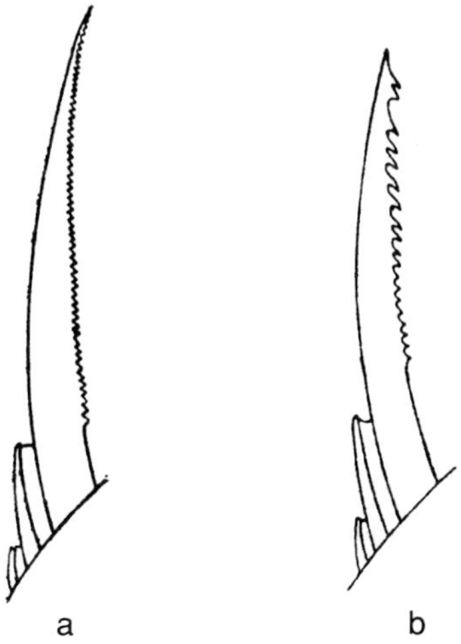

Fig. 14: The first unbranched rays of the dorsal fin of *Carassius carassius* **(a)** and *Carassius auratus* **(b).**

Table 11. Meristic data for *Carassius carassius*.

Locality	Author	Branchiostegal Rays (n)				Vert. (n)	*l.l.*	*Sp.br.*
		Du	*Db*	*Au*	*Ab*			
Germany	SMOLIAN 1920	III–IV	14–21	III	5–8	–	31 $\frac{7-8}{5-6}$ 35	–
Russia	DRJAGIN 1949	III–IV	14–21	II–III	6–8	31–33	28 $\frac{6-8}{6-7}$ 37	22–33
Poland	STAFF 1950	III–IV	14–21	II–III	6–8	–	28 $\frac{6-8}{5-6}$ 37	23–33
Poland	SKÓRA 1961	III	16–18	III	7	31–34	33 $\frac{6-8}{5-6}$ 36	6–34
Poland	GĄSOWSKA 1962	III–IV	14–21	II–III	5–8	–	32 $\frac{6-8}{6-7}$ 35	23–34
Belorussia	ZHUKOV 1965		15–19		6–8	28–37	32–35	23–29
Poland	BIAŁOKOZ 1979	III–IV	15–19	II–III	6–8	31–34	32 $\frac{6-8}{6-7}$ 34	24–35
Poland	WITKOWSKI 1984	II–IV	16–19	III	5–7	–	30 $\frac{7-8}{6-8}$ 34	–
Poland	SZCZERBOWSKI 1995	III–IV	16–19	II–III	5–8	30–33	32 $\frac{7-8}{7-8}$ 35	22–28

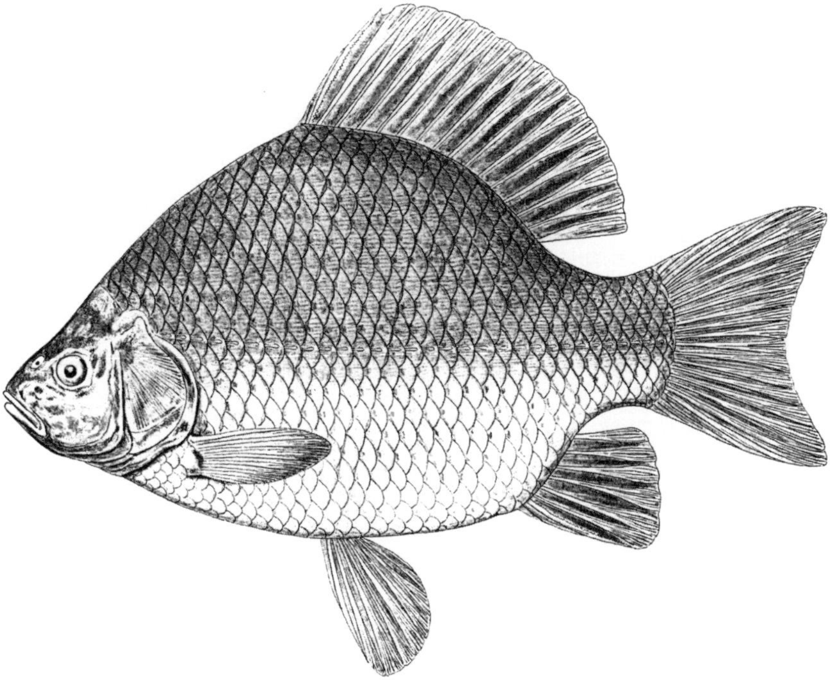

Fig. 15: Crucian carp – *Carassius carassius* (from ANTIPA 1909).

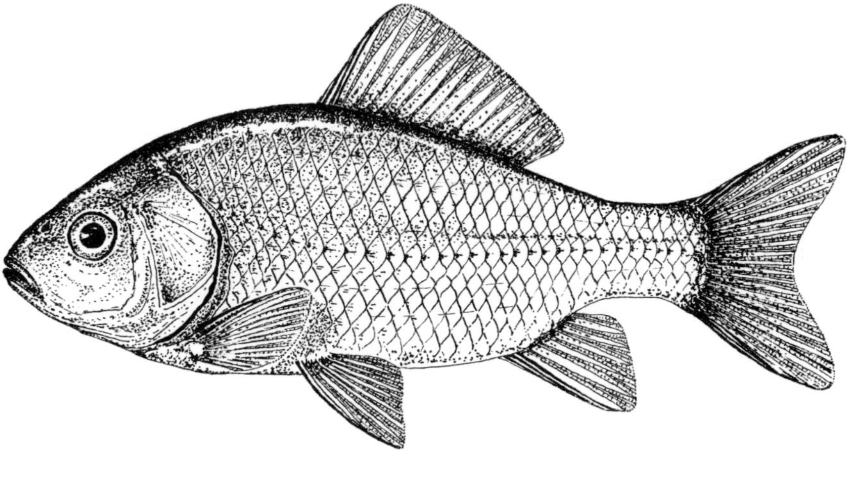

Fig. 16: *Carassius carassius* morpha *humilis* (from BRYLIŃSKA *et al.* 1991).

Table 12: Morphometric data for *Carassius carassius* (average values).

		Dvina R. ZHUKOV 1965	Dnieper R. ZHUKOV 1965	Novo-Marievskii Liman KIRILLOV 1972	Biebrza R. WITKOWSKI 1984	Nameless-Marsh SZCZERBOWSKI 1995
% of *Sl*	*H*	56.84	46.58	39.93	51.96	35.5
	h	17.99	15.72	-	17.15	13.9
	pD	55.93	53.11	51.36	54.71	52.7
	poD	19.30	21.11	-	20.28	24.5
	lpc	15.55	16.91	16.95	16.91	17.6
	lD	41.93	36.00	34.34	37.50	30.3
	hD	21.99	18.63	16.67	21.51	17.7
	lA	13.48	11.72	11.55	11.58	9.1
	hA	17.55	15.47	14.61	17.27	15.1
	lP	19.23	17.67	18.00	18.18	19.3
	lV	22.99	20.18	20.62	21.82	20.8
	P-V	25.17	24.05	-	25.54	22.6
	V-A	29.61	27.55	-	29.09	24.8
	lc	27.17	27.61	28.36	27.34	30.1
% of lc	*Oh*	19.11	20.55	20.95	20.14	21.0
	prO	39.60	31.88	-	34.44	25.6
	poO	48.85	49.65	-	50.89	50.8
Sl (mm)		18.24	-	-	17.13	17.3
n		16	30	-	28	84

The caudal fin is short, making the fish appear longer. It is slightly concave and rounded at the edges, like the other fins. The caudal fin length is 15.7 % to 21.4 % of the body length, on the average; the corresponding percentages for the shortest caudal fins are 15.0 % to 19.1 %. As percentages of standard length (*Sl*), the predorsal (*pD*) distance is 48.0 % to 58.0 %; the base of the dorsal fin (*lD*), 29.9 % to 40.3 %; the post dorsal length (*poD*), 15.5 % to 21.8 %, and the caudal length (A-C), 19.3 % to 24.8 %. The upper edge of the dorsal fin is convex. The length of the intestinal tract in adults is 2.5 times the body length (SKÓRA 1961). The body is golden copper in color and darker dorsally. It is light gold ventrally. The peritoneum is light. The *humilis* form is frequently dark red. There is a large, rounded black spot at the base of caudal fin. The fins, especially the dorsal and caudal, are reddish. It is known that changes in coloration are brought about by changes in the clarity of the water (WAŁECKI 1864).

Karyotype: Originally, the chromosome number of the crucian carp was reported to be 2n = 94 (MAKINO 1939). Studies conducted by LABAT *et al.* (1983) showed that 2n = 100 and that the number of branches (NF) was 164. The chromosomes have been placed in three groups:

20	–	metacentric (m),
44	–	submetacentric (Sm) and subtelocentric (St),
36	–	acrocentric (a) and telocentric (t).

Similar results were obtained by VASIL'EV and VASIL'EVA (1984) and VASIL'EV (1985), who reported the following formulas for crucian carp chromosomes according to the authors cited:

HAFEZ *et al.*	2n–100:	20 m + 44 Sm + 36 a
RAICU *et al.*	2n = 50:	20 m + 12 Sm + 18 St.a
VASIL'EV (1985)	2n = 100:	48 m Sm + 52 St.a

Hence, the crucian carp is tetraploid, although the occurrence of isolated diploid populations is not excluded (e. g. RAICU *et al.*).

Protein specifity: According to VALENTA *et al.* (1977 b), five different transferrins were separated from crucian carp serum. The mobilities of the different transferrins were different from those obtained from other crucian carp populations, goldfish, and other cyprinid species. Investigations of the isoenzyme patterns and the polymorphism of lactate dehydrogenase (LDH) of crucian carp showed that there are subunits with different electrophoretic mobility, which are genetically controlled at the B^1, B^2, A^1, A^2, and C loci. The locus B^2 is polymorphic and has two different alleles for each transferrin (VALENTA *et al.* 1977a). The isoenzyme patterns of LDH are tissue specific. In all tissues examined, the activity at the A^2 and B^1 loci was lower than that at A^1 and B^2. In liver tissue, the polymorphic locus B^2 was expressed together with B^1 and C^1.

Osteological characteristics: Only a few detailed studies of the osteology of the crucian carp have been undertaken. DMITRIEVA (1957) described the most important bones of the head (Fig. 17). The information in the literature concerns mainly the number of vertebrae and the formula of the pharyngeal teeth, which is 4-4 in 96 % of the fishes and 3-4 or 4-3 in the rest. MIASKOWSKI (1959a) compared selected cyprinid fishes and stated that the number of vertebrae in the crucian carp is 33.06. RUDZIŃSKI and SKÓRA (1963) analyzed the effects of crossing the crucian carp with the common carp and reported that the number of vertebrae in the crucian carp ranged from 31 to 34 (Fig. 18). Four of the anteriormost vertebrae (Fig. 19) were drawn by MIASKOWSKI 1959b. The number of intramuscular bones is 77 (LIEDER 1957).

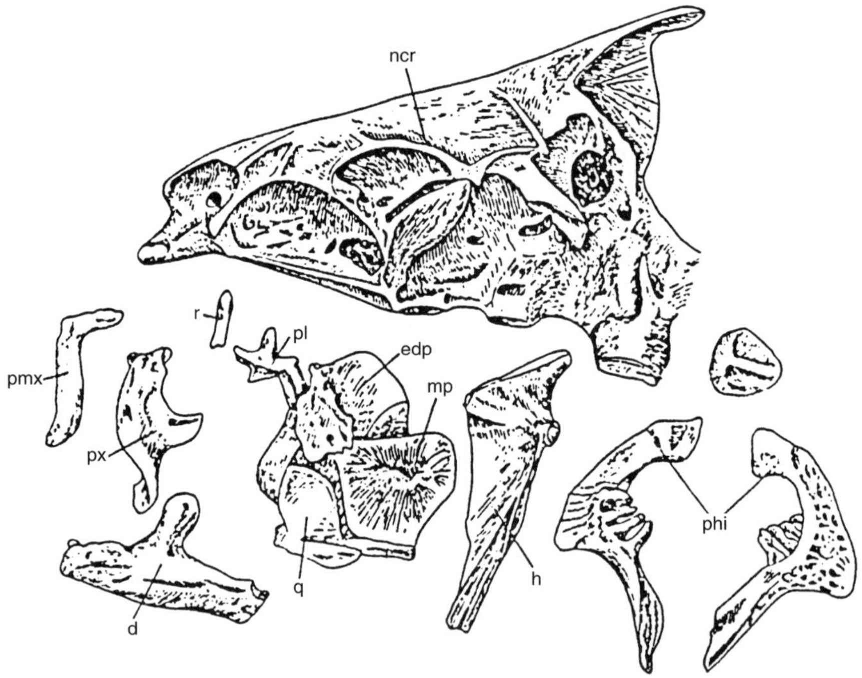

Fig. 17: The most important bones of the crucian carp head.
ncr – neurocranium, **r** – rostrum, **pmx** – praemaxillare, **px** – maxillare, **d** – dentale, **pl** – palatinum, **edp** – endopterygoideum, **mp** – mesopterygoid, **q** – quadratum, **h** – hyomandibulare, **phi** – os pharyngeum inferior

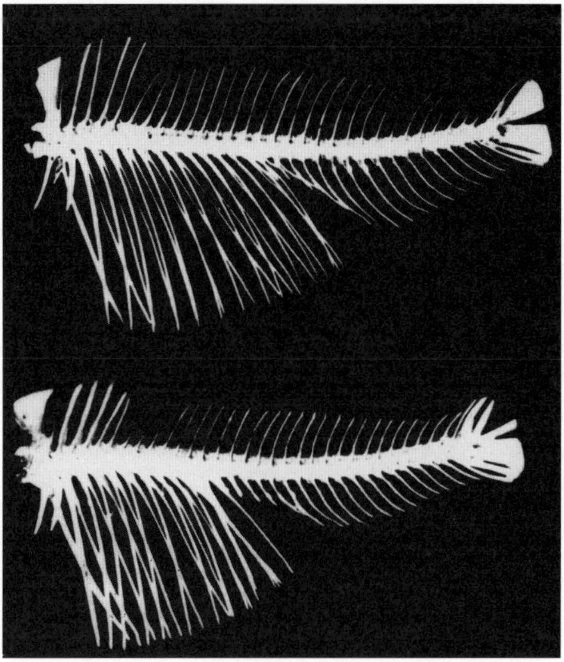

Fig. 18: Part of the skeletons of the crucian carp (above) and a hybrid (below) of the crucian and common carps (RUDZIŃSKI and SKÓRA 1963).

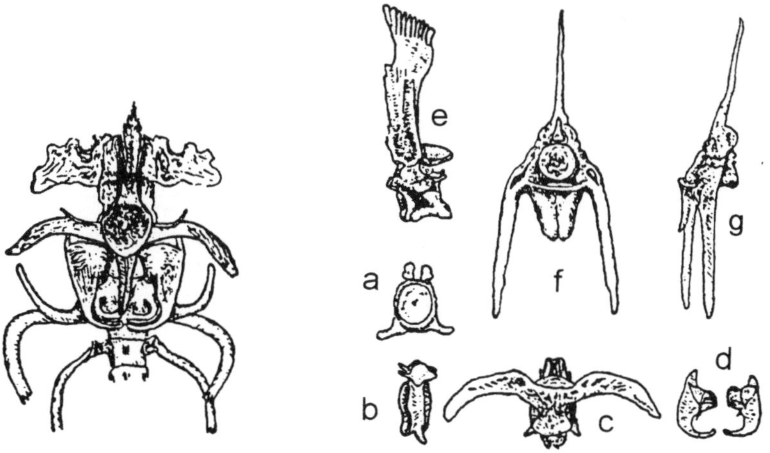

Fig. 19: The first four vertebrae of the crucian carp.
a – Frontal vertebra, **b** – Lateral view of the frontal vertebra, **c** – 2nd and 3rd vertebrae fused, **d** – Processi of the 3rd vertebra, **e** – Lateral view of the 2nd and 3rd vertebrae, **f** – 4th vertebra, back view, **g** – Lateral view of 4th vertebra.

Sexual dimorphism: It is generally accepted that crucian carp males and females do not differ very much externally. However, some authors (BERG 1949, SCHÄPERCLAUS 1953, CHIGARZH 1958) pointed out that within the same length groups of a population, females are slightly heavier due to the increased ovary weight during reproduction. This is reflected by the coefficient of condition, which averages 2.17 for males and 2.24 for females. In the cases discussed by SCHÄPERCLAUS (1953), CHIGARZH (1958), and SZCZERBOWSKI *et al.* (1997 b), females were slightly larger and lived longer than males under the same environmental conditions.

Variations: The crucian carp is distributed over a large area. This results in considerable variation in the habitat conditions and therefore also in morphology, leading to the development of different ecological races (Tables 11 and 12). The form *humilis* is an example, which shows the geographic variations described by BERG (1949) and CHIGARZH (1958) under the following formula: *Du* 3–4, *Db* 14–21, *Au* 2–3, *Ab* (5) 6–8, scales along the *l.l.* 32–35 (36).

Age and size variability: The characteristics of certain parts of the body undergo slight changes with growth. The most prominent of these include the number of gill rakers, which increases with fish length as shown in Figure 20. In fishes up to 100 mm in body height (*H*), it increases more rapidly than the length (LIBOSVARSKY 1964). In larger fishes, the length and height growth is nearly proportional.

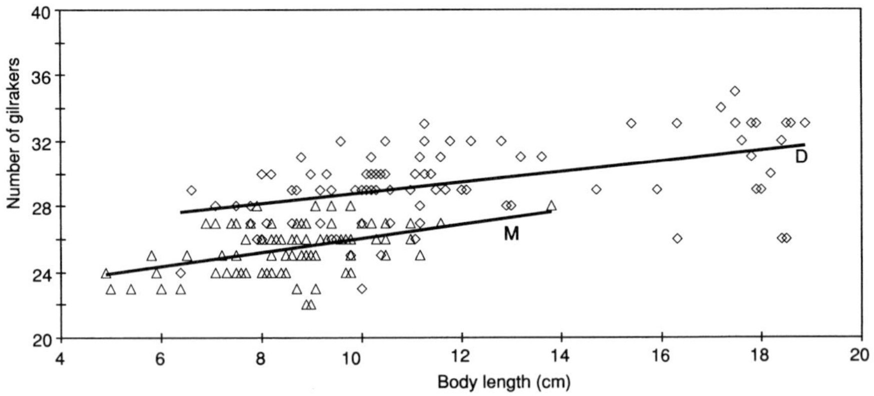

Fig. 20: Relationship between *Carassius carassius* length and gill raker numbers (SZCZERBOWSKI 1995).
D – *Carassius carassius* morpha *tipica* , **M** – *Carassius carassius* morpha *humilis*.

Subspecies

No subspecies of *Carassius carassius* are presently recognized, although WAŁECKI (1889) described one subspecies as *C. vulgaris crassior*, and ROULE (1914) described another, *C. vulgaris lacustris*.

Hybrids

Crucian carp hybrids were already known in the 17th century (LEONHARDT 1904). HECKEL (1835) named the hybrid of the crucian and common carps *Cyprinus kollari*, and HECKEL and KNER (1858) called it *Carpio kollari*. ANTIPA (1909) described the hybrid as *Cyprinus carpio* x *Carassius vulgaris*. According to BĂNĂRESCU (1964), this hybrid is characterized by the following features: *Du* 4, *Db* 17–20 (21), *Au* 3, *Ab* 4–5, Scales along the *l.l.* 35–38 (39), *V* 33–36 (37). Pharyngial teeth may be arranged in one, two, or three rows, their general formulae being 1.4–4.1, 1.1–4.1, 1.4–4.1.1, 1.4–4, or 4–4.1.1. Hybrids are found in natural waters as well as in artificial ponds.

Attempts have also been made to obtain hybrids with certain characteristics. NIKOLIUKIN (1952) studied crosses between the crucian carp with various species of Cyprinidae, including the roach, chub, rudd, tench, gudgeon, white bream, bream, vimba, and common carp. RIABOV (1979) mentioned that there have been as many as 21 attempts to crossbreed the crucian carp with other species. It was found that only the hybrids of the crucian and common carp provided satisfactory results. They lived up to seven years. Hybrids of the crucian carp with other species were characterized by a high mortality, and some of the fishes developed gynogenetically. NIKOLIUKIN (1952) described the morphological features of the *C. carassius* x *C. carpio* hybrid as *Du* 3–5 usually 4, *Db* 16–21 usually 18–19, *Au* 3, *Ab* 5 (6), *l.l.* 35–38 usually 35–36, *Sq. sup.* 6–8, and *Sq. inf.* 6–7. Similar data were reported by SKÓRA (1962). The number of scales was slightly lower than that of the common carp. The pharyngial tooth formulae were highly variable (Table 13), ranging from 1.4–4.1 to 2.4–4.1.1. The number of gill rakers was greater than that of the common carp

Table 13: Variations in the pharyngeal tooth arrangement (PhF, NIKOLIUKIN 1952).

Fish	Formula of *PhF*	n
Wild carp	1.1.3-3.1.1 (rarely 1.2.3-3.2.1)	-
Common carp	1.1.3-3.1.1	50
Common carp x crucian carp	1.4-4.1, 1.1.4-4.1, 1.4-4.1.1, 1.1.4-4.1.1, 1.4-4.2, 2.4-4.2, 1.1.3-3.1.1, 1.4-4, 1.4-3.1.1, 2.4-4.1, 4-4.1.1, 2.4-4.1.1	100
Crucian carp x wild carp	1.4-4.1, 4-4.1, 1.1.4-4.1, 1.4-4, 1.1.4-4.1.1, 4-4, 2.4-4.1, 1.1.4-4	60
Crucian carp	4-4	60

Table 14: Variations in the gill raker (*Sp.br.*) numbers (NIKOLIUKIN 1952).

Fish / *Sp.br.* / n	23	24	25	26	27	28	29	30	31	32	$\bar{x}$	n
Common carp	2	7	21	12	7	1	-	-	-	-	25.4	50
Common carp x crucian carp	-	1	3	13	13	6	1	-	-	-	26.6	37
Crucian carp	-	-	-	3	7	4	7	3	1	-	28.1	25
Crucian carp x wild carp	-	-	-	1	11	21	11	4	1	1	28.3	50
Wild carp	3	12	20	22	11	2	2	-	-	-	25.6	72

Table 15: Meristic data for hybrids *Cyprinus carpio* x *Carassius carassius* (NIKOLIUKIN 1952).

Character		Common carp	Common carp x crucian carp	Crucian carp	Crucian carp x wild carp	Wild carp
% of *Sl*	*pD*	52.9	51.4	53.3	49.8	50.5
	poD	19.4	20.2	19.4	21.0	19.0
	lpc	18.5	19.2	16.1	18.4	18.7
	H	38.1	38.1	48.0	33.4	32.2
	h	14.1	14.2	15.7	13.4	12.7
	lc	32.0	29.4	28.7	29.1	29.6
	Oh	6.0	5.7	6.2	6.4	6.2
	poO	15.0	14.9	14.1	14.0	13.8
	lD	36.9	36.4	37.8	36.2	37.9
	hD	16.2	18.0	20.8	18.9	15.9
	lA	9.4	10.0	12.7	10.2	10.1
	hA	15.2	15.8	17.0	16.3	16.1
	lP	19.8	18.0	18.3	19.2	19.3
	lV	16.8	18.7	22.0	19.7	18.2
% of *lc*	*Oh*	18.8	19.6	21.7	22.1	20.6

(Table 14), being close to that of the crucian carp. The morphometric features were usually intermediate between those of the common and crucian carp (Table 15). The head length and body height in relation to the body length are also intermediate between the two parental species. The angle between the ribs and horizontal body axis is also intermediate. The swim bladder is similar to that of the crucian carp, with its caudal chamber being longer and more voluminous than that of the common carp (RUDZIŃSKI and SKÓRA 1963). NIKOLIUKIN (1952) crossed a ♂ hybrid, ♀ *C. carassius* x ♂ *C. carpio*, with a ♀ crucian carp, and a ♀ hybrid, ♀ *C. carassius* x ♂ *C. carpio*, with a ♂ mirror carp and obtained numerous progeny. The pharyngial tooth arrangements of the progeny were similar to those of the common carp. Only in some individuals was their arrangement 1.1.4–4.1.1 or 1.4–4.1. The fishes were fully scaled and possessed longer barbels than the F_1 hybrids.

Detailed observations of hybrids between crucian and common carp were reported by SKÓRA (1968). Back-crosses were made between ♀ hybrids, (♀ *Cyprinus carpio* x ♂ *Carassius carassius*), and ♂ *Cyprinus carpio* and ♂ *Carassius carassius*. The body shape and scales of these hybrids were similar to those of the hybrids of the first generation. Most fishes had two pairs of barbels, of which the posterior ones were longer. They were longer than those of first generation hybrids. The scales of the lateral line had variable diameters and were irregularly distributed. The ventral fins were partially deformed. The number of rays in the fins varied considerably. Back-crosses of hybrids with male crucian carps had fewer gill rakers, and their number was less variable than in the hybrids with male common carps. The number of pharyngial teeth in the back-crossed hybrids was similar to those in the crucian carp: 4–4, and that of the hybrids with a male the common carp was similar to those of the common carp: 1.1.3–3.1.1 or 1.2.3–3.1.1, 1.1.4–3.1.1. These back-crossed hybrids were a dark coppery colour, darker than that of the first generation hybrids. They are dark dorsally and light gold ventrally. The progeny of the hybrid with the male common carp grew twice as fast as those resulting from the cross of the hybrid with the male crucian carp.

LIEDER (1957) described a number of other specimens, citing KIRPICHNIKOV and PROBST to report that the number of scales on the lateral line of the hybrids was the same as those of the parents. In contrast, the colouration was considerably different. Until reaching a length of 5 to 6 cm, the hybrids were reddish yellow but darker dorsally. The colouration of older fishes was similar to that of the crucian carp. Hybrids with common carp mothers were the same colour as the common carp.

The growth rate of the hybrids was usually lower than that of the common carp but much more rapid than that of the crucian carp (NIKOLIUKIN 1952, RUDZIŃSKI and SKÓRA 1963). In the third year of life, the hybrids attain 26.6 and 27.7 cm in *Tl* and 323.8 to 382.7 g in weight (Table 16), while common carp belonging to natural populations attained 1.3 and 1.8 kg. LIEDER (1957) was able to accelerate hybrid growth by decreasing the stocking density. The largest hybrid grew to 39.5 cm and 1330 g at the age of 5 to 6 years. Back-crossed hybrids reached 28 g during the first

Table 16: Length – weight relationship in hybrids *Cyprinus carpio* x *Carassius carassius* (Nikoliukin 1952).

Fish / Age class	*Tl* (cm) 1	2	3	4	*W* (g) 1	2	3	4
Crucian carp	6.7	10.6	13.0	15.6	-	-	-	-
Common carp x crucian carp	10.1	21.4	26.6	31.2	17.1	181.0	323.8	496.8
Crucian carp x wild carp	10.7	25.5	27.7	-	17.4	231.9	382.7	-
Wild carp	14.6	31.0	37.8 ♂ 38.7 ♀	48.0 ♂ 50.0 ♀	54	416	1323 ♂ 1806 ♀	2106 ♂ 2721 ♀

autumn and 493 g after two years. In contrast, common carp grew to only 477 g within the same time period. However, mortality of the hybrids was higher.

Most hybrids were sterile, especially males (Gomel'ski *et al.* 1985). Some females developed gonads, but these were strongly reduced. Rapidly growing individuals attained sexual maturity during the second year of life, and slowly growing ones, during the third. Among F_1 hybrids, males accounted for 87 % of the population.

Hybrids were more resistant to diseases and poor environmental conditions than the common carp. Probst (1953) found that they were considerably more resistant to carp dropsy than the common carp.

Hybrids of common and crucian carps were not of commercial interest. This was mostly due to a number of unsatisfactory features. Usually, the hybrids were characterized by slower growth, possessed more bones (Lieder 1957), and attained sexual maturity earlier than the common carp. Some culturists alleged that the flesh quality was also worse, although Nikoliukin (1946) stated that hybrid flesh was very tasty, similar to that of the crucian carp.

Distribution

The crucian carp is native throught most of Eurasia, except the northwestern and southwestern parts of Europe including Norway, northern Finland, Scotland, western France, the Iberian Peninsula, Italy, and the southern and western parts of the Balkan Peninsula. In Asia, it is present in the basins of rivers in the Siberian parts of the Arctic Ocean watershed as far east as the Lena River, but absent from China, Korea, and Japan. In Western Asia, it is absent from rivers in the catchment area of the Aral Sea, Anatolia, and eastern Transcaucasia. Among the rivers in the watershed of the Aegean Sea on the Balkan Peninsula, it seems to be native only to the Maritza (Drensky 1951). It is difficult to determine the exact limits of its native range in western Europe. It is native to France only in the Northeast, in the drainage area of the Rhine. In England, it is believed to be native only to the river systems of the eastern part from Yorkshire to Suffolk. It is nowhere native in Italy, although the "form" *humilis* was recorded from Sicily by Heckel as early as 1840 (Fig. 21).

The crucian carp has suffered a great decline in abundance during the past few decades, at least in the basins of the lower and middle Danube in Slovakia, Hungary, and Romania. It has gradually been replaced by the more competitive Prussian carp. At the beginning of the twentieth century, it was quite abundant throughout all of Romania. During the 1930s, it began to become rarer in fishponds, natural ponds, and other standing water bodies in eastern Romania, but it remained abundant in the shallow lakes on the floodplain of the Danube, in the Danube Delta, and everywhere north and west of the Carpathians. In the early 1950s, the Prussian carp was introduced north and west of the Carpathians and began to

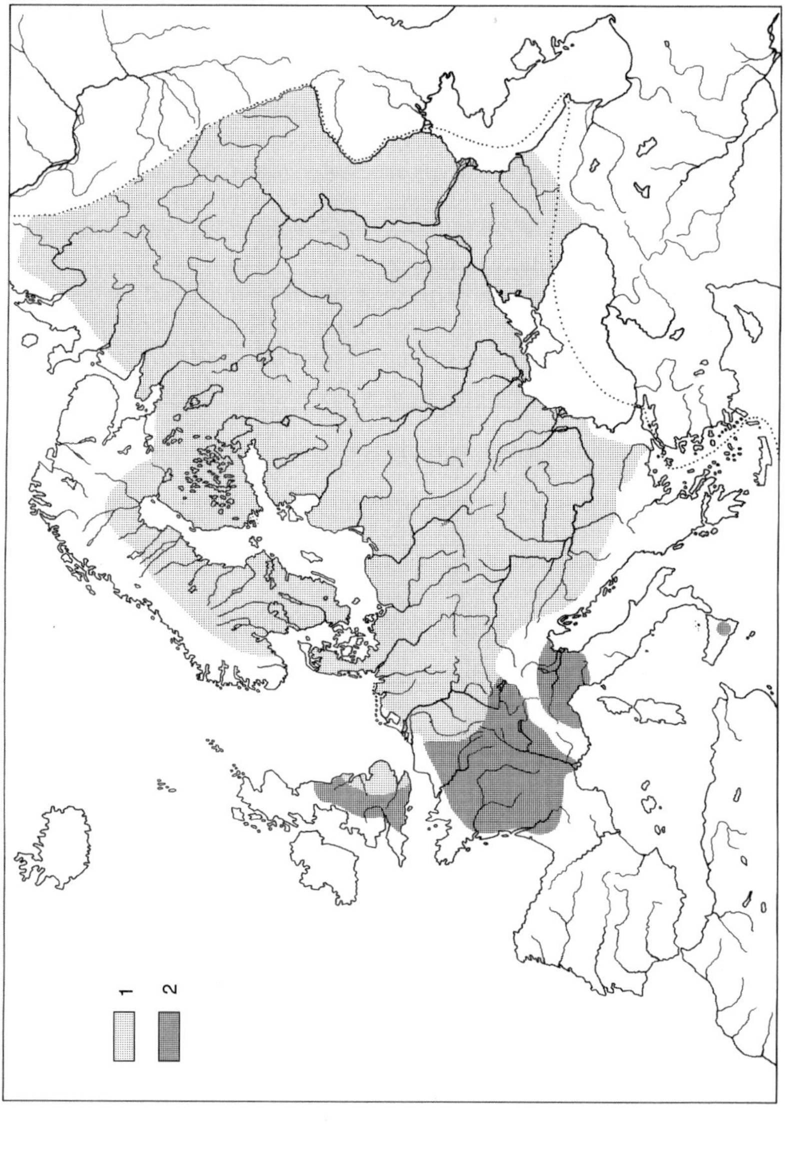

Fig. 21: Geographical distribution of *Carassius carassius* in Europe.
1 – Original area of distribution, 2 – Successful introductions in recent centuries.

replace crucian carp there, as well. During the early 1960s, it was still quite abundant in all lakes of the Danube Delta, but it was already rare in some shallow lakes on the floodplain of the Danube. Twenty years later, it was already very rare in the Danube Delta, as well. Presently, only small, isolated colonies survive, mostly in isolated water bodies.

One cause of this drastic decline in abundance is pollution of the water with nutrients, which results in an almost total destruction of the submersed vegetation, in which the crucian carp lives, feeds, and spawns. However, this decline in abundance began many years before fertilizers came into general use in agriculture. Another important cause of the decline in abundance is competition from the Prussian carp (LELEK 1987).

Introductions

The crucian carp has been introduced from its native range into most of France, Italy, and Great Britain north to Sunderland and including eastern and central Wales but excluding Scotland (MAITLAND 1972). Many publications have reported the introduction of the crucian carp to the Iberian Peninsula and even to other continents, such as North America, Australia, Northern India, and South Africa. Most of these reports actually refer to the Prussian carp, and recent publications have shown that this is the only species of its genus present on the Iberian Peninsula, in North America, and in Australia (LEE *et al.* 1980). However, the report on occurrence of the crucian carp in northern India (KULLANDER *et al.* 1999) actually refers to this species.

Ecology

The crucian carp inhabits standing freshwater bodies, such as lakes, slow flowing rivers, and all kinds of small ponds, even temporary ones. It prefers waters with low transparency, up to 2 m in Secchi depth, and a depth of not more than 3.5 m, at which the temperature near the bottom does not drop below 19.0 °C in summer. These are usually warm waters with plenty of aquatic vegetation, which are frequently called crucian carp waters. The crucian carp is characterized by its low oxygen requirements. In winter, it is able to survive oxygen deficiencies and can live in water in which no oxygen can be detected analytically. It can also survive at temperatures below 0 °C. It may even survive for a few days with its integument frozen. If water is lacking, this fish is able to bury itself in the mud and survive until water again becomes available (RYBKIN 1958).

Habitat: Shallow waters overgrown with plants are typical habitats for the crucian carp, whether they are located in small ponds or near the shores of large lakes. The crucian carp inhabits these areas during spring and summer. It moves to deeper waters during the winter.

Migrations: The crucian carp does not migrate. It only moves in spring from deeper water into shallower, warmer places, where it feeds and reproduces. It is a slow-moving fish. When scared, it moves along the bottom and may bury itself in the mud or beneath the remains of aquatic vegetation. In autumn, it moves into deeper water to spend the winter.

Hardiness: The crucian carp is one of the most hardy fishes. It can survive for several months in anoxic water at low temperatures by taking advantage of anaerobic metabolism (BLAZHKA 1960, HOLOPAINEN and HYVÄRINEN 1985). Because of this ability, it is often the only fish species present in shallow ponds of northern and eastern Europe that are regularly free of oxygen during winter (CERNY 1971, NIKOLSKI and SHUBNIKOVA 1974). It survives well for three weeks in a suspension containg 200 mg of suspended solids per l. It is resistant to variations in water temperature. In winter, it can survive for a few months at temperatures close to 0°C. In summer, the lethal temperature is as high as 38.5 °C, although temperatures within a range of 34.6 to 36.0 °C cause disturbance. The optimal summer temperature is about 27 °C.

The crucian carp is also resistant to pH change and can live at a pH in the range from 4.0 to 10.5. The optimal pH is 5.0 to 8.5 (HELLAWELL 1989). The fish is resistant to phenol and survives for 146 to 172 hours at a phenol concentration of 800 mg/l (ŁUKJANIENKO 1974). Crucian carp fry are even more resistant to phenols than adults. Crucian carp fry survive at a concentration of 25 mg/l, while the adults die after 128.7 hours, on the average.

Feeding habits: The crucian carp has been classified as a benthophage (ZAWISZA and CIEPIELEWSKI 1973), although the species has a varied diet. It is omnivorous (USPENSKAJA 1953, PREJS 1973). Its diet changes with the season and according to environmental conditions. The intestinal tracts of juvenile fishes contain rotifers and planktonic crustaceans. Adults may also feed on plankton, but some of them become typical phytophages. NOVOSEL'CEVA and NOVOSEL'CEV (1972) showed that in stomachs of fishes aged 4+ and 5+, zooplankton constituted up to 79.5 % of the food. Among this zooplankton, cladocerans were most important, with individual species accounting for the following percentages of the stomach contents: *Daphnia cristata* 42.3 %, *Chydorus sphaericus* 12.1 %, *Bosmina longirostris* 4.5 %, and *Leptodora kindti* 5.0 %. Cyclopidae were present with 14.2 %. Benthic fauna represented 20.5 % of the food mass, and among these organisms, nematodes accounted for 10.8 %, Chironomidae for 7.6 %, and Ostracoda for 1.4 %. Some fishes feed on blue-green algae, which accounted for up to 75 % of the food mass. The prey size increases with the size of the fish (PASZKOWSKI *et al.* 1989). Older fishes feed mainly on insect larvae and molluscs. In Lake Warniak, the proportion of plant components in the food of the crucian carp was the smallest in April, when it accounted for 1 % of food weight. It was greatest, constituting 11 %, in August and September (PREJS 1973). Detritus is also found in the digestive tracts of older fishes. The crucian carp practically stops feeding in winter. Most intensive feeding is observed during summer, from June through August, when it feeds for almost 24 hours a day. The diet is also most varied during this period. In spring and autumn, foods of animal origin predominate. SZLAUER (1971) stated that the crucian carp feeds on animals most effectively at night, when it feeds mainly on cladocerans and less intensively on insect larvae.

Longevity: The maximal life span of the crucian carp has been estimated to be 9 to 10 years.

Growth: The growth rate of the crucian carp may vary considerably because it is influenced by environmental conditions and population density (IVLEV 1961, NIKOLSKY 1963, BERG 1949). It was found that the crucian carp grows more rapidly in waters with a lower fish stock density. Populations of crucian carp from relatively large lakes with rich assemblages of other fishes and moderate abiotic conditions have particularly low population densities and are characterized by a predominance of very large, deep-bodied individuals, usually varying in length from 20 to 40 cm and averaging about 1 kg in weight (BERG 1949, HAMRIN 1979, HOLOPAINEN and PITKÄNEN 1985). A considerable amount of data is available on the growth of this fish (HEUSCHMAN 1938, SCHÄPERCLAUS 1953, ZAWISZA and ANTOSIAK 1961, BRYLIŃSKA *et al.* 1991). Generally, the growth rate is low. Under normal conditions, individuals at an age of four years have already attained 6.4 to 15.1 cm (Table 17) and 6 to 150 g (Table 18). In artificial ponds in which the stock density was lower and the competition for food less than in natural waters, the crucian carp attained 15.5 to 25.8 cm in *Tl* and 320 to 402 g in weight by the same age.

In Central Europe, fishes longer than 30 cm and heavier than 1 kg are very rare. There are great differences in the growth rates of individual populations. In Lake Skopy (BIAŁOKOZ 1977), the crucian carp reached 11 cm during the tenth year of life, while in Lake Przerwanki, it had already attained the same length during the third year of life. SCHÄPERCLAUS (1953) reported an example of rapidly growing crucian carps reaching this length during the second year of life, while SKÓRA (1961) observed this length during the first year of life among individuals raised in ponds. It is generally known that the crucian carp grows rapidly in relatively warm regions, while in the northern part of its natural range, in Yakutia, its growth is much slower than it is to the south (SILIN 1983).

The form *C. carassius* var. *humilis* is characterized by a much slower growth. MASLOVSKI (1940) showed that this form reaches only 13.3 cm in *Tl* during the sixth year of life, and its maximal weight is only 88 g. In a small, marshy pond

Table 17: The annual body length (*Sl* in cm) increments for crucian carp at various ages.

Place of research	Author	Sex	Age								
			I	II	III	IV	V	VI	VII	VIII	IX
L. Il'men'	DRJAGIN 1949	♂	-	-	12.7	15.0	18.3	19.4	-	-	-
L. Il'men'	DRJAGIN 1949	♀	-	-	14.1	16.5	19.9	21.0	-	-	-
L. Gir'ial	DRJAGIN 1949		-	11.0	13.3	16.0	17.8	20.2	-	-	-
Ponds	SCHÄPERCLAUS 1953	♀	5.9	17.4	24.2	25.8	-	-	-	-	-
Ponds	SCHÄPERCLAUS 1953	♂	5.9	15.2	19.0	22.6	-	-	-	-	-
Ponds	NIKOLSKI 1956		4.5	11.0	13.5	15.5	18.5	-	-	-	-
Peczora R.	NIKOLSKI 1956		2.6	5.0	7.9	9.3	10.5	12.7	-	-	-
L. Arklickie	ZAWISZA, ANTOSIAK 1961		3.3	6.2	8.9	11.9	15.7	18.1	21.7	-	-
L. Oświn	ZAWISZA, ANTOSIAK 1961		3.2	6.2	9.0	11.7	14.3	18.2	-	-	-
L. Bimbinek	ZAWISZA, ANTOSIAK 1961		3.3	6.0	8.5	12.0	15.7	19.2	-	-	-
L. Przerwanki	ZAWISZA, ANTOSIAK 1961		3.6	7.7	11.8	15.1	19.5	22.1	-	-	-
Ponds	SKÓRA 1961		11.1	18.0	20.0	20.2	20.8	-	-	-	-
Balta Brailei	PAPADOPOL cited by BĂNĂRESCU 1964		6.3-7.8	11.6-13.4	15.0-17.0	18.4	-	-	-	-	-
Dnieper R.	ZHUKOV 1965		3.3	6.7	9.0	12.4	15.4	17.2	-	-	-
Neman R.	ZHUKOV 1965		1.9	4.3	6.4	8.9	11.1	13.5	15.2	17.4	19.1
Drissa R.	PENIAZIU cited by ZHUKOV 1965		3.6	7.2	10.0	13.2	16.4	18.8	20.5	21.0	-
Dnieper R.	KOSTIUCHENKO cited by ZHUKOV 1965		1.9	4.5	6.4	7.8	9.9	-	-	-	-
L. Przerwanki	CIEPIELEWSKI 1967		4.8	7.6	11.8	15.1	17.6	19.8	21.6	23.0	24.4
Nameless-Marsh	SKRZYDŁO 1977		2.4	4.1	5.5	6.6	7.7	8.9	10.9	13.5	-
L. Pogorzelisko	BIAŁOKOZ 1977		3.7	6.1	8.6	10.8	12.2	16.9	19.7	21.2	-
L. Skopy	BIAŁOKOZ 1977		3.0	4.3	5.5	6.4	7.2	8.1	9.1	10.2	10.8

Table 18: The annual body weight (g) increments of crucian carp at various ages.

Place of research	Author	Sex	Age								
			I	II	III	IV	V	VI	VII	VIII	IX
L. Gir'ial	DRJAGIN 1949		53	98	162	205	313	-	-	-	-
L. Il'men'	DRJAGIN 1949		-	113	209	275	395	-	-	-	-
Ponds	SCHÄPERCLAUS 1953	♀	5	118	298	402	-	-	-	-	-
Ponds	SCHÄPERCLAUS 1953	♂	5	77	134	251	-	-	-	-	-
L. Przerwanki	ZAWISZA, ANTOSIAK 1961		-	20	75	150	300	425	-	-	-
Ponds	SKÓRA 1961		56	225	317	320	363	-	-	-	-
Balta Brailei	PAPADOPOL cited by BĂNĂRESCU 1964		13–20.4	69–160	181–246	299	-	-	-	-	-
Dnieper R.	ZHUKOV 1965		1.5	12.6	30	84	163	228	-	-	-
Neman R.	ZHUKOV 1965		0.3	3.2	10.2	28	58	95	140	208	276
Drissa R.	PENIAZIU cited by ZHUKOV 1965		2.2	17	49	108	206	310	402	430	-
L. Przerwanki	CIEPIELEWSKI 1967		6	19	78	138	210	309	383	446	521
Nameless-Marsh	SKRZYDŁO 1977		2	5	8	11	17	26	65	-	-
L. Pogorzelisko	BIAŁOKOZ 1977		-	-	-	39	55	86	271	304	-
L. Skopy	BIAŁOKOZ 1977		-	-	6	9	10	16	22	35	36

Table 19: Changes in numbers and biomass of a crucian carp population (Nikolski and Shubnikova 1974).

Date	n	%	% in relation to preceding year	Biomass		% in relation to preceding year
				g	%	
4.VIII.1961	342	100.0	–	1316.7	100.0	–
4.VIII.1962	217	63.0	63.0	1050.2	80.0	80.0
4.VIII.1963	291	85.0	135.0	1405.9	106.0	132.5
4.VIII.1964	293	85.3	100.4	1414.2	107.0	100.9
4.VIII.1965	213	62.0	72.7	1318.4	100.8	94.2
4.VIII.1966	158	46.0	74.2	979.7	74.0	73.4
4.VIII.1967	34	9.8	21.3	428.9	32.6	44.0
28.VII.1968	20	5.8	59.2	295.7	22.5	69.0
9.VIII.1969	31	9.1	156.9	573.4	43.8	194.7
9.VIII.1970	28	8.2	90.1	505.9	38.5	87.9
4.VIII.1971	6	1.8	21.9	112.8	8.5	22.1
12.VIII.1972	4	1.2	66.7	92.7	7.1	83.5

covering only 0.51 ha in northeastern Poland, this form had attained only 8.9 cm in *Tl* by the seventh year of life. The weight of this fish was only 65 g (SKRZYDŁO 1977).

The most rapid increase in body weight occurs during the spring, and it is fastest if the fishes consume a mixed diet of plankton and benthic organisms.

Population dynamics: The crucian carp is scarce in water bodies with a rich ichthyofauna and abundant predatory species. As the environmental conditions deteriorate, eliminating less resistant species, including predators, the crucian carp

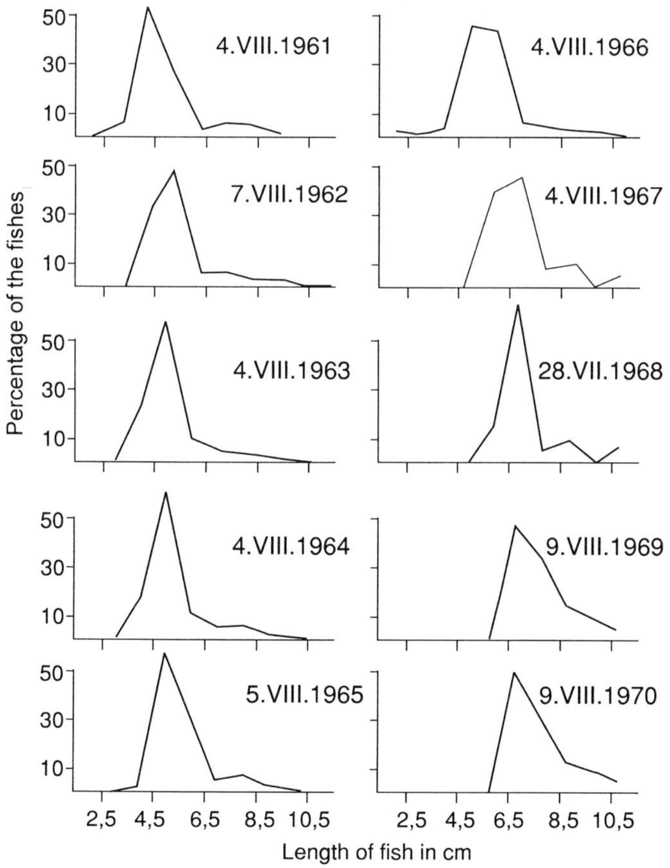

Fig. 22: Growth of *Carassius carassius* in length within a single population during successive years (NIKOLSKI and SHUBNIKOVA 1974).

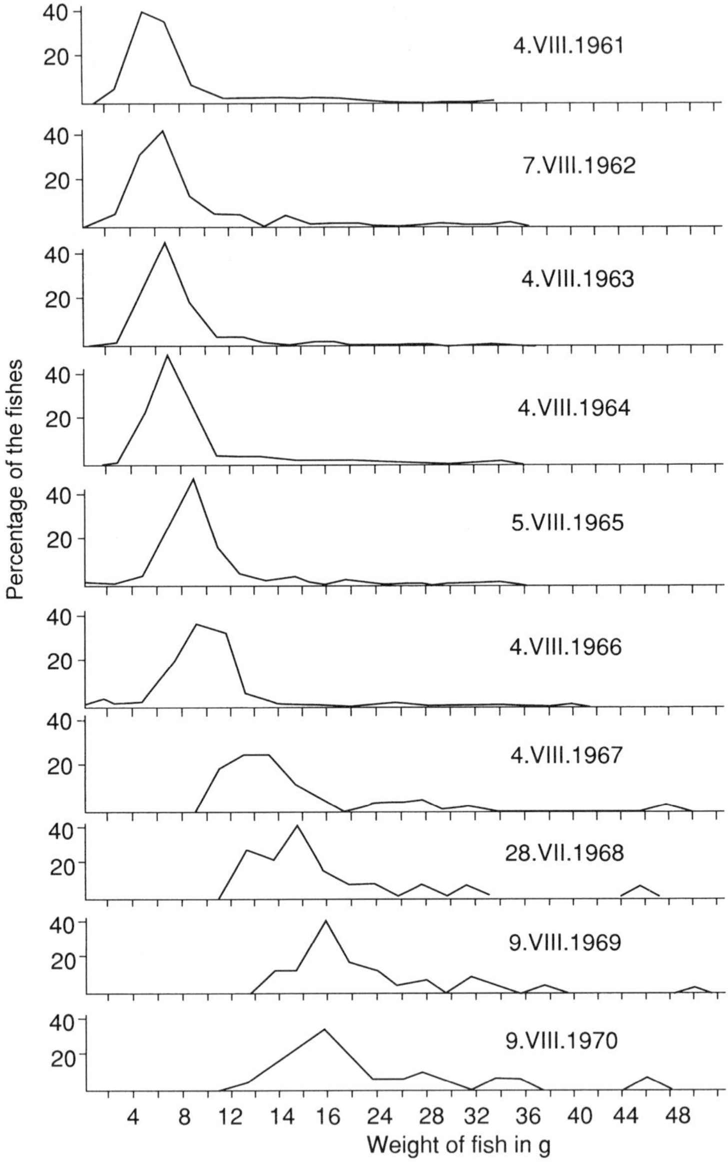

Fig 23: Growth of *Carassius carassius* in weight within a single populations in successive years (NIKOLSKI and SHUBNIKOVA 1974).

increases in abundance, and its population may become very dense when it lives in a monoculture (Zawisza and Antosiak 1961). Nikolski and Shubnikova (1974) showed that population densities can exceed 29,000 individuals per ha. Holopainen and Pitkänen (1985) caught 44,000 crucian carps, averaging 6 cm in length, in a 1.5 ha Finnish pond. Studies by Skrzydło (1977) on the catchable stock of crucian carps that were raised in a monoculture revealed that 56,569 fishes were present per ha, and their total weight was 287.6 kg. Nikolski and Shubnikova (1974) showed that in such ponds, mortality was low during the first five years of life, but it increased during the sixth year (Table 19). As the age of the fishes increased beyond six years, their total mass decreased noticeably, but less rapidly than the total number of fishes present. This is because the increases in the length (Fig. 22) and weight (Fig. 23) of the surviving fishes were considerable. Ciepielewski (1967) studied crucian carp in lakes and found the survival of this fish to be 37.9 %, 26.2 %, and 9.1 % during the sixth, seventh, and eighth year of life, respectively. The average survival rate was 24.4 %. From these figures, it can be calculated that only 1 in every 100 six-year-olds reaches the age of 10 years.

It is generally assumed that the ratio of males to females is 1:1 (Libosvarsky 1963). Deviations from this ratio usually occur because of climatic conditions and the availability of food. In Yakutia, where the water temperature is low and the amount of available food limited, the ratio of males to females is actually 1:1. In lakes with rich food resources, males tend to predominate over females, and the ratio can reach 2:1 or even 3:1 (Silin 1983).

Reproduction biology

Maturity: The crucian carp attains sexual maturity during the third to fifth year of life and usually during the fourth. Males reach sexual maturity a year earlier than females. In the southern part of the range, sexual maturity is attained earlier. For instance, in Romania, a crucian carp may have already matured during its second year (Bănărescu 1964 citing Papadopol), while in the northern part of Russia, it may not be reached until the fifth year. Maturation also occurs earlier in pond culture, as long as food is abundant (Skóra 1961).

Gonads: The fecundity of the crucian carp changes with age, length, and weight. Available data suggest that the fertility of this fish is influenced mainly by its weight. This is represented by the straight line (Fig. 24) plotted by Astanin and Podgornyi (1968). It shows that fishes weighing 40 to 70 g produce about 8,000 eggs, while the absolute fecundity of 190 to 220 g fishes is about 28,000. The absolute fecundity of females 3+ to 7+ years old averages 16,460 eggs, while their relative fecundity is 163 eggs per g of body weight. According to Drjagin (1939), crucian carp 18 to 22 cm in length have a relatively high fecundity, amounting to between 137,000

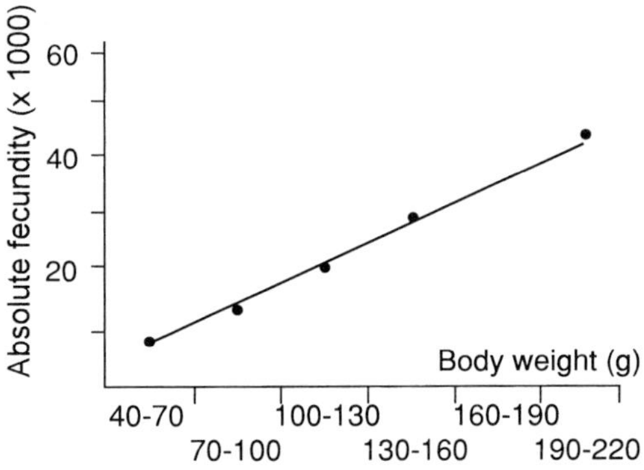

Fig. 24: Relationships between body weight and fecundity of *Carassius carassius* (ASTANIN and PODGORNYI 1968).

and 244,000 eggs. The total egg weight as a percentage of total body weight was found to be 11.7 % to 20.3 %. ZHUKOV (1965) studied this relationship among fishes from 6.5 to 22.8 cm in *Tl* and demonstrated that the weight of the eggs as a percentage of the fish weight was between 0.94 % and 12.0 % (Table 20).

Observations made in Poland (BIAŁOKOZ 1977) showed that the absolute fecundity of an individual could reach 400,000 eggs, but on the average, it was much lower. In Lake Pogorzelisko, the fecundity of fishes 10.0 to 21.0 cm in length and 34 to 320 g in weight varied from 700 to 81,500 eggs. The relative fecundity varied within a wide range from 15,000 to 700,000 eggs per kg of weight. Fishes in Lake Giriial of similar length, 11 to 20.2 cm, and weighing 53 to 313 g had a fecundity of 5,000 to 155,600 eggs, and those in Lake Ilmen weighing 278 to 500 g contained 137,000 to 207,000 eggs. Of theses eggs, only 24,000 to 90,400 were released in the first batch (DRJAGIN 1939).

Spawn: The eggs of the crucian carp are 0.8 to 1.2 mm in diameter just prior to spawning. Those in this size range are located in the caudal part of the ovary; while the eggs in the anterior part are much smaller. This is due to the fact that the crucian carp releases only a portion of its eggs at one time. The smaller eggs are released later in the reproductive season after they have attained 0.2 to 0.7 mm (DRJAGIN 1939). Detailed observations by ASTANIN and PODGORNYI (1968) showed that the diameter of the eggs in crucian carp from the Novo-Marievski Reservoir averaged 0.91 mm in March, 0.98 mm in April, 0.93 mm in May, 0.80 to 0.91 mm in June, and 0.90 mm in August. These dimensions were stable, and the average egg diameter for the whole population was 0.903± 0.004 mm. Crucian carp eggs from this reservoir are 0.07 mm smaller in diameter than goldfish eggs, and much

Table 20: Fecundity of *Carassius carassius*.

Date	*Sl*	Weight g	Weight of eggs g	N of eggs per g	Fecundity		Weight of eggs in % of fish weight
					absolute	relative	
West Dvina - PENIAZIU (ZHUKOV 1965)							
22.IV.1960	13.5	117	1.1	3000	5544	47	0.94
	16.5	237	2.6	2692	15735	66	1.1
	17.0	235	8.5	4300	68000	247	3.6
	17.2	219	6.3	3320	41080	188	2.9
	18.5	325	13.0	2100	61725	190	4.0
	19.0	375	10.0	2630	55730	149	2.7
	22.1	455	17.0	2160	92480	203	3.7
	22.8	550	18.5	2800	111092	200	3.4
Sozh - (ZHUKOV 1965)							
10.V.1960	14.0	127	6.9	-	28360	223	5.4
20.VII.1960	9.9	27	1.05	744	3300	122	1.1
Dnieper - KOSTIUCHENKO (ZHUKOV 1965)							
14.VI.1961	6.5	9.3	0.7	1150	2945	315	7.5
	7.0	9.0	0.7	1550	3521	390	7.7
	7.0	9.2	1.1	1550	4580	498	11.9
	7.2	10.5	1.2	1350	4360	415	11.4
	7.5	12	1.0	1400	3750	312	8.4
	7.7	11	1.05	870	3735	339	9.6
	7.8	15.5	2.0	800	5760	371	12.8
	8.2	16	1.0	1050	3700	231	6.3
	8.4	16	1.3	1200	4490	280	8.2
	9.3	24	2.0	950	7540	314	8.4
	10.1	29	3.5	1080	10980	365	12.0
13.IV.1961	19.8	367	14.5	2940	114800	315	3.9

smaller than eggs of the crucian carp in West Siberia, where the egg diameter is 1.05 to 1.35 mm (KRIVOSHCHEKOV 1953).

Spawning period: The crucian carp spawns during May, June, and July (DRJAGIN 1939, KRYZHANOVSKI 1949). The time of peak spawning can occur from May 20 to July 20, depending on the latitute. Spawning occurs at temperatures above 18 °C.

A number of authors have stated that the crucian carp characteristically spawns in batches, with eggs being released three to five times during the reproductive season (DRJAGIN 1939, BERG 1949, SABANEEV 1959, ZHUKOV 1965, 1988, ASTANIN and PODGORNYI 1968). The egg batches do not change much with the size and age of the fish.

Spawning sites: The crucian carp is an obligatory plant spawner. It spawns in warm, shallow places that are overgrown with aquatic vegetation. The eggs are released on plants, to which they adhere for the entire period of embryonic development.

Mating habits: The percentages of males and females in spawning populations of the crucian carp are similar. However, the number of males present at the spawning sites may be greater at the beginning of the reproductive period than at the end. The spawners become very active before egg ejection, with the males following the females into shallow water overgrown with vegetation. The spawning is rather noisy because the fish are very active. The females release the eggs, and the males eject sperm over them.

Early ontogeny: Egg incubation lasts for 4 to 7 days, depending on the temperature. Their development depends on thermic requirements of 95 to 100 degree-days being met (KRYZHANOVSKI 1949). In a fertilized crucian carp egg kept at 18 °C, the embryo develops according to the following schedule:

Time requirement (hours and minutes)	Stage reached
1:30	2 blastomere stage
1:45	4 blastomere stage
2:00	8 blastomere stage
3:00	16 blastomere stage
6:30	morula stage
8:30	blastula stage

Gastrulation begins about 20 hours after fertilization. Eye spots appear after 50 hours, and thereafter, the embryo forms. The newly hatched larva is almost colorless; it is about 5 to 6 mm in length and weighs 10 mg. Pigmentation is found only in

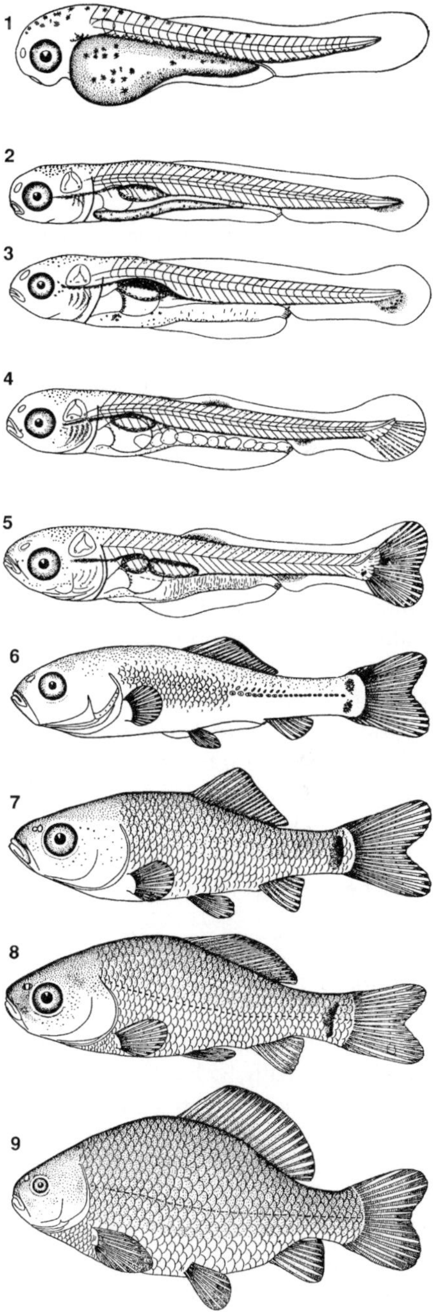

Fig. 25: Selected stages of *Carassius carassius* development (DMITRIEVA 1957). *Sl* in mm: **1** – A (4.0), **2** – B (5.4), **3** – C_1 (6.0), **4** – C_2 (7.6), **5** – D_2 (10.8), 6 – E (14.2), **7** – F (16.0), **8** – H (30.0), **9** – K (190.0).

the eyes, on the top of the head, and on the yolk sac. The larvae possess a well developed respiratory system located in the fin folds and wall of the yolk sac. This is related to the fact that the crucian carp frequently occurs in habitats with low oxygen concentrations. The larva possesses an adhesive gland on the forehead, with which they adhere to aquatic plants and remain immobile until the yolk sac has been resorbed about 10 days after fertilization. Later, they swim upward to gulp air. Swimming larvae have been observed forming small groups only four days after hatching. The larvae begin to search for food at that time. Larvae that have hatched from the first egg batch attain 15 mm by the end of July and develop scales (Fig. 25). Fry 1.5 to 4.0 cm in length are observed in autumn.

Important parasites and diseases

The crucian carp is considered to be a fish most frequently infected with parasites and particularly susceptible to diseases. A list compiled by Bykhovskaia-Pavlovskaia *et al.* (1962) includes 68 organisms reported to be parasites of the crucian carp. Among these, 6 are species of *Myxobolus*, 10 of *Dactylogyrus*, and there are a number of other parasites which attack most cyprinids in standing waters. A more recent compilation (Bayer 1985, Lom and Dykova 1989) contains fewer parasites and indicates that the genera *Dactylogyrus* and *Gyrodactylus* are of particular importance. From among the protozoans, 21 species have been listed as parasites of the crucian carp. The most pathogenic ones are *Ichthyophthirius multifiliis*, *Myxidium rhodei*, *Myxobolus dispar*, *Trichodina reticulata*,, and *Trypanosoma carassi*. Bănărescu (1964) mentioned four protozoan parasites of the crucian carp in Romania belonging to the genera *Myxobolus* and *Trichodina*. He also reported parasitic Trematoda, including *Diplostomum spathaceum*. Among the Monogenea, there are four species of *Dactylogyrus* and two of *Gyrodactylus*. The Copepoda are represented among the parasites by *Ergasilus sieboldi*.

Economic importance

The crucian carp is not produced on a large scale. A few are caught in small lakes and bogs, usually by local people. It is also cultured in ponds in order to increase the effectiveness of carp production. Hence, it is very difficult to determine annual catches of this fish. Official FAO statistics state that in the period from 1986 to 1995, the crucian carp catches in the world ranged from 93,291 to 557,541 tons. According to these statistics, the highest yields in Europe were recorded in the Czech Republic, where 49 t were produced in 1994. These data are highly questionable. They provide no information on Eastern Europe, where the crucian carp is very common and has been widely introduced into new reservoirs. For instance, in Botkin'ski Dam Reservoir, crucian carp catches from 1966 through 1973 ranged from 13 to 58 tons (PUSHKIN 1976). Crucian carp catches in Poland from 1956 to 1984 ranged from 30 to 114 tons (SZCZERBOWSKI *et al.* 1995). The annual catches in the Danube Delta and on the floodplain of the lower Danube in Romania ranged between 2,000 and 3,000 tons at the beginning of the twentieth century (ANTIPA 1909). These had decreased to 300 to 2,500 tons by the 1950s and 1960s, when the species was already replaced in many shallow lakes of the floodplain by the Prussian carp (BĂNĂRESCU 1964). Nevertheless, the crucian carp continued to be the most abundant species in catches from lakes in the Danube Delta until the middle or end of the 1970s. It presently has no economical importance anywhere in the drainage area of the middle and lower Danube. Therefore, it can be roughly estimated that catches of the crucian carp in Europe is several thousand tons. The FAO statistics on crucian carp catches in mainland China, Taiwan, Japan, and the Republic of Korea must certainly refer to the Prussian carp.

The flesh of the crucian carp has many small intramuscular bones. In spite of this, it is quite popular on the market. Its flesh is very tasty and has a high nutritive value. The caloric value of its flesh is 810 kcal/kg, which is less than the 1189 kcal of goldfish flesh and the 1103 kcal of wild common carp flesh. The caloric value of the whole crucian carp is 437 kcal per kg of total fish weight (SUKHOVERKHOV 1951).

Literature

ANTIPA, G., 1909: Fauna ichtiologica a Romaniei. Bucuresti. **XVI.** – Academia Romana. 294 pp.

ASTANIN, L. P., and M. I. PODGORNYI, 1968: [Special characteristics of fecundity of *Carassius carassius* (L.) and *C. auratus gibelio* (Bloch).]. – Vopr. Ikhtiol. **8**(2): 266–273. (in Russian).

BĂNĂRESCU , P., 1964: Pisces – Osteichthyes. – Fauna Republicii Populare Romine 13: 1–961.

BAYER, O. N., 1985: (Key to the parasitic fauna of freshwater fish in the USSR.). Izd. "Nauka", Leningrad. 385 pp. (in Russian).

BERG, L. S., 1909: (Fishes of Amur Basin.). – Zap. Akad. Nauk (8), po fizikomat. otd. **24**(9): 270 pp. (in Russian).

BERG, L. S., 1949: (Freshwater fishes of the USSR and the neighbouring countries.). Vol. 2. I. Izd. Akad. Nauk SSSR, Moscow – Leningrad. pp. 496–925. (in Russian).

BIAŁOKOZ, W., 1977: [The effect of selected individual and population characters on the absolute and relative fecundity of bream (*Abramis brama* L.), crucian carp (*Carassius carassius* L.) and guppy (*Lebistes reticulatus* Peters).]. – Bibl. ART, Olsztyn. 62 pp. (in Polish).

BIELZ, E. A., 1853: Übersicht der lebenden Fische Siebenbürgens. – Verh. u. Mitth. Siebenb. Ver. Naturwiss. **4:** 172–185.

BLAZHKA, P., 1960: [Notes on the biology of common crucian carp, *Carassius carassius* (L.) morpha *humilis* Heckel.]. – Zool. Zh. **39**(9): 1384–1389. (in Russian).

BRYLIŃSKA, M., E. BRYLIŃSKI, J. TERLECKI, W. BIAŁOKOZ, J. RADZIEJ, B. MŁYNIEC, R. BARTEL, W. KOPIEJEWSKA, M. TADAJEWSKA, T. KRZYWOSZ, and K. GORYCZKO, 1991: (Freshwater fishes of Poland.). PWN, Warsaw. 429 pp. (in Polish).

BYKHOVSKAIA-PAVLOVSKAIA, I. E., A. W. GUSEV, M. W. DUBININA, N. A. IZIUMOVA, T. S. SMIRNOVA, I. L. SOKOLOVSKAIA, G. A. SHTEIN, S. S. SHUL'MAN and U. M. EPSHTEIN, 1962: Key to the parasitic fauna of freshwater fish in the USSR. Izd. Akad. Nauk SSSR. Moscow – Leningrad. 776 pp. (in Russian).

CERNY, K., 1971: Growth and mortality of crucian carp *Carassius carassius* (L.) morpha *humilis* Heckel 1840, in the natural pond Mansfeldova in the Central Elbe Basin. – Vést. Cesk. Sp. Zool. **35:** 251–260.

CHIGARZH, I., 1958: [Comments on crucian carp systematics (*Carassius carassius* m. *humilis* Heckel, 1840).]. – Vopr. Ikhtiol. **11:** 136–141 (in Russian).

CIEPIELWSKI, W., 1967: Growth and survival of an isolated population of crucian carp (*Carassius carassius* L.) in a small unexploited lake. – Rocz. Nauk Rol. **90**(2): 239–248. (in Polish).

DMITRIEVA, E. N., 1957: (Morpho-ecologic analysis of two species of crucian carp.). – Tr. Inst. Morf. Zhivt. **16:** 102–170. (in Russian).

DRENSKY, P., 1951: (Fishes of Bulgaria.). Izd. B'lgar. Akad. Nauk., Sofia. 340 pp. (in Bulgarian).

DRJAGIN, P. A., 1939: (Spawning by batches of Cyprinidae.). – Izv. VNIORKH **21:** 81–119. (in Russian).

DRJAGIN, P. A., 1949: (Crucian carp. Commercial fishes of the USSR.). Pishchepromizdat. Moscow: 443–448 (in Russian).

DYBOWSKI, B., 1862: Versuch einer Monographie der Cyprinoiden Livlands. Dorpat; 217 pp.

DYBOVSKI, B., 1877: (Fishes of the Amur River system.). – Izv. Sib. otd. IRGO **8**(1–2): 1–22. (in Russian).

FITZINGER, L. J., 1832: Prodrom. Faune Austriae. Wien. 334 pp.

GOMEL'SKI, B. I., O. V. EMEL'IANOVA, and A. V. REKUBRATSKI, 1985: (Production and some biological properties of amphiploidal hybrids of crucian carp and common carp.). – Dokl. Akad. Nauk. SSSR. **301**(5): 1210–1213. (in Russian).

GRATZIANOV, V. I., 1907: (A review of the fish in Russian empire). Moscow. 567 pp. (in Russian).

HAMRIN, S., 1979: Biomass of tench, perch, crucian carp and pike in a small, eutrophic lake in southern Sweden. – Jyväs. yliop. biol. lait. tied. **19:** 57–67.

HECKEL, J., 1835: Annalen des Wiener Museum. **I**:, Vienna, 223 pp.

HECKEL, J., 1837: Ichthyologische Beiträge zu den Familien der Cottiden, Scorpeaeniden, Gobioiden und Cyprinoiden. Ann. Wiener Mus. Naturgesch., **2:** 143–164, pls. 8, 9.

HECKEL, J., and R. KNER, 1858: Die Süsswasserfische der Östreichischen Monarchie mit Rücksicht auf die angränzenden Länder. Leipzig. 388 pp.

HELLAWELL, J. M., 1989: Biological indicators of freshwater pollution and environmental management. Elsevier Applied Science, London and New York. 546 pp.

HEUSCHMAN, O., 1938: Karausche, *Carassius carassius* L. und Giebel, *Carassius auratus gibelio* (Bloch). – Z. f. Fischerei Bd **36:** 249–285.

HOLOPAINEN, I. J., and H. HYVÄRINEN, 1985: Ecology and physiology of crucian carp (*Carassius carassius* L.) in small Finnish ponds with anoxic conditions in winter. – Verh. internat. Verein. Limnol. **22:** 2566–2590.

HOLOPAINEN, I. J., and A. K. PITKÄNEN, 1985: Populations size and structure of crucian carp (*Carassius carassius* L.) in two small, natural ponds in Eastern Finland. – Ann. Zool. Fennici. **22:** 397–406.

IVLEV, V. S., 1961: Experimental ecology of the feeding of fishes. Yale Univ. Press, New Haven. 302 pp.

KESSLER, K. F., 1856: (Natural history of the Kiev scientific region.). Ryby, Kiev. 98 pp. (in Russian).

KIRILLOV, F. N., 1972: (Fishes of Yakutia.). Nauka, Moscow. 360 pp. (in Russian).

KRIVOSHCHEKOV, G. M., 1953: (Crucian carp of West Siberia.). – Tr. Barabinsk. otdel. VNIORKH **6**(2): 71–124. (in Russian).

KRYZHANOVSKI, S. G., 1949: (Ecological and morphological regularities of the development of cyprinids, misgurnes and catfishes.). Izd. Akad. Nauk SSSR. Moscow – Leningrad. pp. 5–332. (in Russian).

KULLANDER, S. O., F. FANG, B. DELLING, and E. AHLANDER, 1999: The fishes of the Kashmir Valley. In: L. NYMAN (Ed.), River Jhelum, Kashmir Valley – Impacts on the aquatic environment. Swedmar, Stockhom: 99–167.

LABAT, R., R. HAFEZ, and E. QUILLIER, 1983: Cytogenetic studies in some species of cyprinid fishes. – Rocz. Nauk Rol., s. H **100:** 89–93.

LEE, D. G., C. R. GILBERT, C. H. HOCULT, E. R. JENKINS, D. E. MCALLISTER, and J. R. STAUFFER, Jr. (Eds.), 1980: Atlas of North American freshwater fishes. North Carolina State Museum of Natural History, Raleigh. North Carolina, 854 pp.

LELEK, A., 1987: The Freshwater Fishes of Europe. Vol. 9, Threatened Fishes of Europe, p. 168–170. AULA-Verlag, Wiesbaden.

LEONHARDT, E., 1904: Die Bastarde der deutschen Karpfenähnlichen Fische. Neumann-Neudamm, Leipzig/Radebeul. 58 pp.

LEŚNIEWSKI, P. E., 1837: (National fisheries or natural history of local fish.). Warsaw. 360 pp. (in Polish).

LIBOSVARSKY, G., 1963: Zum Geschlechtsverhältnis bei der Karausche *Carassius carassius* (L.). – Zool. Anz. **170**(5–6): 191–197.

LIBOSVARSKY, J., 1964: Über die Veränderungen der Körperhöhe bei der Karausche (*Carassius carassius* L.). – Z. Fischerei Hilfswiss (N.F.) **12**(1/2): 1–12.

LIEDER, U., 1957: Die Ergebnisse der in Jahre 1956 durchgeführten Karpfen – Karauschenkreuzungen. – Zeitschr. f. Fischerei **6**(1–7): 283–299.

LINNAEUS, C., 1758: Systema Naturae per Regna tria Naturae, secundum classes, ordines, genera, species, cum characteribus, differentiis, synonymis, locis. X. ed. Holmiae. 316 pp.

LOM, J., and I. DYKOVA, 1989: (Protozoological parasites of commercially exploited fish.). Cesky rybarsky svaz., Prague. 315 pp. (in Czech).

ŁUKJANIENKO, W. I., 1974: (Fish toxicology.). PWRiL, Warsaw. (in Polish).

MAITLAND, P. S., 1972: Key to British Freshwater Fishes. Freshwater Biological Association. Publ. 27. 139 pp.

MAKINO, S., 1939: The chromosomes of the carp *Cyprinus carpio*, including those of some related species of Cyprinidae for comparison. – Cytologia **9:** 430–440.

MASLOVSKI, A. D., 1940: (Materials to study crucian carp in the lakes of the Don River basin.). – Tr. Donec. gidrobiol. A. I. Charkov, pp. 145–163 (in Russian).

MEN'SHIKOV, M. I., and A. I.REVNIVYKH, 1937: (Notes on the biology of commercially exploited fish and lakes of the Vagoisky region, Omskoi area.). – Tr. Biol. inst. Permsk. Univ. **7**(3–4): 143–175. (in Russian).

MIASKOWSKI, M., 1959a: Variabilitätsstudien an der Wirbelsäule von Cypriniden. – Arch. Fischereiwiss. B. **10:** 31–52.

MIASKOWSKI, M., 1959b: Variabilitätsstudien an den Halswirbeln, dem Weberschen Apparat und dem Schlundgebiß von Cypriniden. – Arch. Fischereiwiss. B. **10:** 119–144.

NIKOLIUKIN, N. I., 1946: Experimental Progeny Obtained from Intergeneric Hybrids of Cyprinidae. – Compl. Rend. (Doklady) Acad. Sc. **51**(9): 737–740.

NIKOLIUKIN, N. I., 1952: (Inter-species hybridisation of fish.). Saratovskoe Obl. Gos. Izdat. 312 pp. (in Russian).

NIKOLSKI, G., 1956: (Detailed ichthyology.). PWN, Warsaw. 624 pp. (in Polish).

NIKOLSKY, G. V., 1963: The ecology of fishes. Academic Press, London. 352 pp.

NIKOLSKI, G.W., and N. G. SHUBNIKOVA, 1974: (The character of some quantitative relations and biological indices in a population of crucian carp *Carassius carassius* L.). – Vopr. Ikhtiol. **14**(4): 581–588. (in Russian).

NILSSON, S., 1832: Prodromus Ichthyologiae Scandinaviae. Berlingska Tryckeriet, Lund, 32 pp.

NORDMANN, A. N., 1840: Observations sur la fauna pontique. In: A. DEMIDOFF (Ed.), Voyage dans la Russie meridionale et la Crimée. B. Paris, 756 pp.

NOVOSEL'CEVA, R. I., and G. E. NOVOSEL'CEV, 1972: Feeding of goldfish acclimated in Lake Karasove (Bol'soi Solevecki Ostrov). – Trudy Sevniorkh. **6**: 90–93. (in Russian).

PASZKOWSKI, C. A., W. M. TON, and I. J. HOLOPAINEN, 1989: An experimental study of body size and food size relations in crucian carp, *Carassius carassius*. – Env. Biol. Fish. **24**(4): 275–286.

PREJS, A., 1973: Feeding of introduced and autochthonous non-predatory fish. – Ekologia Polska **21**(30): 465–505.

PROBST, E., 1953: Die Beschuppung des Karpfens. – Münchener Beiträge zur Abwasser-, Fischerei- und Flußbiologie **1**: 150–227.

PUSHKIN, Iu. A., 1976: (Characteristics of a commercial stock of crucian carp and the dynamics of its exploitation in Botkin.). – Rybokhoz. Izuts. Vnutr. Vod. **17**: 47–56. (in Russian).

REISINGER, I., 1830: Specimen Ichthyologiae sistems Pisces aquarum dulcium Hungariae. Budae, Typog. R. Univers. Hungaria.

RIABOV, I. N., 1979: (Hybridisation of different representatives of the subfamilies of Cyprinidae family.). – Vopr. Ikhtiol. **19**(6): 1025–1041. (in Russian).

ROULE, L., 1914: Traité de la Pisciculture et des pêches. Paris, 734 pp.

RUDZIŃSKI, E. and S. SKÓRA, 1963: Die Beerbung einiger Elternmerkmale durch den Mischling Karpfkarausche. – Acta Hydrobiol. Kraków **5/4**: 343–352.

RYBKIN, N. I., 1958: (Ability of crucian carp to bury in the bottom.). – Ryb. Khoz. **8**: 34–37. (in Russian).

SABANEEV, L. P., 1959: (Life and catches of freshwater fish.) Gos. izd. sel. lit. Ukrain. SSR, Kiev. 667 pp. (in Russian).

SCHÄPERCLAUS, W., 1953: Die Züchtung von Karauschen mit höchster Leistungsfähigheit. – Z. f. Fisch. **1/2**: 19–69.

SILIN, B. V., 1983: (Details of the species status of crucian carp in Yakutian water bodies.). – Vopr. Ikhtiol. **23**(2): 186–192. (in Russian).

SKÓRA, S., 1961: Karausche (*Carassius carassius* L.) aus der Teichwirtschaft Golysz. – Acta Hydrobiol. Kraków **3**(2-3): 91–112.

SKÓRA, S., 1962: Die Karpfkarausche und ihre Zucht in Teichen. – Acta Hydrobiol. Kraków **4**(3–4): 245–266.

SKÓRA, S., 1968: Auswirkung der Rückkreuzung zwischen Mischlingen von Karpfen (*Cyprinus carpio* L.) und Karauschen (*Carassius carassius* L.) auf Wachstum und Variabilität der Nachkommenschafts-Populationen. – Acta Hydrobiol. Kraków **10**(1–2): 239–257.

SKRZYDŁO, A., 1977: (Water chemistry, food resources and ichthyofauna of a disappearing forest pool in Stary Dwór forest district.). Bibl. ART, Olsztyn. 55 pp. (in Polish).

SMOLIAN, K., 1920: Merkbuch der Binnenfischerei. Fisch. Gmbh., Berlin. 1228 pp.

STAFF, F., 1950: (Freshwater fishes of Poland and the neighbouring countries.). TeiM, Warsaw. 286 pp. (in Polish).

SUKHOVERKHOV, F. M., 1951: Economic importance of Prussian carp in pond fish culture. - Ryb. Khoz. **4:** 36–37. (in Russian).

SZCZERBOWSKI, A., 1995: [Variability of biotaxonomic characters in crucian carp, *Carassius carassius* (Linnaeus, 1758).]. IRS, Olsztyn. 56 pp. (in Polish).

SZCZERBOWSKI, J. A., B. ZDANOWSKI, A. KRÜGER, W. DEMBIŃSKI, K. GORYCZKO, K. LOSSOW, R. BARTEL, M. STUDNICKA, and J. ZAMOJSKI, 1995: Inland Fisheries in Poland. IRS, Olsztyn. 544 pp.

SZCZERBOWSKI, J. A., and A. SZCZERBOWSKI, 1996: (The Carassius.). IRS, Olsztyn. 175 pp. (in Polish).

SZCZERBOWSKI, A., M. J. ŁUCZYŃSKI, Z. ZAKES, and M. SZKUDLAREK, 1997a: Dwarf crucian carp (*Carassius carassius* L.) population in a forest pond. - Arch. Ryb. Pol. **5:** 155–166.

SZCZERBOWSKI, A., M. J. ŁUCZYŃSKI, Z. ZAKĘŚ, and M. SZKUDLAREK, 1997b: Maturation and growth of a stunted form of crucian carp, *Carassius carassius* (L.), in natural and controlled conditions. - Pol. Arch. Hydrobiol. **44**(1–2): 171–180.

SZLAUER, B., 1971: [Ingestion of phytophilic fauna by crucian carp, *Carassius carassius* (L.).]. - Zesz. Nauk. WSR, Szczecin **35:** 85–103. (in Polish).

USPENSKAJA, V.D., 1953: (Feeding of crucian carp in dam reservoirs of the Khazma River basin.). -Tr. vsesoiuzn. gidrobiol. Ob. **5:** 347–364. (in Russian).

VALENTA, M., A. STRATIL, and L. KÁLAL, 1977a: Polymorphism and heterogeneity of transferrins in some species of the fish family Cyprinidae. - Anim. Blood Grps biochem. Genet. **8:** 93–109.

VALENTA, M., V. ŠLECHTA, V. ŠLECHTOVA, and L. KÁLAL, 1977b: Genetic polymorphism and isoenzyme patterns of lactate dehydrogenase in tench (*Tinca tinca*), crucian carp (*Carassius carassius*) and carp (*Cyprinus carpio*). - Anim. Blood Grps biochem. Genet. **8:** 217–230.

VARPAKHOVSKI, N., 1886: (A review of the ichthyofauna of Kazan District.). - Zap. Akad. Nauk, L. **2**(3): 1–70. (in Russian).

VASIL'EV, V. P., 1985: (Evolutionary karyology of fish.). Izd. Nauka, Moscow, 300 pp. (in Russian).

VASIL'EV, V. P., and E. D. VASIL'EVA, 1984: (Goldfish, *Carassius carassius,* a diploidal or a triploidal species.). - Dokl. AN SSSR **283**(1): 228–230. (in Russian).

WAŁECKI, A., 1864: (Systematic review of native fish.). Warsaw. 115 pp. (in Polish).

WAŁECKI, A., 1889: (Comments on our ichthyofauna.). - Pam. Fizjogr. Warszawa **9**: 275–295. (in Polish).

WITKOWSKI, A., 1984: (Analysis of the ichthyofauna in the Biebrza River basin.). - Cz. I. Acta Univers. Wratislaviensis, No. **646**. Wroclaw. (in Polish).

ZAWISZA, J., and B. ANTOSIAK, 1961: [Rate of growth of crucian carp (*Carassius carassius* L.) in lakes near Wegorzewo.]. - Rocz. Nauk Rol. **77-B-2:** 527–548. (in Polish).

ZAWISZA, J., and W. CIEPIELEWSKI, 1973: [Changes of the autochthonous ichthyofauna due to the introduction of carp (*Cyprinus carpio* L.) and bream (*Abramis brama* L.).]. - Ekologia Polska **21**(28): 423–444.

Zhukov, P. I., 1965: (Fishes of Belorussia.). Nauka i Tekhnika, Minsk. 414 pp. (in Russian).
Zhukov, P. I., 1988: (Guide to the ecology of freshwater fish.). Nauka i Tekhnika, Minsk. 310 pp. (in Russian).

Cyprinus Linnaeus, 1758

E = carp; F = carpe; G = Karpfen; R = sazan, karp

Petru Bănărescu, Vlastimil Baruš and Milan Peňáz

Synonyms

Cyprinus Linnaeus, 1758, X: 320 (Type species: *Cyprinus carpio* Linnaeus, 1758)
Mesocyprinus Fang, 1936: 699 (Type species: *Cyprinus micristius* Regan, 1906)

Etymology: See etymology of the species.

Description

Typical cyprinid fishes, rather sluggish, heavy-bodied, free swimming, usually with two pairs of barbels. Eye situated along or above the body axis. Pharyngeal teeth of masticatoriid form (Fig. 26), arranged in three lines, usually 1.1.3–3.1.1, rarely 1.2.3–3.2.1. The upper jaw is protractile; the gill membranes are attached to the isthmus. Gill rakers are not fused. The dorsal fin is elongated (III–IV, 11–22); the last

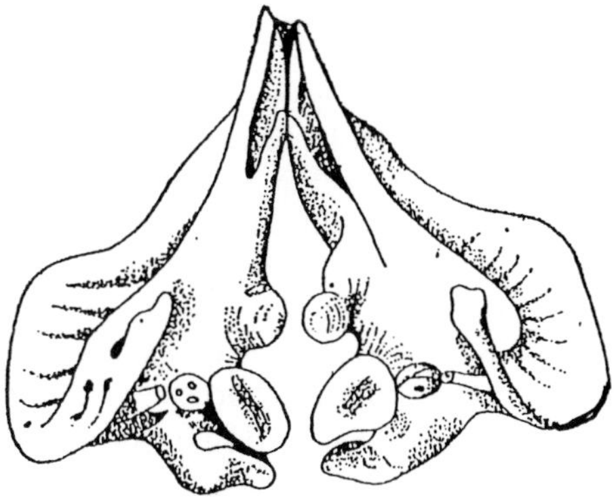

Fig. 26: Pharyngeal teeth of the wild common carp, *Cyprinus carpio carpio*, from the Danube River. After Bănărescu (1964).

unbranched dorsal ray is serrated along its posterior margin. The anal fin is short with 5 or 6 branched rays and a serrated, unbranched ray. The origin of A is anterior to the vertical line passing through the end of the D. The lateral line is continuous along the central part of body; the scales are large, with 26 to 45 in the lateral line.

The bony elements on the surfaces of the neurocranium and operculum are characterized by a distinct sculpture. The supraorbital sensory canal joins the postorbital canal in the region of the os pteroticum. The intestine is relatively long and can reach two and a half to three times the body length. Its relative length increases with age. The peritoneum is silvery to whitish. The brain is characterized by a strong development of the lobus facialis and the lobi vagales as well as by a median and posterior divergence of the two halves of the tectum opticum, which are connected by a thin membrane, beneath which the valvula cerebeli is visible (Fig. 27).

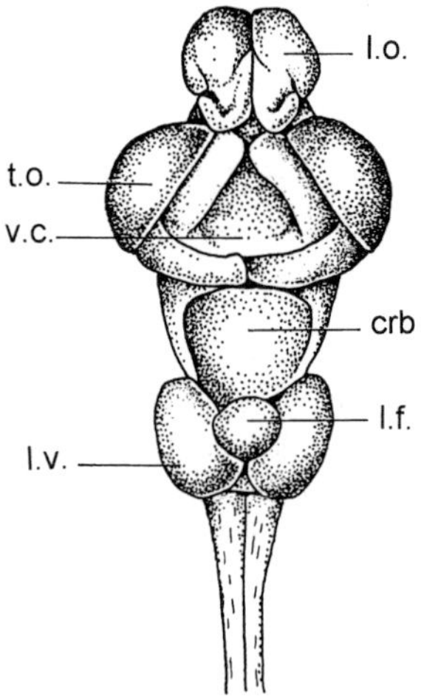

Fig. 27: The brain of *Cyprinus carpio,* as presented in an unpublished thesis by P. Bănărescu.
l.o. = lobi olfactorii, **t.o.** = tectum opticum; **v.c.** = valvula cerebeli; **crb** = cerebelum; l.f. = lobus facialis; **l.v.** = lobi vagales.

Fossil forms of this genus are known from the Neogene in Siberia, the Pliocene in China, and the Holocene in Europe.

The taxonomy within this genus is controversial. Fang (1936) divided the genus *Cyprinus* into two subgenera: *Cyprinus* sensu stricto with the type species *Cyprinus* (*C.*) *carpio* (L.) and *Mesocyprinus* Fang, 1936, with the type species *Cyprinus* (*M.*) *micristius* (Regan, 1906). Bănărescu (1964) considered the genus *Procypris* Lin, 1933, as well as the subgenera *Cyprinus* and *Mesocyprinus* to be synonyms, an opinion which can no longer be accepted. Chinese authors, such as Wu (1977), Wang (1979), Zhou and Chu (1986), and Zhou (1989), continue to recognize the two subgenera proposed by Fang (1936) and consider the genus *Procypris* to be distinct.

Opinions on the number of species in the genus *Cyprinus* vary greatly. NICHOLS (1943), studying original material, recognized five species, of which three originated in Chinese territory: *C. carpio* L. from Tungting Lake, Hunan; Shantung; Swatow; Fukien; the vicinity of Canton, and Hainan; *C. fossicola* Richardson, 1864, from Canton; *C. micristius* Regan, 1906, from Yunnan; *C. pellegrini* Tchang, 1933, from Tunghai, Yunnan; and *C. rabaudi* Tchang, 1930, from Katin and Fontou in Szechwan. The last two of these were reported from literature citations only. Concerning *C. pellegrini*, a carp without barbels, NICHOLS (1943) remarked that it is hard to determine whether this form is a "good" species or an abnormal form of *C. carpio*. About the nominal taxon, *C. yunnanensis,* he mentioned that it was described by TCHANG (1933) from the same locality as *C. pellegrini* as a carp with only a small pair of maxillary barbels, making it intermediate between *C. carpio* and *C. pellegrini*. Justified doubts have been raised concerning the validity of both *C. pellegrini* and *C. yunnanensis*. BERG (1949) recognized the existence of only three species within the genus *Cyprinus*, namely *C. carpio, C. micristius,* and *C. rabaudi*. In contrast, BĂNĂRESCU (1964) reported that there are five species in the genus, of which *C. carpio* is native to Europe and the Far East, while the other four are exclusively East Asian: *C. micristius* Regan, 1906; *C. rabaudi* Tchang, 1930; *C. pellegrini* Tchang, 1933; and *C. merus* Lin, 1933, from the West River in Kwangsi and the Fu River near Wuchow in Kwangsi. Actually, *C. rabaudi* and *C. merus* belong to the genus *Procypris*.

The taxonomic disagreement on the number of species and subspecies in the genus *Cyprinus*, illustrated only for southern China, is apparent from the data compiled by ZHOU and CHU (1986) in Table 21. The data were obtained in Yunnan Province, which is drained by rivers in four catchment areas, three of which are East Asian and one, the Lankangjiang or Mekong, is Southeast Asian. This review is based on the opinions of several authors who have reported the number of species to be between 9 and 13, depending upon the number of subspecies they recognize. ZHOU and CHU (1986) reported that only one nominal taxon, *C. crassilabris* Chu et Hwang, 1977, is a synonym of *C. carpio chilia* Wu, 1977.*

At this point, it should be mentioned that BALON (1995a, b) reported the possibility of the Chinese carps being derived from introduced European fishes and of many species of *Cyprinus* recognized by Chinese authors being merely domesticated or feral forms. A different opinion was expressed by KOTELATT (1997), who introduced the hypothesis of radiation in southern China and adjacent areas, especially in lakes of Yunnan and Myanmar (= Burma), accompanied by specialization for occupying different feeding niches, which is at least as reasonable as the hypothesis of BALON (1995a, b). Ongoing DNA analysis may confirm or refute these

---

*Actually, in China are recognized 11 species within the subgenus *Cyprinus* and additional five species within in the subgenus *Mesocyprinus* (Yue Peigi et al., 2000: Fauna Sinica. Osteichthyes, Cypriniformes III. Science Press, Beijing, China, pp. 1–661, in Chinese).

Table 21: Comparison of the systematic positions of species and subspecies in the genus *Cyprinus* in Yunnan Province, China, according to various authors. The information provided by ZHOU and CHU (1986) was modified for this table.

CHEN and HWANG 1977	WANG 1979	ZHOU and CHU 1986
C. carpio rubrofuscus	*C. carpio*	*C. carpio rubrofuscus*
C. carpio chilia	*C. chilia*	*C. carpio chilia* *Syn. C. crassilabris*
C. micristius micristius	*C. micristius*	*C. micristius micristius*
C. micristius fuxianensis	*C. fuxianensis*	*C. micristius fuxianensis*
C. yilongensis	*C. yilongensis*	*C. yilongensis*
C. crassilabris	*C. crassilabris*	
C. megalophthalmus	*C. megalophthalmus*	*C. megalophthalmus*
C. longipectoralis	*C. longipectoralis*	*C. longipectoralis*
C. pellegrini pellegrini	*C. pellegrini*	*C. pellegrini*
C. pellegrini barbatus	*C. barbatus*	*C. barbatus*
C. yunnanensis yunnanensis	*C. yunnanensis*	*C. yunnanensis*
C. yunnanensis daliensis	*C. daliensis*	*C. daliensis*
C. ilishaestomus	*C. ilishaestomus*	*C. ilishaestomus*

phylogenetic hypotheses. Further insights into these relationships may be derived from parasitological data presented by NIE (1995), who studied parasitic worms from six lakes in the highlands of Yunnan Province, Lakes Fuxian, Xingyun, Tunghai, Erhai, Lugu, and Chenghai, and reported only the common carp, *C. carpio* L., as a host.

The above review of opinions on the systematics and number of species in the genus *Cyprinus*, does not, in fact, substantially influence the taxonomic stability of the wild European common carp, *Cyprinus carpio* L. In the following description of this species, we therefore employed a well supported and easy to survey system of intraspecific classification recognizing three subspecies, as proposed by KIRPICHNIKOV (1967).

Affinities

The closest relatives of *Cyprinus* are *Procypris* Lin, 1933, with the type species, *P. merus* Lin, and *Paraprocypris* Fang, 1936, with the type species, *P. papillososlabiatus* Fang (= *Cyprinus rabaudi* Tchang, 1930). *Procypris* differs from *Cyprinus* by its pharyngeal tooth formula of 4.3.2 - 2.3.4, slightly hooked teeth, and papillose lips. Its two species inhabit East Asia. More distant relatives are *Carassioides* Oshima, 1926, a monospecific genus in East Asia, and *Carassius* Jarocki, 1822 (p. 1).

Distribution

The present range of this genus is disjunct, encompassing Europe and Anatolia as well as Eastern and parts of Southeastern Asia. It is absent from the Arctic Ocean watershed in Siberia, where it lived prior to the Ice Age. For its range in Europe, see the chapter on *C. carpio*. It is native throughout continental East Asia from the Amur to the Song Koi River. It is now present in Japan and the other East Asian islands, where it may have been introduced. In Southeastern Asia, it is present in but perhaps not native to the drainage areas of the Mekong and Salween Rivers. The main species, *C. carpio*, is widely distributed throughout East Asia. Four other species have been described as endemic to the upper section of the Xijiang or Pearl River in Yunnan province, China: *C. yilongensis*, *C. pellegrini*, *C. yunnanensis*, and *C. ilihaestomus*. A fifth one, *C. micristius*, lives in the Changjiang and Xijiang drainage areas.

Five other species have been described as endemic to the Lankangjiang River, as the uppermost part of the Mekong is called. These are *C. barbatus*, *C. longipectoralis*, *C. daliensis*, *C. megalophthalmus*, and *C. crassilabris*. *C. intha* is known from the Salween Basin in Burma. It has been suggested that the five species from the Lankangjiang River are actually feral forms of domesticated *C. carpio*. However, it is most difficult to accept this hypothesis in the case of *C. intha*.

Cyprinus carpio (Linnaeus, 1758)

E = Common carp, carp, Wild carp, F = carpe commune, carpe,
G = Wildkarpfen, Karpfen, R = sazan

VLASTIMIL BARUŠ, MILAN PEŇÁZ and KLAUS KOHLMANN

Synonyms*

Cyprinus carpio Linnaeus, 1758: 320 (Europe)
Cyprinus rex cyprinorum Bloch, 1782: 107, 193 (near Torgau, Dresden, Germany)
Cyprinus nudus Bloch, 1784: 178 (Silesia, Poland)
Cyprinus rex Walbaum, 1792: 24 (near Torgau, Dresden, Germany)
Cyprinus carpio var. *caspius* Walbaum, 1792: 17 (mouth of Volga and Don Rivers)
Cyprinus viridi-violaceus Lacépède, 1798: 547 (China, from a Chinese drawing)
Cyprinus rubro-fuscus Lacépède, 1798: 547 (China)
Cyprinus nigro-auratus Lacépède, 1798: 547 (China)
Cyprinus viridescens Lacépède, 1803: 541, 543 (Rouen, France)
Cyprinus rondeleti Shaw, 1804: 194 (Lyon, France)
Cyprinus carpio Pallas, 1811: 290 (Onon, Argun, and Shilka Rivers)
Cyprinus macrolepidotus Hartmann, 1828: 183 (Lake Constance, Switzerland)
Cyprinus flavipinnis Cuvier et Valenciennes, 1828: 71 (Java, from a drawing by KUHL and VAN HASSELT)
Cyprinus vittatus Cuvier et Valenciennes, 1828: 72 (according to a drawing)
Cyprinus carpio var. *lacustris* Fitzinger, 1832: 333 (Neusiedler See)
Cyprinus elatus C. Bonaparte, 1832 – 1841, Fig. 3 (Lago Sabatino, Italy)
Cyprinus hungaricus Heckel, 1836: 222 (Neusiedler See)
Cyprinus haematopterus Temminck et Schlegel, 1842: 189 (Nagasaki)
Cyprinus melanotus Temminck et Schlegel, 1842: 190 (Japan)
Cyprinus conirostris Temminck et Schlegel, 1842: 191 (Japan)
Cyprinus nordmannii Valenciennes, in Cuvier et Valenciennes, 1842: 66 (market in Odessa, from Dniester and Bug Rivers)
Carpio vulgaris Rapp, 1845: 141 (Lake Constance, Germany – Switzerland – Austria)

---

*A comprehensive analysis of the synonymy of the nominal species based on European and Western and Central Asian fishes was recently published by KOTTELAT (1997).

Cyprinus acuminatus Richardson, 1846: 289 (Canton)
Cyprinus fossicola Richardson, 1846: 229 (Canton)
Cyprinus atro-virens Richardson, 1846: 287 (Canton, according to a drawing)
Cyprinus flammans Richardson, 1846: 288 (Canton, according to a drawing)
Cyprinus hybiscoides Richardson, 1846: 289 (Canton, according to a drawing and one specimen)
Cyprinus sculponeatus Richardson, 1846: 290 (Canton, according to a drawing)
Cyprinus chinensis Basilewski, 1855: 227 (Tien-tsin, northern China)
Cyprinus obesus Basilewski, 1855: 228 (Tien-tsin)
Cyprinus carpio var. *gibbosus* Kessler, 1856: 357 (Dnieper near Ekaterinoslav)
Cyprinus bithynicus Richardson, 1856: 372 (Apollonikis Lake, Asia Anatolia)
Cyprinus carpio Heckel et Kner, 1858:54 (Danube Basin, former Austrian Empire)
Cyprinus acuminatus Heckel et Kner, 1858: 58 (Danube, Neusiedler See, Lake Balaton)
Cyprinus hungaricus Heckel et Kner, 1858: 60 (Hungary; Neusiedler See, Lake Balaton, Tisza River)
Cyprinus regina, Heckel et Kner, 1858 : 60 (Hungary, former Austrian Empire)
Cyprinus carpio var. *mürgo* Dybowski, 1869: 950 (rarely in the Onon River)
Cyprinus carpio var. *haemoptera,* Martens, 1874 : 401 (Shanghai)
Cyprinus flavipinnis Karoli, 1882: 179 (Canton)
Cyprinus carpio Berg, 1909: 68 (Amur River)
Cyprinus carpio forma *typica* Antipa, 1909: 115(lower Danube in Romania)
Cyprinus carpio var. *gibbosus,* Atipa, 1909: 116 (lower Danube in Romani191a)
Cyprinus carpio var. *hungaricus* Antipa, 1909: 116 (lower Danube Basin)
Cyprinus carpio var. *oblongus* Antipa, 1909: 117 (lower Danube Basin)
Cyprinus hybiscoides Chaudhuri, 1911: 13 (Talifu Lake, Yunnan)
Cyprinus carpio anatolicus Hanko, 1924: 150 (Pursak River near Eski – Chehir and Kötske – Kissik, Turkey)
Cyprinus carpio Berg, 1932: 533 (ex parte: Europe, Caucasus, and Aral Sea Basin)
Cyprinus carpio Svetovidov, 1933: 259 (Sea of Azov and the Black, Caspian, and Aral Seas; parts of Europe; Lake Issyk-Kul)
Cyprinus carpio viridiviolaceus Svetovidov, 1933: 259 (South China south of Canton, Indochina)
Cyprinus carpio viridiviolaceus natio *haematopterus* Svetovidov, 1933: 260 (Amur Basin, Japan, China, southward to Fou-zhou)
Cyprinus carpio viridiviolaceus natio *haematopterus* monstrositas *hybiscoides* Svetovidov, 1933: 260 (Java)
Cyprinus carpio morpha *aralensis* Spiczakow, 1935: 428 (Aral Sea near Vozrozhdeni'a and Lazareva Islands)
Cyprinus carpio fluviatilis Pravdin, 1945: 11 (floodplain lakes of the Volga River near Saratov)
Cyprinus carpio Berg, 1949: 831 (Europe and Northern Asia)

Cyprinus carpio haematopterus Berg, 1949: 843 (Amur River Basin, Japan, Korea and China south to Fuzhien)
Cyprinus carpio var. *communis* Verma, 1970: 207 (Bhopal, India)
Cyprinus carpio carpio Mišík, 1958: 55 (Danube Basin in Slovakia)
Cyprinus carpio haematopterus Mišík, 1958: 55 (Amur River Basin)
Cyprinus carpio carpio Banarescu, 1964: 474 (Danube Basin in Romania)
Cyprinus carpio carpio Kirpichnikov, 1967: 39 (Europe, Transcaucasia)
Cyprinus carpio aralensis Kirpichnikov, 1967: 39 (Aral Sea Basin)
nomen praeoccupatum: *C. acuminatus* Richardson, 1846: 289 (Canton) – sec. Berg (1949)
Cyprinus carpio haematopterus Kirpichnikov, 1967:39 (Far East, China, Korea, Japan)
Cyprinus carpio viridiviolaceus Kirpichnikov, 1967: 39 (Vietnam)
Cyprinus carpio var. *communis*, Verma, 1970 : 207 (Bhopal, India)
Cyprinus carpio carpio Balon, 1974: 7 (Danube Basin in Slovakia)
Cyprinus carpio haematopterus Balon, 1974: 7 (Amur River Basin and China)

Holotype

No holotype was designated. Linnaeus (1758) probably described this species using specimens from a pond culture, designating Europe as the terra typica. Presuming that all domestic European forms originated as wild carps from the Danube, specimens from the Danube River can be considered as typical (Mišík 1958a, b, Balon 1974, 1995a, b).

Etymology

The generic name originated as latinized Greek. The specific name comes from the colloquial Celtic name for the fish. "Kyprinos" or "Kyprianos" was the name given to this fish by Aristotle (384 – 322 B.C.; 1862) and was probably derived from "Cypris", a secondary name for the Goddes of Love, Aphrodite, perhaps because even then the great fertility of the carp was known. Later the name was latinized to "Cyprinus", probably by Pliny. The specific name, however, probably originated later and appeared as "carpa" for the first time in the works by Cassiodorus (A.D. 490 – 585; 1626) in the sixth century (Steffens 1958; Balon 1974, 1995a, b).

Description

Du 2–4, *Db* 18–22 (mean 19.4), *Au* 2–3, *Ab* 4–5 (4.9), *Cu* 4–8, *Cb* 16–18 (17.0), *Cu* 4–8, *Pu* 1, *Pb* 14–19 (16.2), *Vu* 2, *Vb* 7–9 (8.0) for populations of wild carps from the Slovakian section of the Danube (MIŠÍK 1958a). *Du* 2–4, *Db* 15–23 (mean 18.8), *Au* 2–4, *Ab* 3–7 (5.1), *Cu* 4–8, *Cb* 13–20 (16.8), *Cu* 4–8, Pu 1, *Pb* 13–16 (15.0), *Vu* 1, *Vb* 5–9 (7.8) for populations in Lake Velence, Hungary, reared in ponds in Germany (STEFFENS 1964a).

L.l. 32–41, *Squ. sup.* 5–6, *Squ. inf.* 5–6 – for European populations, according to BERG (1949);

or *l.l.* 34–40 (mean 37.8), *Squ. sup.* 5–7, *Squ. inf.* 5–7; according to MIŠÍK (1958a);

or *l.l.* 34–39 (mean 36.4), *Squ. sup.* 5–7, *Squ. inf.* 5–7; according to STEFFENS (1964a).

Sp.br. on the outside of the first gill arch, 21–29, usually 23–26 (BERG 1949); or 22–28, mean 25.1 (MIŠÍK 1958a); or, for the right gill arch at age K_1, 18–26, mean 21.4; at K_2, 21–25, 23.2; at K_3, 23–27, 24.6; and for the left gill arch at age K_1, 19–24, mean 21.3; at K_2, 19–26, 22.7; at K_3, 23–27, 24.8, according to STEFFENS (1964a).

Sp.br. on the inside of the first gill arch, 27–34, mean 31.7 (SVETOVIDOV 1933, for European populations); or 29–36, mean 31.9 (MIŠÍK 1958a); or for the right gill arch at age K_1, 24–31, mean 27.0; at age K_2, 27–31, 29.1; at age K_3, 29–33, 30.2; and for the left gill arch at age K_1, 25–31, mean 26.9; at age K_2, 27–31, 28.6; and at age K_3, 28–32, 30.2, according to STEFFENS (1964a).

D.ph. 1.1.3–3.1.1, rarely 1.2.3–3.2.1 (BERG 1949, MIŠÍK 1958a), or usually 1.1.3–3.1.1, and rarely 1.3–3.1.1, 1.3–3.1, 1.2.3–3.1.1 (STEFFENS 1964a).

Vert. 36–38, most frequently 37 (BERG 1949); for migratory populations from the Volga River basin, 35–38, mean 36.1; and for resident populations, 35–39, mean 37.1 (ZUSSER 1938), or 35–37, mean 36.4 (STEFFENS 1964a). The number of vertebrae may be constant at 38 (MIŠÍK 1958a, BALON 1958a, b, 1995a, b), and most frequently it corresponds to the number of scales occurring in the *l.l.*

The variability in the meristic features of European wild carp populations was also apparent from the less extensive data of earlier authors (HECKEL et KNER 1858, SIEBOLD 1863, HERMAN 1887, ANTIPA 1909, BĂNĂRESCU 1964). The following variations in selected meristic features of populations of wild carps, *Cyprinus carpio,* from different parts of their European, Central Asian, and Far Eastern ranges were reported by BERG (1949): *Du* 3–4, *Db* (15)16–21(22), *Au* 3, *Ab* 5(6), *l.l.* (32) 33–40 (41), *Sp.br.* on the exterior edge of the first gill arch 17–29, *Vert.* 32–39. Only the last two values, the number of gill rakers on the exterior edge of the first gill arch and the number of vertebrae, exceed those known for European populations.

The common carp is a relatively large fish which attains 1000 to 1050 mm in *Sl* and 13 kg in weight. However, specimens 16 kg or more have been reported (BERG 1949, KAZANCHEEV 1981). One "wild carp" weighing 32 kg was caught in the Lago di Garda near Lugano in 1886 (BERG 1949). SABANEEV (1892) reported the capture of wild carps in Russia near the end of the 19th century, one with a weight of 55 kg caught 80 km from Taganrog, Krivaia Kosa, and another, 68 kg, from the Voroniezh River in the Lebedian District. Apparently, the larger of these specimens was considerably deformed, having a shape like "a one meter block almost 70 cm broad".

The same author also mentioned that wild carps weighing 16 to 24 kg are not rare in numerous large and small rivers of southern Russia, but those in the Volga and Dniepr Rivers only rarely attain a weight of 16 kg. BĂNĂRESCU (1964) recorded a specimen from the Romanian Danube Basin that was 1100 mm in *Sl* and 27 kg in weight. He remarked that carps exceeding 10 kg are exceptional. BLAHÁK (1994) reported a trophy specimen of 14 kg and 930 mm in *Tl* from the Danube arm at Medvedovo, Slovakia, captured in 1969. HOLČÍK (1970) determined its age at 14+ years. Specimens 420 to 530 mm in *Tl* and with a mean weight of 2.5 kg are considered to be the most abundant among wild carps caught in the Danube, Volga, Kura, Ural, and Artek River Systems as well as in the basins of other European rivers (BĂNĂRESCU 1964, KAZANCHEEV 1981).

According to STEFFENS (1958), *Cyprinus carpio* can be distinguished from *C. micristius* Regan, 1906, from its type locality in Yunnan, China, by its longer dorsal fin. It differs from *C. rabaudi* Tchang, 1933, from Katin and Fontou, Szechwan, in its larger scales, reflected by its number of *l.l.* from 32 to 41 compared to 45 in *C. rabaudi*. NICHOLS (1943) expressed doubts concerning the validity of a species, *C. pelegrini* Tchang, 1933, from Tunghai, Yunnan, which lacks barbels, and *C. yunnanensis* Tchang, 1933, from Yunnan, which possesses only one pair of small maxillary barbels. It is impossible to distinguish *C. carpio* from other nominal taxa from China and Vietnam, because data made available by WU (1977), MAI DING (1978), and ZHOU & CHU (1986) were not used for comparison with European populations of the wild common carp.

Morphology: Body elongated and torpedo-shaped (Fig. 28). The dorsal profile at the junction of the head and body is nearly straight without the prominent notch

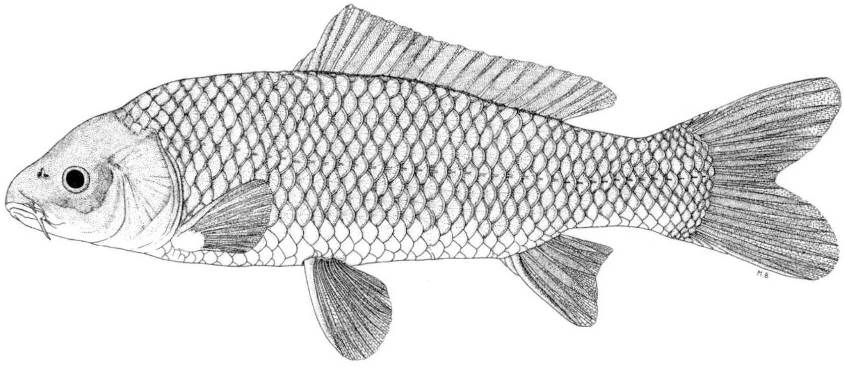

Fig. 28: Wild carp, *Cyprinus carpio carpio*, 379 mm in *Sl*, from the Danube River, Slovak Republic, October 13, 1960. MIRIAM BARADLAI, Bratislava.

Table 22: Selected morphometric characters, mainly as mean values, in wild carp populations, *Cyprinus carpio,* from different parts of the range.

Locality	**Danube R.** **n = 100**	**Southern Bug R.** **n = 11**	**Psel R.** **n = 13**	**Dnieper-Petrovsk** **n = 36**	**Kakhovsk Res.** **n = 29**	**Aral Sea** **n = 100**
Author	**MIŠÍK (1958a)**	**DEMENTEV and SHCHERBUKHA (1974)**				**NIKOLSKIY (1934)**
Sl (in mm) range/mean	359-635 465.50±0.09	- 398.2±2.17	- 404.7± 1.82	- 372.7±0.41	- 363.4±0.93	- -
% of *Sl*						
lc	22.7-26.7 24.18±0.09	- 24.45±0.39	- 24.48±0.26	- 23.11±0.14	- 24.27±0.16	- 24.51±0.08* 24.45±0.09**
prO	7.6-10.8 9.26±0.06	- -	- -	- -	- -	- -
Or	2.9-3.9 3.30±0.05	- -	- -	- -	- -	- -
io	8.9-11.7 10.25±0.06	- -	- -	- -	- -	- 10.00±0.05* 9.79±0.7**
poO	11.2-13.8 12.25±0.06	- -	- -	- -	- -	- -
pD	42.1-48.1 44.58±0.15	- 44.85±0.32	- 48.85±0.51	- 44.44±0.31	- 46.08±0.22	- 46.76±0.13* 46.33±0.17**
H	24.6-30.9 27.69±0.15	- 29.50±0.55	- 33.23±0.54	- 29.30±0.23	- 33.17±0.43	- 27.55±0.16* 27.65±0.12**
laco	13.8-22.0 17.16± 0.20	- -	- -	- -	- -	- -
lpc	18.1-23.2 20.26±0.10	- 19.35±0.31	- 18.17±0.29	- 18.41±0.19	- 18.34±0.22	- -
h	11.1-13.6 12.36±0.05	- 12.15±0.41	- 12.48±0.07	- 11.82±0.12	- 12.52±0.03	- 12.59±0.06* 12.92±0.06**
P-V	20.7-26.4 23.00±0.12	- 24.49±0.46	- 22.85±0.29	- 22.77±0.20	- 22.66±0.19	- -
V-A	27.5-33.1 29.86±0.12	- 30.05±0.32	- 28.24±0.31	- 28.91±0.22	- 28.86±0.16	- -
lD	35.8-44.9 40.46±0.16	- 39.45±0.42	- 40.24±0.60	- 38.81±0.30	- 39.48±0.25	- -
lA	7.3-10.4 8.87±0.06	- 9.25±0.40	- 9.09±0.15	- 8.79±0.08	- -	- -
lP	15.6-20.0 18.73±0.11	- 19.05±0.34	- 18.25±0.33	- 18.07±0.21	- 17.96±0.12	- 16.80±0.07* 17.53±0.12**
lV	14.7-19.9 16.97±0.11	- 17.65±0.45	- 16.93±0.22	- 16.08±0.18	- -	- -
hD	14.0-18.8 16.62±0.12	- 12.05±0.26	- 12.40±0.22	- 12.41±0.25	- 10.76±0.14	- -
hA	14.3-17.5 16.01±0.08	- 13.15±0.29	- 12.62±0.21	- 12.44±0.18	- -	- -

* upper numbers = females, ** lower numbers = males

Table 22: (cont.)

Locality	**Delta Ili R.** **n = 100**	**L. Balkhash** **n = 100**	**L. Issyk-Kul** **n = 55**	**L. Bolon'** **n = 50**	**L. Udyl** **n = 60**	**Amur R.** **n = 60**
Author	**Burmakin (1956)**	**Burmakin (1956)**	**Turdakov (1956)**	**Gromov (1973)**		
Sl (in mm) range/mean	- 291.80±0.21	- 302.50±0.16	- -	- 422.0	- 424.0	- 518.0
% of *Sl*						
lc	- 26.56±0.08	- 26.66±0.09	- 24.00±0.19	21.5-25.5 23.2±0.1	23.0-25.3 24.0±0.1	21.6-28.5 23.3±0.1
prO	- 9.79±0.05	- 9.97±0.05	- -	7.5-9.7 8.6±0.1	7.9-9.5 8.5±0.1	7.6-10.5 8.8±0.1
Or	- 4.22±0.03	- 4.23±0.03	- 3.10±0.005	2.3-4.7 3.9±0.1	3.5-6.2 4.0±0.0	2.6-4.5 3.7±0.1
io	- 9.95±0.04	- 10.31±0.04	- -	8.1-10.0 9.0±0.1	8.8-10.9 9.6±0.1	7.5-8.5 8.8±0.1
poO	- 13.29±0.05	- 13.77±0.05	- 12.10±0.09	10.0-15.3 11.7±0.3	11.3-13.3 12.2±0.1	8.6-13.5 11.3±0.1
pD	- 47.53±0.14	- 47.44±0.12	- 44.70±0.29	42.2-50.3 46.3±0.2	46.0-53.0 47.8±0.2	43.5-51.5 46.4±0.2
H	- 30.15±0.17	- 32.79±0.19	- 26.20±0.25	25.5-34.0 29.7±0.3	27.5-36.0 30.8±0.2	26.5-35.5 29.8±0.2
laco	- -	- -	- 16.50±0.16	- -	- -	- -
lpc	- -	- -	- 20.10±0.15	16.0-20.9 19.2±0.1	17.4-21.5 19.5±0.1	15.5-20.5 18.3±0.1
h	- 12.09±0.05	- 12.41±0.05	- 11.50±0.08	10.4-14.8 12.1±0.1	11.8-14.0 12.6±0.1	10.5-13.5 12.2±0.1
P-V	- 24.42±0.13	- 24.23±0.10	- 22.50±0.10	20.6-24.6 22.9±0.1	21.5-26.6 23.5±0.1	21.5-25.5 23.2±0.1
V-A	- 28.20±0.12	- 28.47±0.13	- 28.20±0.22	- -	- -	- -
lD	- 39.34±0.10	- 39.41±0.10	- 38.10±0.25	34.0-43.6 39.2±0.3	35.0-42.0 38.8±0.2	35.5-42.5 39.3±0.2
lA	- 8.90±0.10	- 8.78±0.05	- -	7.6-9.8 8.9±0.1	7.4-11.0 8.6±0.1	8.5-11.5 9.0±0.1
lP	- 19.70±0.11	- 19.44±0.11	- -	15.2-21.3 18.8±0.2	15.8-21.2 18.7±0.1	15.5-20.5 18.2±0.1
lV	- 16.96±0.09	- 16.45±0.10	- -	14.4-19.1 17.0±0.1	14.4-18.2 17.0±0.1	14.5-18.5 16.6±0.2
hD	- 14.91±0.10	- 14.42±0.11	- -	9.6-14.8 12.6±0.1	11.2-14.6 13.2±0.1	10.5-14.5 12.5±0.1
hA	- 14.73±0.08	- 14.58±0.09	- -	9.8-15.8 13.2±0.1	10.3-15.3 13.3±0.1	10.5-14.5 13.0±0.1

that is typical of domestic and feral carps. The cycloid scales are large and regular in shape; their caudal edges are marked by dark pigment giving the body a mesh-like pattern. The colour is brown to dark brown on the dorsum, dark golden laterally, and light golden with orange mottling on the ventral part. The anterior spines on the anal fin and the area around the anus are yellow-orange. The other fins are dark brown. The lateral line crosses the mid-lateral part of the body. It is complete but barely visible (BERG 1949, BĂNĂRESCU 1964, STEFFENS 1958, 1964a, BALON 1974, 1995a, b).

A complete set of morphometric data on the Danube wild carp (n = 87) was presented by MIŠÍK (1958a). Females were larger than males, reaching 496.5 ± 11.17 mm in *Sl* compared to 447.5 ± 6.95 mm. The data on the remaining characters are compiled in Table 22.

In % of standard length, Sl – anterior barbel length (*lba*) 0.9–2.5 (mean 1.7), posterior barbel length (*lbp*) 2.2–5.4 (3.6), head depth (*hc*) 18.5–23.4 (20.5), preventral distance (*pV*) 42.1–47.3 (44.7), preanal distance (*pA*) 70.8–77.0 (73.5), length of dorsal caudal lobe (C_1) 21.2–27.4 (24.1), length of ventral caudal lobe (C_3) 20.8–26.9 (24.2).

In % of head length, lc – snout length (*prO*) 32.6–41.3 (mean 38.7), anterior barbel length (*lba*) 4.1–10.8 (7.2), posterior barbel length (*lbp*) 9.5–22.6 (15.2), internasal width (*ina*) 19.9–26.3 (22.8), orbit diameter (*Or*) 12.0–17.3 (14.3), interorbital width (*io*) 38.4–46.1 (42.5), postorbital distance (*poO*) 47.5–54.5 (50.6), head depth (*hc*) 78.0–94.3 (84.5), head width (*lac*) 55.3–73.0 (62.6).

In % of caudal peduncle length, lpc – caudal peduncle depth (*hpc*) 63.9–83.3 (71.9), caudal peduncle width (*lapc*) 32.6–50.8 (40.7), minimum body depth (*h*) 52.7–69.9 (61.3).

BALON (1974) described the morphometric features of dwarf wild carps reared in aquaria (n = 7). The fin ray counts of starved carps are within the same range as those of wild carps inhabiting the Danube (MIŠÍK 1958a) or reared in ponds (STEFFENS 1964a). Counts of scales in the lateral line (*l.l.*) and the number of gill rakers (*Sp.br.*), which are significant characters for intraspecific taxonomy of the wild carp, are also within the range of normal variability. On the other hand, starving specimens exhibit distinct changes in body proportions and variation in one meristic character. The pharyngeal tooth formula (*D.ph.*) and arrangement in dwarf carps are unusual 1.2.1–3.1, 1.3–3.1, or 1.2.1–1.2.1. The head length (*lc*) as a percentage of *Sl* is considerably greater: 29 to 33 with a mean of 30 %. Also greater as percentages of *Sl* was the first pair of barbels (*lba*), which vary from 3.3 to 5.2 with a mean of 4.2 %; the eye diameter (*Oh*), ranging between 6.7 and 9.0 and averaging 7.4 %; the preventral distance (*pV*), varying from 50 to 52 with a mean of 50 %; and the caudal fin lobes, the upper one (lC_1) varying from 24 to 37 and averaging 30 % and the lower one (lC_3) ranging between 25 and 38 and averaging 30 %. There are also more ventral fin rays, which number 18 to 22 and average 20. Dwarf carps are also characterized by a slightly greater postorbital distance (*poO*), ranging from 14 to 15 with a mean of 14 %, a greater head depth (*hc*), varying from 19 to 25 and averaging 22 %; a greater predorsal distance (*pD*), from 48 to 55 and averaging 50 %; greater maximum body depth (*H*), from 22 to 35 with a mean of 30 %; greater length of the longest pectoral fin (*lP*), from 19 to 25 with a mean of 21 %; and longer ventral (*lV*), 18 to 22 and averaging 20 %; dorsal (*lD*), 18 to 22 and averaging 20 %; and anal (*lA*), 18 to 20 and averaging 19 % fin rays, with the data given as percentages of *Sl*. No proportions are considerably less in the dwarf fishes except the interorbital width (*io*), which varies from 45 to 49 with a mean of 46 %, and the postorbital distance (*poO*), which ranges from 66 to 79 with a mean of 73 %, shown as percentages of head length. The snout length (*prO*), which ranges from 32 to 35 with a mean of 33 %; head depth (*hco*), varying from 66 to 79 with a mean of 73 %; head width (*lac*), from 53 to 63 with a mean of 58 %; and length of the dorsal fin base (*lD*), from 34 to 38 with a mean of 36 %, are somewhat less than the relative values for wild carps.

Neither wild nor domestic pond carps have as many irregular teeth as the experimental carps (BALON 1974), though 15 % of the domestic pond carps show similar irregularities (STEFFENS 1964a), which are absent in wild carps. Similar differences in morphometric characters were observed by WUNDER (1949) among starved domestic carps. However, RUDZINSKI (1961) reported that wild Danube carps reared in Polish ponds had smaller heads than the domestic carps.

The gape of the mouth, a character that RUDZINSKI (1961) and STEFFENS (1964a) used successfully to distinguish wild and domestic carps, also proved to be a useful feature to characterize wild European carp populations. The value of the index obtained by dividing 10 times the mouth gape in cm^2 by the *Sl* in cm equals 1.14 (RUDZINSKI 1961). It is 1.09 or 1.24, depending on whether the carps had consumed supplemented or natural diets, respectively (STEFFENS 1964a). It was only 0.40 (BALON 1974) for wild carps that had been deprived of food. Domestic carps always have greater values for this index, which range from 1.91 to 2.30 (RUDZINSKI 1961, STEFFENS 1964a). Even more pronounced differences are encountered among the mouth-gape indices calculated by dividing 10 times the mouth gape in cm^2 by *lc* in cm. Its value is 4.46 or 5.57 for wild carps, depending upon whether they received supplemented or natural diets, respectively, and 8.12 for domestic carps living under the same conditions (STEFFENS 1964a). It is only 1.30 for wild carps that have been deprived of food (BALON 1974). The values of the mouth-gape index calculated by dividing 10 times the mouth gape in cm^2 by the fish weight in g do not show any notable differences. They amount to 0.08 to 0.10 for both wild and domestic carps (STEFFENS 1964a). Only wild carps that were deprived of food show greater values for this index, equaling 0.18 (BALON 1974). From these findings, it is clear that wild carps always have a smaller gape than domestic forms. RUDZINSKI (1961) and STEFFENS (1964a) believed that the enlargement of the mouth in domestic carps is related to changes in feeding habits and probably resulted from artificial selection.

The intestinal length in relation to *Sl* in K_3 wild carps from Polish pond cultures was 1.97 to 2.30:1 (RUDZINSKI 1961). In K_1 to K_3 carp reared in Germany, it ranged from 1.91:1 to 2.30:1 (STEFFENS 1964a). Both authors found that the intestines of wild carps were generally 15 to 20 % shorter than those of domestic carps. The mean index calculated by dividing the intestinal length by *Sl* in cm was 2.11 for wild carps and 2.64 for domestic ones (STEFFENS 1964a). It was 1.17 for wild carps that had been deprived of food (BALON 1974). The corresponding values for the index calculated by dividing 10 times the intestinal length in cm by the wet body weight in g are 3.02, 2.55 (STEFFENS 1964a), and 8.33 (BALON 1974), respectively. That the intestine is longer in domestic carps is probably due to an adaptation to utilize additional food items. The number of intestine coils of carp 4 to 6 cm in total length averaged six (KLUST 1939/1940, STEFFENS 1962); in wild carps deprived of food, it varied from four to seven (BALON 1974, 1995a, b).

RUDZINSKI (1961) and STEFFENS (1964a) also documented a marked reduction in the size of the posterior chamber of the swim bladder in the domestic carp. The average ratio of the volumes of the anterior and posterior swim bladder chambers is 16:9 (RUDZINSKI 1961) or 20.3:13 (STEFFENS 1964a) for wild carps and 4:1 (RUDZINSKI 1961) or 9:1 (STEFFENS 1964a) for domestic carps. STEFFENS (1964a) calculated an index expressing the relationship between the weight of the fish in g and the total volume of the swim bladder in ml. This ranged from 14.8 to 20.0 with a mean of 16.9 for wild carps (K_2–K_3) and 16.4 to 31.4 with a mean of 21.0 for domestic carps of the same age. The differences in the size and volume of both swim bladder chambers are probably explained by the general body structure of this species. The apparent reduction of the posterior swim bladder chamber volume may be due to the greater relative weight of the head in domestic carps (STEFFENS 1964a).

Karyotype: *Cyprinus carpio,* like other species of the cyprinine genera *Aulopyge, Barbus* s.str., and *Carassius,* originated through tetraploidy, as discussed in reviews by RÁB and COLLARES-PEREIRA (1995). This section is confined to a discussion of data pertaining to *C. carpio,* mainly of wild origin.

The karyotype is 2n = 100 somatic chromosomes. It is characterized by a remarkably large number of *a* chromosomes. There are two NORs. Male diakinesis was studied by OHNO *et al.* (1967), and bivalent meiotic chromosomes were identified. No sex chromosomes were found in the carps. The haploid karyotype characteristics are known for carps from the lower Danube River: 12m + 12sm, st + 26a (RAICU *et al.* 1972), and from the middle Danube: 14m + 18sm + 1NORsm + 11st + 6a (KRUPKA *et al.* 1989). Karyotype of the wild common carp, *C. carpio*, from the Danube is identical with those of the cultured domestic Rumanian, Hungarian and Ukrainian forms of the carp (RAICU *et al.* 1972) as well of the French ones (HAFEZ *et al.* 1978). Similarly, no substantial differences exist between the karyotypes of European carps and some Chinese subspecies (ZHAN and SONG 1980, RÁB 1995).

RÁB (1994) stated that there is no difference between the karyotype of carps from wild populations in the Danube and that of wild carps from the Amur River.

Protein specificity: The genetics of wild carp populations is not well known. The available literature was reviewed by DE LIGNY (1969) and KIRPICHNIKOV (1987). Publications devoted to this topic are scarce, and most report the results of studies on domestic carps. The most detailed, thorough, and exhaustive review dealing with biochemical genetics of carps, including some information on wild populations, is that of PAAVER (1983).

The genetic variability of carps is relatively great, and it depends on the heredity of the population. A summary of published data shows that a total of 44 protein enzyme products of structural loci have been analyzed, and 21 of them were found to be more or less polymorphic. The data for different populations vary. In his in-depth study, BRODY *et al.* (1979) analyzed the products of 33 loci in carps, mostly domestic races and landraces, and discovered polymorphism of 6 % to 18 %, depending on the population. This corresponds to the average variability generally common in fishes.

Similar data on the wild carp are scarce. They were summarized by PAAVER (1983), who analyzed populations in different geographical regions. He studied populations in the Danube Delta, Volga River, and Lake Paleostomi as well as the European subspecies, *C. c. carpio*; the subspecies in the Amur drainage area, *C. c. haematopterus*; and that in Lake Yaskhan, probably *C. c. aralensis*. He found differences not only between the wild and domestic carps but also among the wild carp populations. In general, the variability of the domestic carp is greater because of its origin. A greater than usual protein variability is a sign that the individual population arose by a crossing of different stocks.

The differences among populations of wild carps were found at different protein loci. European populations of the subspecies *C. c. carpio* are characterized by a prevalence of the dominant allele A for myoglobin approaching monomorphism, the slow allele for esterase-1, the rapid allele for phosphoglucomutase (PGM) almost to a monomorphic level, the three allele system for liver-specific lactate dehydrogenase-1 (LDH-C1), and monomorphism for the slow esterase locus 2. Moreover, in the Danube River, there are feral specimens of domestic carp together with hybrids of this morph with varieties of wild *C. carpio carpio*, including individuals with characteristics of domestic carps, including mirror scaling.

The Amur wild carp, *C. c. haematopterus,* is characterized by the prevalence of myoglobin allele a, a recessive "zero" allele, as well as the rapid allele for esterase-1 and the slow allele for PGM. It shows a high variability of esterase-2 and both liver-specific LDHs, LDH-C1 and C2. The variability of the transferrin locus, with as many as 23 variants, was also found to be much greater than that of European populations, which have 4 to 6 variants. The occurrence of rare alleles for esterase-1 and PGM is also more frequent.

Osteological characteristics: Like other species of Osteichthyes, the skeleton of the wild carp provides firm support for the body, which has a typical torpedo-like shape. The body is conspicuously wider than those of most other cyprinids. The osteology of *C. carpio,* based mainly on studies of domestic forms, was reviewed by URBANOWICZ (1956), who provided an evaluation of early information and a rich set of illustrations. The brief characterization of the skeleton by STEFFENS (1958) is summarized as follows:

The skeleton consists of three main parts: the skull, the vertebral column with the ribs, and the fin bones. The skull has a neurocranium with a set of bones firmly connected to each other and concealing the brain and the viscerocranium consisting of a row of paired visceral arches. Lower, intermediate, and upper jaws, together with small connecting bones, form the skeleton of the carp snout. The second visceral arch forms the hyoideum. Posterior to this are four gill arches with 25 to 30 gill rakers along their internal surfaces (FIEBIGER 1918–1919) and with gill filaments on their external side. The fifth gill arch is rudimentary, and like the lower pharyngeal bone, it bears no gills but is armed with five pharyngeal teeth arranged in three rows. These paired bones are opposed by similar, paired upper pharyngeal bones. These bear no teeth. They originate from the upper parts of the four pharyngobranchial gill arches and form a horny, hard chewing plate, which is situated on a process of the basioccipital. Pharyngeal teeth are continuously replaced throughout the whole life. The skull also consists of some paired dermal bones. These include the plates located around the eyes, the circumorbitalia, and they form the gill cover, which consists of four bones, namely the main operculum, preoperculum, interoperculum, and suboperculum.

The bones of the trunk are mainly the vertebrae with their lateral processes bearing ribs. The total number of amphicoelous vertebrae is frequently 36 or 37. However, it can fluctuate over a wide range. WUNDER (1934, 1949) was able to distinguish 3 cervical, 11 thoracic, 6 lumbar, and 16 to 17 caudal vertebrae in the spinal column. The vertebrae are initially cartilaginous and ossify later on. The spherical, elastic vestiges of the chorda are retained between vertebrae, which are joined to each other by a thin ligament. The spinal cord passes through dorsal, neural arches of the vertebrae, while the main blood vessels run through the ventral, hemal arches. However, no arches form along the column in the trunk below the vertebral processes that bear ribs. The skull and spinal column are joined firmly to each other by the first three vertebrae. Only in the middle of the body are the vertebrae more flexible. The flattened cervical vertebrae and the last caudal one are different in structure. The first vertebra is very small; the second one has a strongly elongated upper arch. The processes of the last vertebra, the urostyle, are fan-like and support the rays of the caudal fin. As in other fishes, the pectoral bone, or sternum, is absent from the carp.

Below the first vertebra is a set of small movable ossicles, the so called Weberian apparatus, connecting the swim bladder with the otic capsule and serving to transmit pressure changes.

The pectoral girdle, which articulates with the neurocranium and supports the pectoral fins, consists of three bones, the capsula, coracoid, and mesocoracoid, and five radials connected to fin rays. The bones of the pelvic girdle are embedded in the ventral musculature and do not articulate with the axial skeleton.

The intermuscular bones, very unpleasant to most fish consumers, are secondary supporting elements that are formed by the ossification of myosepta, which serve to connect the muscle segments.

The head of the wild carp is about 10 to 15 % shorter than that of the domestic carp (GRODZINSKI 1961), and its mouth opening is considerably smaller (see description above, RUDZINSKI 1961, STEFFENS 1964a, BALON 1974). The smaller and finer bones of its skull, especially the premaxillary and dentale, are correspondingly different. The relative shortness of the head is also related to the slightly narrower neurocranium (RUDZINSKI 1961).

There are also some differences between the wild and domestic carp in the axial skeleton of the trunk. The upper processes of the vertebrae, called the spinae neurales, pleurapophyses, and pleural ribs, are always shorter in the wild carp (RUDZINSKI 1961). RUDZINSKI (1961) also showed that all fins of the wild carp are smaller, and their fin rays are shorter. The total weight of all fins after removal from a wild carp is about 20 % less, relative to the total weight of the fish, than that of the domestic carp fins.

Additional data concerning the body form and anatomy, obtained from the wild carp population in Lake Velence, are provided by STEFFENS (1964a).

Sexual dimorphism: No fully acceptable hypotheses have been proposed concerning sexual dimorphism of the wild carp. BORZENKO (1926), TURDAKOV (1956), and BERG (1949) found no sexual dimorphism in the meristic and morphometric features. BURMAKIN (1956) also found no differences in the meristic characters. However, in the population from the Ili River, he found sexual differences in the predorsal and *P-V* distances and in the body width. In the population from Lake Balkhash, there were differences in the head length, body depth, interorbital width, and predorsal distance. NIKOLSKIY (1934) found sexual dimorphism in the minimum body depth and length of the pectoral fin (Table 23).

Probably the most detailed investigation of sexual dimorphism was conducted by MIŠÍK (1958a), who examined members of the Danube wild carp population numbering 56 males and 30 females. He attributed existing differences between the standard length of males, 447.5 ± 6.95 mm, and that of females, 498.5 ± 11.17 mm, to their different ages during sexual maturation. Most of the specimens he examined were taken from a spawning school captured in the Lesser Danube River, a Slovakian tributary of the Danube. The females of this population are noted (BALON 1957) for delayed sexual maturation. The average preanal distance is greater in females than in males because of the size reached by the ovaries, which swell as they ripen, rounding off the ventral profile of the abdomen, especially near the

Table. 23: Characters expressing the sexual dimorphism in the wild carp populations of different geographical origin. According to Mišík (1958 a) and others (mean values of characters expressed as % of *Sl*).

Character	Danube R. (Mišík 1958 a)		Aral Sea (Nikolskiy 1934)		L. Balkhash (Burmakin 1956)		Ili R. (Burmakin 1956)	
	males	females	males	females	males	females	males	females
% of standard length, *Sl*								
lc	23.98±0.12	24.53±0.17	24.45±0.09	24.51±0.08	26.22±0.10	27.08±0.12	26.39±0.11	26.69±0.11
io	10.12±0.07	10.50±0.08	9.79±0.07	10.00±0.05	10.16±0.06	10.46±0.06	9.99±0.05	10.26±0.07
poO	12.21±0.08	12.33±0.11	-	-	13.11±0.07	13.93±0.07	13.29±0.07	13.29±0.07
pD	44.39±0.18	45.00±0.23	46.33±0.17	46.76±0.13	47.06±0.15	47.80±0.16	47.03±0.20	48.05±0.17
pA	73.18±0.13	74.27±0.23	-	-	-	-	-	-
H	27.46±0.19	28.10±0.27	27.55±0.16	27.65±0.12	32.59±0.24	31.11±0.30	29.81±0.25	30.49±0.19
laco	16.27±0.16	18.70±0.39	-	-	-	-	-	-
h	12.41±0.08	12.27±0.14	12.92±0.06	12.59±0.06	12.46±0.07	12.37±0.08	12.04±0.07	12.13±0.07
lA	9.04±0.08	8.53±0.11	-	-	8.83±0.07	8.73±0.09	9.12±0.15	8.68±0.14
lP	18.96±0.15	18.27±0.18	17.53±0.12	16.80±0.07	19.84±0.12	19.16±0.18	20.07±0.16	19.33±0.14
lV	17.32±0.13	16.29±0.16	-	-	16.69±0.11	16.14±0.17	17.25±0.11	16.67±0.12

anal opening. This distance in the specimens examined depended upon the degree of maturation of the females caught at the spawning sites. The body width is the dimorphic character most significant statistically. The body width of female wild carps increases more rapidly during gonad maturation than that of the males, and this occurs to a greater degree than in other cyprinids. This probably has resulted from an adaptation to the kind of habitat in which a deviation from the characteristic cylindrical body shape during the prespawning period would be detrimental. BALON (1995a, b) wrote that the retention of a torpedo-like shape is possibly necessary to overcome strong currents and reach innundated areas at higher elevations for spawning.

The fins of males are longer and deeper than those of females. These differences are not significant statistically except of the cases of the lengths of the anal and ventral fins. The head width as a percentage of the head length is 61.79 ± 0.42 % in males and 64.28 ± 0.58 % in females, while the width of the caudal peduncle is 39.49 ± 0.45 % in males and 42.88 ± 0.75 % in females. The greater values in the females are statistically significant. There is also a difference in the body form of the females, which is more cylindrical and wider than that of the males (MIŠÍK 1958a).

The length of the pectoral fin as a percentage of the *P-V* distance is 83.75 ± 0.98 % in males and 77.30 ± 1.02 % in females. The greater distance in the males is related to differences between both sexes in the length of the pectoral fin and the *P-V* distance. However, these differences are not statistically significant. Some minor sexual dimorphism is also evident in the head length, snout length, predorsal distance, length of the upper caudal lobe, depth of *A*, interorbital width as a percentage of head length, and minimum body depth as a percentage of the length of the caudal peduncle. These and perhaps other differences are rather small and can be detected only statistically. They are therefore insignificant and of no practical value for sex determination. The greater body width of the females during the spawning period is clearly a sexually dimorphic character. In addition, males can be recognized during the spawning period by the milt they release when pressed slightly.

Shortly before and during spawning, the males can be safely distinguished by the breeding tubercles, which appear as very small granules on the scales above or, more frequently, below the lateral line, on the caudal peduncle, and on the head. They are especially noticable on the anterior parts of opercula and praeopercula and under the eyes. Tubercles even occur along the inner surfaces of the first pectoral and pelvic fin rays as well as along the unbranched rays of the dorsal, anal, and caudal fins. Breeding tubercles may also be present on some females, but only on the head. No female displayed them on scales or fins (MIŠÍK 1958a).

An interesting kind of sexual dimorphism was observed among the wild carp from Lake Balkhash. This is the greater body depth (*H*) relative to the *Sl* in males, as reported by BURMAKIN (1956) for specimens that had not finished spawning.

Because BURMAKIN (1956) did not consider body width, it is not known whether it increases during sexual maturation, as it did among the population in the Slovakian part of the Danube (MIŠÍK 1958a). However, DEMENTEV and SHCHERBUKHA (1974) evaluated 15 meristic characters of a population in Kakhovsk Reservoir (males = 53, mean Sl 504.4 ± 0.51 mm; females = 53, mean Sl 507.3 ± 0.50 mm) and found differences only in the body depth, which is greater in females. The depth of males was 27.43 ± 0.27 % of *Sl*, while the corresponding depth of females was 29.01 ± 0.26 %. TARONOVA (1959) analyzed 14 characters of the wild carp population from the Ili River delta and found that only the depth of the anal fin shows sexual dimorphism to a significant degree. That of the males was 14.2 ± 0.19 % of *Sl*; the anal fin depth of females was 10.89 ± 0.27 %. For populations from the Ili River delta and Lake Balkhash, TARONOVA (1959) reported values that differed from those of BURMAKIN (1956). For males and females, her mean value of H was 27.33 ± 1.09 % and 28.20 ± 0.04 % of *Sl*, respectively, and the body width was 15.70 ± 0.19 % and 16.79 ± 0.26 %, respectively.

After comparing data on the sexual dimorphism of wild carps from different geographical regions, MIŠÍK (1958a) concluded that sexual differences are not associated with specific morphological characters. However, such differences can be manifested to a different degree by different characters alone or in combination. Only the body width and the lengths of *P* and *V* can be considered sexually dimorphic characters in most populations (Table 23).

Variations: Complete information on variations in meristic and morphometric characters are only available for the population of wild carp in the Danube (MIŠÍK 1958a). Additional data from the literature are sufficient only for a preliminary evaluation of the variations in the characters. This, of course, is also important for intraspecific taxonomy.

Meristic characters that have been investigated include variations in the numbers of branched rays in *D* and *A*, scales in *l.l.*, *Sp.br.*, and *Vert.* (Table 24). The number of branched rays in *D* (*Db*) varies over a wide range from 14 to 22 (BERG 1949, MIŠÍK 1958a). From the available mean values for this character in populations from different geographical regions, a tendency toward a slightly increased count of the rays in the dorsal fin of the European, Central Asian and East Asian populations could be detected. The number is moderately lower in East Asian populations. Unfortunately, enough data are not available for an evaluation of the number of branched fin rays in *A* (*Ab*), which varies in the range of four to six with five being most common.

The maximum range of the lateral line scale count is 33 to 41 among European, Central Asian, and East Asian populations (MIŠÍK 1958a, SVETOVIDOV 1933, KIRPICHNIKOV 1967); however, even lower numbers are encountered among populations of wild carp from Vietnam. The variation in the mean counts of these scales for populations in Europe and Central and East Asia is 35.60 to 39.60. Al-

Table 24: Some selected meristic characters of wild common carps, *Cyprinus carpio*, from different parts of its range according to various authors.

Author, locality	Character – Range/Mean				
	Db	*Ab*	*l.l.*	*Sp.br.*	*Vert.*
Mišík (1958 a) Danube R.	18–21 (22) 19.36	4–5 4.95	(34–35)36–39(40) 37.36	22–28 25.05	38 –
Kirpichnikov (1967) Don R.	– 19.17	– –	– 38.85	– –	– 36.79
Kirpichnikov (1967) Don R.	– 18.99	– –	– 38.66	– 26.49	– 36.29
Dementev & Shcherbukha (1974) Southern Bug R.	18–21 19.50	5–6	36–39 38.30		
Psel R.	19–21(22–23) 20.30	5	38–41 38.30		
Dnieper Estuary	(16–17)19–21 19.10	(4)5(6)	(34) 35–39 36.90		
Kakhovsk Res.	(16–17)18–22(23) 19.73	5	(33–35)36–39 36.28		
Kirpichnikov (1967) Volga R.	– 19.24	– –	– 38.77	– 24.78	– –
Svetovidov (1933) Europe	17–21 18.10	– –	37–41 38.40	– 24	– –
Kirpichnikov (1967) L. Tarapavani	– 18.55	– –	– 39.08	– 23.85	– 35.95
Borzenko (1926) Kura R.	16–22 –	(4)–5 –	35–40 37.90	23–29 25.60	– –
Kirpichnikov (1967) Aral Sea, Amu-darya R.	– 18.34	– –	– 36.10	– 25.60	– 35.99

Table 24: (cont.)

Author, locality	Character – Range/Mean				
	Db	*Ab*	*l.l.*	*Sp.br.*	*Vert.*
Dryagin (1936)	16–21	–	35–39	21–26	–
Tchu R.	18.78	–	37.35	23.5	–
Kirpichnikov (1967)	–	–	–	–	–
L. Issyk-Kul	18.47	–	37.50	25.70	36.19
Turdakov (1956)	–	–	–	–	–
L. Issyk-Kul	19.40	–	36.0	24.40	–
Kirpichnikov (1967)	–	–	–	–	–
L. Balkhash	19.07	–	38.87	26.88	36.20
Burmakin (1956)	–	–	–	–	–
Estuary of Ili R.	19.40	–	37.84	25.77	36.38
Nikolskiy & Evtukhov (1940)	–	–	–	–	32–38
Ili R.	19.3	–	36.50	25.5	34.5
Taronova (1959)	–	–	–	–	–
Ili R.	18.93	4.43	36.55	25.36	32.05
L. Balkhash	–	–	–	–	–
	17.60	4.05	35.60	27.60	32.87
Nekrashevich (1934)	17–21	–	35–40	22–28	34–38
L. Balkhash and Tshany	19.1	5.6	37.8	24.8	36.1
Kirpichnikov (1967)	–	–	–	–	–
Amur R.	17.82 or 17.84	–	38.06	21.30	36.53
Nikolskiy (1956)	–	–		–	–
Amur R.	18.2	–	37.3	20.1	36.6
Berg (1949)	(14)15–20(21)	5	37–39 (40)	17–25	34–37
Amur R.	–	–	–	–	36.2

Table 24: (cont.)

Author, locality	Character – Range/Mean *Db*	*Ab*	*l.l.*	*Sp.br.*	*Vert.*
Gromov (1973)	18–22	–	35–40	19–24	34–36
Amur R.	19.3	3.7	38.0	21.8	35.0
L. Bolon′	15–22	–	36–40	20–25	35–37
	18.3	3.8	38.1	22.6	35.8
L. Udyl	16–20	–	36–40	–	–
	18.2	–	37.8	–	–
L. Dalaynur	–	–	–	–	–
	17.3	–	38.1	21.7	–
Urshun R.	–	–	–	–	–
	19.4	–	39.6	24.1	–
Cend Aiuush (1967)	15–20	5–6	–	23–28	–
L. Buir Nur	16.5	5.7	–	24.1	–
Kirpichnikov (1967)	–	–	–	–	–
Korea (1 specimen only)	18	–	35	–	–
Svetovidov (1933)	17–20	–	(34)35–39(40)	17–25	–
China, Japan, Amur Basin	18.1	–	37.3	20.0	–
Kirpichnikov (1967)	18–21	–	33–39	18–21	–
China	–	–	–	–	–
Kirpichnikov (1967)	–	–	–	–	–
North Vietnam	20.33	–	32.60	20.75	33.40
Czhan Din Chzhong (1967)	–	–	–	–	–
North Vietnam	19.86	–	32.12	18.10	33.87

though the differences may be considerable, they are not statistically significant. A moderate decrease in the mean count of lateral line scales was recorded for populations in the Danube moving from west to east, although there were many exceptions. The wild carps from Vietnam differ in this character from all populations mentioned above; the mean number of its *l.l.* scales is 32. However, it should be noted that data on its individual populations are scarce (KIRPICHNIKOV 1967, CHZHAN DIN CHZHONG 1967).

The importance of gill raker counts, *Sp.br.*, for the taxonomy of *Cyprinus carpio* was pointed out by SVETOVIDOV (1933). However, he counted gill rakers on the internal side of the first gill arch, while most other investigators counted those on the outside. Because most literature refers to the gill rakers on the outside of the first gill arch, only these are considered in this chapter. The range of variation for this character is 18 to 29 in European and Central Asian populations, while it is 17 to 28 among those in East Asia, which is similar to that of Vietnamese populations of wild carp (Table 24). The mean values for this character vary within the ranges of 21.42 to 27.60 for European and Central Asian populations, 20.0 to 24.1 for East Asian, and 18.10 to 20.75 for South Chinese and Vietnamese wild carps. In spite of a considerable overlap in the individual ranges as well in the mean values for individual, geographically limited populations, there is a clear, statistically significant tendency for the *Sp.br.* count on the external edge of the gill arch in wild carp populations to be lower in East and South Asia than in Central Asia and Europe, especially in the Danube.

KIRPICHNIKOV (1967) also used the number of vertebrae (*Vert.*) to distinguish the wild carp populations. The range of this character reaches 32 to 38 for European, Central Asian, and East Asian populations, and the mean value for individual populations ranges between 34.50 and 36.79. There are no significant geographical differences in this character. Only among the Vietnamese wild carp populations is the mean value of this character, 33.40 to 33.87, significantly lower than those of the others. However, these mean averages were calculated for only a few specimens from two populations that were examined.

In our opinion, of all the meristic characters discussed, only the numbers of branched rays in *D*, scales in the *l.l.*, and *Sp.br.* on both the internal and external sides of the first gill arch might be of importance for the intraspecific classification of *Cyprinus carpio.*

As MIŠÍK (1958a) reported, there is a remarkable paucity of data on the morphometric characters in the available literature, and since he wrote this, few additional data have been obtained (TARONOVA 1959, PAVLOV 1964, PAPANDOPOL and CAUTIS 1965, DEMENTEV & SHCHERBUKHA 1974). This is documented in Table 22, which shows that the mean lengths of the specimens compared are very different. Thus, MIŠÍK (1958a) could only report that a certain degree of similarity between the wild carp populations from the Danube and those from Lake Issyk-kul and the Aral Sea exists, based on an analysis of morphometric characters. The

Danube populations differ in morphometric characters among themselves more than they differ from populations in Lake Balkhash and the estuary of the Ili River. However, this finding might result from an artefact in the data resulting from different standard lengths (*Sl*) of the specimens from the different localities. The body forms of specimens from the two Asian population somewhat resemble those of younger carps. It was possible to demonstrate some differences between the mean values of characters compared, such as predorsal distance and maximum body depth; however, these probably have no taxonomic significance because they remain within the natural ranges of variation and may be characteristic only of local ecomorphs. The necessity of comparing the morphometric characters of specimens of wild carp of nearly equal *Sl* and *Tl* in collections from different water bodies was discussed by DEMENTEV and SHCHERBUKHA (1974).

Generally, the maximum body depth (*H*) is an important character for evaluating the genetic purity in the Danube and among all other wild carp populations. This was reported by MIŠÍK (1958a, b, c) and experimentally confirmed by TUČA (1958), BASTL (1961) and TUČA and BASTL (1961). Both authors made extraordinary efforts to characterize the exterior body shape of the wild carp in the Danube. Earlier authors also gave special attention to the maximum body depth. It was used by HECKEL (1836) and HECKEL and KNER (1858) to distinguish three forms: *Cyprinus carpio* with a body depth equal to 33 % of the standard length, *C. acuminatus* with a depth 36 to 40 % of the *Sl*, and *C. hungaricus* with a depth of 25 to 29 % of *Sl*.

ANTIPA (1909) considered HECKEL's species to be varieties and distinguished a new variety, *oblongus,* which had a maximum body depth of 24 to 27 % of *Sl*. KESSLER (1856) had given a wild carp from the Dnieper the name *C. carpio* var. *gibbosus;* its body depth was 36 to 40 % of *Sl*. Other authors described additional morphs until BERG (1949) grouped all of them into three main taxa: *typica* with a depth 33% of the standard length, *hungaricus* with a depth from 25 % to 28 % of the *Sl*, and *elatus* with a depth of 36 % to 40 % of *Sl*. The domestic common carp is reported to have a body depth from 31 % to 39 % of its length (BRYLIŃSKA 1986).

BALON (1974, 1995a, b) stated that while the Slovakian wild carp studied were mostly within the morphometric range of *hungaricus*, some specimens fell into the range of *typica*. Since the body depth reflects the condition, sex, and possibly whether the fish inhabited a lotic or lentic habitat, the use of this character even to distinguish morphs is unwarranted. On the other hand, a great body depth is a clear sign of domestication and probably reflects both the conditions in the pond and a conscious selection; therefore, if deep-bodied forms of carp, such as *elatus, gibbosus,* and *acuminatus,* occur together with low-bodied wild forms, they must be feral forms with domestic ancestors. The characteristic maximum body depth of typical carps from the Danube ranges from 24.6 % to 30.9 % with a mean of 27.7 % of *Sl* (MIŠÍK 1958a, b, c; TUČA 1958, BASTL 1961, MIŠÍK & TUČA 1965). Such individuals have completely scaled bodies and other characteristics, which were described above. This mean value of *H* (Table 22) is exceeded, according to DEMENTEV and

SHCHERBUKHA (1974) and BURMAKIN (1956), only by the wild carp populations from the Psel River and Kakhovsk Reservoir, where the mean averages are 33.23 % and 33.17 %, respectively; from the Ili River, where it is 30.15 %; and from Lake Balkhash, in which it is 32.79 % of *Sl*. A wild origin of all these populations is not suspected. TARONOVA (1959) investigated the same populations of wild carp from the Ili River (n = 100) and Lake Balkhash (n = 50) and found that for specimens with a standard length ranging from 237 to 267 mm, the body height, *H*, was 27.33 % ± 1.09 % and 28.20 % ± 0.04 % of *Sl*, respectively, and the corresponding body width was 15.70 % ± 0.19 % and 16.67 % 0.26 % of *Sl*. These findings agree well with results reported by MIŠÍK (1958a). In our opinion, it is important to remember that a low body shape is a genetically determined feature of wild carp populations within their original ranges. This was confirmed experimentally in ponds (TUČA 1958, BASTL 1961, RUDZINSKI 1961, STEFFENS 1964a).

BALON (1995a, b) pointed out that the barbels of both pairs on the wild Danube carps are longer than those of Russian wild carps (BERG 1949). However, MIŠÍK (1958a) noted a substantial variation in a way that allowed the carps to be grouped into long and short barbel morphs. This distinction, if it is valid at all, has no distinct importance for the regional classification of wild carp populations.

KIRPICHNIKOV (1967) mentioned differences in other morphometric characters, including the eye diameter as a percentage of the head length, *lc*, and the lengths of the barbels (see above) and intestine. These are differences among wild carp populations of European, East Asian, and Southeast Asian, particularly Vietnamese, origin. Except in the case of the European carp, differences have been determined only by examination of single specimens, so the evidence is insufficient, and the conclusions are not convincing. This does not mean that they could not be verified after a larger number of specimens are examined. The differences found could then be effectively used to distinguish geographically limited populations.

The greater strength, mobility, and viability of wild carps are demonstrated by some physiological characteristics. The earliest information on this was obtained during studies of specimens from the population in Lake Velence that were reared in German ponds (STEFFENS 1964a). Their erythrocyte counts were significantly higher in the wild carps than they were in domestic forms also raised in ponds. For 50 wild carps, the range was 0.86 to 2.88 million per mm^3 with a mean of 1.86: s = ±0.28, and v = 15.1%. For 52 domestic carps, the corresponding values were 0.94 to 2.29 million per mm^3 with a mean of 1.51: s = ±0.30, v = 19.9 %. A similar tendency was also found for the hemoglobin content. Among the same fish specimens, the blood of the wild carps contained more hemoglobin than that of the domestic carps in the ponds. The range of the wild carps was 4.16 to 13.12 g/100 ml of blood with a mean of 9.17: s = ±1.62, v = 17.7 %, while the corresponding values for the domestic carps were 4.54 to 12.02 g/100 ml of blood with a mean of 7.56: s = ±1.57, v = 20.0 %, respectively. The blood sugar level was 16 % to 26 % higher than that of the domestic form. The wild carps also had a higher total protein content in the blood serum. The range for 19 K_3 (three-year-old) carps was 2.98 to 4.55 with a mean of 3.80 g/100 ml serum: s = ±0.44 g/100 ml serum, v = 11.6 %; for 12 K_4 carps, the range was 3.33 to 4.38 with a mean of 3.90 g/100 ml serum: s = ±0.36 g/100 ml serum, v = 9.1 %. Corresponding values for 22 domestic K_3 carps were 2.28 to 4.38 with a mean of 3.17 g/100 ml serum: s = ±0.63 g/100 ml serum, v = 19.9 %. Furthermore, the wild carps had a lower water content in the muscles, a greater fat content in individual organs, more glycogen in the liver, and more vitamin A in the eyes, intestine, and liver. Additional data concerning these and other characters were provided by STEFFENS (1964a) and KIRKOVÁ (1990). The wild carps are more mobile, stronger, and nimbler because their muscles are better vascularized and do not fatigue as quickly as those of domestic carps (BALON 1995a, b). Therefore, the wild carp is, or was, also better able to overcome river currents.

Age and size variability: The variability of morphometric characters during the adult and senescent periods of the life history, particularly that of the wild Danube carp, was studied by MIŠÍK (1958a). After statistical evaluations of morphometric data recorded especially for the purpose, he was not able to determine any significant changes related to the age of the fish. He also assumed that the standard length of the fish plays a much more important role in the changes in morphometric characters.

The relationship of the morphometric characters to the growth velocity seems very interesting. BALON (1957), analyzing specimens of wild carp from the Lesser Danube, a tributary of the Danube, which had been studied by MIŠÍK (1958a), was able to identify three distinct groups according to their growth rates. MIŠÍK (1958a) subsequently compared mean values for these morphometric characters and found tendencies for some of them to either increase or decrease according to the growth velocity. In one case (Table 25), the head length, predorsal and preventral distances, body depth and width, and length of the dorsal fin, all as percentages of *Sl*, as well as the head depth and width as percentages of the head length and the depth and width of the caudal peduncle and minimum body depth as percentages of the caudal peduncle length tend to be greater when the growth rate is slower. In contrast, the lengths of the caudal peduncle, *A*, C_3, and *P*, as well as the height and depth of *A*, all as percentages of *Sl*; the length of the upper barbels, distance between the anterior margins of the nasal openings, and the eye diameter as percentages of the head length; and length of *P* as a percentage of the *P-V* distance, all tend to decrease with the growth rate.

A significant increase of the body depth and width with decreasing growth rate is considered to be the most important finding by MIŠÍK (1958a), who considered it to be a specific morphoecological adaptation of the wild Danube carp to the strong currents in the stream habitat and a surplus of food. After considering the changes in the other morphometric characters, he assumed that the faster growing forms of the carp are limnophilic, while the slower growing ones are somewhat rheophilic.

Table 25: Mean values of morphometric features in a population of Danube wild carps, in relation to the growth rate (Mišík 1958a).

Growth	slow	medium	fast
n	2	39	8
% of *Sl*			
lc	23.5	24.0	24.4
pro	8.8	9.4	9.6
lb_1	2.0	1.7	1.7
lb_2	4.1	3.6	3.9
Oh	3.4	3.4	3.4
io	9.9	10.2	10.3
poO	12.0	12.1	12.5
hc	19.6	20.4	20.9
pD	42.9	44.4	44.6
pV	44.2	44.5	45.1
pA	73.2	73.4	73.5
H	27.3	27.5	28.2
laco	14.7	16.7	17.6
lpca	21.4	20.4	20.1
h	12.4	12.4	12.5
P-V	22.4	23.0	23.1
V-A	29.7	29.9	29.4
lD	39.9	40.6	41.0
lA	9.3	8.9	8.7
lC_1	25.4	24.2	24.3
lC_3	25.0	24.3	24.1
lP	20.1	18.7	18.4
lV	18.0	17.0	17.1
hD	18.3	16.5	17.2
hA	16.7	16.0	15.8
% of head length, *h*			
pro	35.1	39.1	39.1
lb_1	8.5	7.4	7.1
lb_2	17.6	15.0	16.0
ina	24.1	23.0	22.6
Oh	14.5	14.4	13.7
io	42.2	42.5	42.0
poO	51.1	50.3	51.1
hc	83.7	84.8	84.9
lac	60.8	61.7	63.3
% of caudal peduncle length, *lpc*			
hpc	68.5	71.7	71.9
lapc	34.0	40.4	41.1
h	57.8	60.9	62.2
% of distance *P-V*			
lP	90.0	81.4	79.7
% of distance *V-A*			
lV	63.2	56.8	58.4

Table 26: Mean values of morphometric characters in a population of Danube wild carps, *C. c. carpio,* in successive length groups (MIŠÍK 1958a) and those from a population from Kakhovsk Reservoir (DEMENTEV and SHCHERBUKHA 1974).

Length group (*Sl*, mm)	**Danube R.**							**Kakhovsk Res.**		
	350	**400**	**450**	**500**	**550**	**600**	**650**	**363**	**491**	**662**
n	1	25	26	21	8	4	1	29	103	37
% of Sl										
lc	25.4	24.5	24.0	24.3	24.2	23.3	22.7	24.3	22.6	22.0
pro	9.6	9.4	9.3	9.4	9.5	8.4	8.5	-	-	-
lb_1	1.9	1.7	1.6	1.9	1.9	2.0	-	-	-	-
lb_2	4.0	3.6	3.6	3.5	3.8	4.5	3.9	-	-	-
Oh	3.5	3.5	3.4	3.4	3.4	3.5	3.2	-	-	-
io	10.3	10.3	10.2	10.3	10.3	9.8	10.2	-	-	-
poO	13.0	12.2	12.2	12.4	12.3	12.1	12.0	-	-	-
hc	21.1	20.3	20.5	20.9	20.7	20.1	19.1	-	-	-
pD	47.1	44.8	44.6	44.6	44.5	43.6	44.1	46.1	42.8	42.1
pV	45.9	44.8	44.4	45.1	45.0	44.6	43.3	-	-	-
pA	74.6	73.3	73.5	73.9	74.2	73.5	73.2	-	-	-
H	29.5	28.0	27.4	27.5	27.7	27.1	24.6	33.2	28.5	28.8
laco	15.3	16.0	16.8	18.5	18.1	17.9	18.6	-	-	-
lpca	20.1	20.6	20.6	19.8	19.5	20.2	20.5	18.3	18.2	18.5
h	12.2	12.4	12.4	12.4	12.3	12.5	12.3	12.5	11.6	11.6
P-V	22.8	22.6	22.8	23.9	23.5	23.7.	23.1	22.7	22.8	23.4
V-A	30.1	29.5	30.0	30.2	30.4	29.9	30.9	28.9	29.0	29.1
lD	39.6	40.7	39.9	40.7	41.3	41.1	40.5	39.5	38.5	38.4
lA	8.9	9.0	8.8	9.0	8.4	8.7	9.4	-	-	-
lC_1	25.6	24.4	24.4	23.6	23.3	23.4	-	-	-	-
lC_3	26.2	24.7	24.0	23.6	23.4	23.2	-	-	-	-
lP	19.5	19.0	18.6	18.7	18.1	18.2	-	18.0	17.2	16.3
lV	18.1	17.4	16.9	17.0	16.2	16.3	16.4	-	-	-
HD	18.4	16.9	16.8	16.5	16.4	16.4	15.8	10.8	11.0	10.3
HA	16.7	16.2	16.1	16.0	15.7	15.8	16.7	-	-	-
% of head length *h*										
pro	37.8	38.7	38.7	39.0	39.3	36.3	37.5	34.0	34.4	35.2
lb_1	7.7	7.1	7.0	7.6	7.7	8.7	-	-	-	-
lb_2	15.8	14.8	15.1	14.5	16.0	19.4	17.1	-	-	-
Ina	23.0	23.1	22.6	23.1	22.2	23.7	25.3	-	-	-
Oh	13.8	14.4	14.4	14.2	14.3	15.1	14.3	14.9	14.0	12.7
Io	40.8	42.3	42.6	42.7	42.4	42.3	45.0	-	-	-
poO	51.5	50.2	50.7	50.8	50.8	51.8	52.8	49.0	49.4	50.6
Hc	79.4	82.9	85.7	85.7	84.1	86.4	84.5	-	-	-
Lac	56.4	60.2	62.2	64.4	63.9	68.2	73.0	-	-	-
% of caudal peduncle length *pdc*										
hpc	73.5	71.3	71.4	73.7	72.5	70.9	63.9	-	-	-
lapc	38.3	38.6	39.5	43.2	44.3	39.3	47.0	-	-	-
H	60.7	59.8	60.5	63.3	63.3	61.9	60.0	-	-	-
of distance *P-V*										
lP	85.4	84.1	82.1	79.1	77.1	76.7	-	-	-	-
% of distance *V-A*										
lV	60.2	60.0	56.5	57.0	53.6	55.1	53.1	-	-	-

Mišík (1958a) evaluated the variability of the morphometric characters in seven groups of wild Danube carp within the length range of 350 to 650 mm and concluded that these characters in the adults in his collection were rather stable and showed no tendency to change as the fishes grew in length (Table 26). In one of the analyses, however, he noted a slight decrease in the head length, interorbital width, praedorsal distance, body depth, and the length of both caudal lobes, all as percentages of *Sl*, as well as in the length of *P* as a percentage of the *P-V* distance and the length of *V* as a percentage of the *V-A* distance with increasing body length among the males. In contrast, slight increases in the length of the dorsal fin base as a percentage of *Sl* and in the height and width of the head as a percentage of the head length were observed among the same specimens.

Among females, Mišík (1958a) reported a moderate decrease in the head length, predorsal distance, and the length of the lower caudal lobe, all as percentages of *Sl*, and somewhat smaller increases in the head height as a percentage of the head length and the length of *P* as a percentage of the *P-V* distance. Other morphometric characters showed a tendency to increase with the body length, including the length of *D* as a percentage of *Sl*, the distance between the nostrils and head width as percentages of the head length, and the width of the caudal peduncle and minimum body depth as percentages of the caudal peduncle length.

Similar results were obtained by Nikolskiy (1934), who compared the morphometric characters of juvenile and adult wild carps from the Aral Sea and found the most significant differences to be a shorter head length and predorsal distance as percentages of *Sl*. A less pronounced decrease in the preorbital distance as a percentage of the head length was also noted. Dementev and Shcherbukha (1974) found similar tendencies in the changes of these characters with length among the wild carp populations from Kakhovsk Reservoir (Table 26). In our opinion, these facts do not contradict the statements cited above concerning less pronounced sexual dimorphism among wild carp populations.

While in most other cyprinid species, an increase in the relative body depth with increasing body length has been observed, illustrating positive allometry, in the wild carp, the depth decreases, illustrating negative allometry. This seems to be peculiar for the species. The same relationship was also observed among wild carps from the Aral Sea (Nikolskiy 1934). The data presented by Tuča (1958) provided additional evidence for this relationship. He found that the mean maximum body depth was 32.5 % of *Sl* in juvenile specimens 50 mm in *Sl*, while Mišík (1958a) reported that among adults of the same population with a *Sl* of about 500 mm, the mean maximum body depth was only 27.5 % of *Sl*. A less obvious relationship of this kind has frequently been observed among wild carps from pond cultures. It was expressed by the ratio of body length to body width. For two-year-old wild Danube carps reared in a Slovakian pond, this ratio, expressed as an index, ranged from 2.94 to 3.13 with the mean 3.04. Corresponding index values were obtained experimentally by STEFFENS (1964a), who calculated the following indices for wild carps originating in Lake Velence and cultured in a German pond: for fed K_1 2.80 to 3.28 (mean 3.01), K_2 2.70 to 3.27 (2.97), K_3 2.89 to 3.44 (3.19); for unfed carps K_2 2.69 to 3.21 (2.97), K_3 2.89 to 3.44 (3.19). Toward the end of the four year period of culture, the index reached a mean value of 3.37.

RUDZINSKI (1961) calculated the same index for wild carps from the Danube that were cultured in Polish ponds, some of which received feed. He calculated the following mean indices for those that were fed and those that were not: for $K_{2,}$ 2.89 and 2.98, respectively, and for $K_{3,}$ 2.93 and 3.13, respectively. Similarly, CHYTRA *et al.* (1962) calculated the corresponding mean indices for wild Danube carps reared in South Bohemian ponds: $K_{1,}$ 2.90; $K_{2,}$ 3.05; and $K_{3,}$ 3.24. Finally, for the wild form in the Neusiedler See, UNTERÜBERBACHER (1964) calculated indices for the K_2, $K_{3,}$ and K_4 carps, which were 2.72, 3.11, and 3.36, respectively. This brief review of the literature demonstrates that this index is very useful for distinguishing carps of wild populations from the domestic form, which always has a greater body depth (MIŠÍK 1958a, b, TUČA 1958, CHYTRA *et al.* 1962, RUDZINSKI 1961, STEFFENS 1964a).

The importance of the body width has been discussed in the section on sexual dimorphism, but the experimental data obtained by STEFFENS (1964a) should be given some special attention. Under the pond culture conditions discussed above, he calculated ratios of body length to body width for K_1 carps receiving feed that ranged from 4.84 to 6.00 with a mean of 5.51. For K_2 carps, the corresponding values were 5.14 to 6.29 with a mean of 5.61, and for $K_{3,}$ they were 4.60 to 6.38 with a mean of 5.29. For carps that did not receive feed, the values for the K_2 ranged from 5.00 to 6.43 with a mean of 5.70, and those for the K_3 were 5.00 to 6.14 with a mean of 5.54. RUDZINSKI (1961) and CHYTRA *et al.* (1962) reported the body width as a percentage of *Sl.* RUDZINSKI (1961) calculated a mean for this percentage of 16.7 % for K_2 carps that received no feed, and 17.8 % for unfed K_3. For the K_2 carps that had been fed, he found a corresponding percentage of 17.9 %, while the percentage for fed K_3 was 18.1 %. CHYTRA *et al.* (1962) reported the mean values for K_1, $K_{2,}$ and K_3 to be 18.3, 17.6, and 16.6 %, respectively. STEFFENS (1964a) concluded that the differences between the wild and domestic carps from ponds are, within a given age group, clearly manifested by the relation of body length to body width. For all experimental groups of wild carp, this relation ranged from 5.29 to 5.70 with a mean of 5.53. For all groups of domestic carp, they ranged from 4.51 to 5.25 with a mean of 4.86.

Additional information on how the mass of the body and individual organs changes with the age and size of the fish, as determined from studies on the wild Danube carp population, was provided by STEFFENS (1964a).

Subspecies

Although the wild carp and its numerous domestic forms have been cultured and exploited for centuries, the subspecific systematics of this taxon and its division into subspecies are still subject to much discussion. Preliminary data on the intraspecific systematics of *C. carpio* were published by SVETOVIDOV (1933), who distinguished two subspecific categories on the basis of three meristic characters: the number of *Sp.br.* on the interior edge of the first gill arch, the number of scales in the *l.l.* and the number of branched rays in the dorsal fin. These two groups are the European and East Asian carp. After engaging in experimental systematics of the carp, KIRPICHNIKOV (1943) revised and considerably expanded his earlier findings (KIRPICHNIKOV 1967). He concluded that there are at least four subspecies of the wild carp: the European and Transcaucasian nominotypic subspecies, *C. carpio carpio* (L.); the Central Asian *C. carpio aralensis* Kirpichnikov, 1967; the East Asian *C. carpio haematopterus* (Temminck et Schlegel, 1842); and the South Chinese and Vietnamese *C. carpio viridiviolaceus* (Lacépède, 1798). To the meristic characteristics used by SVETOVIDOV (1933) to distinguish these subspecies, KIRPICHNIKOV (1967) added the number of vertebrae and three morphometric characters, the eye diameter, length of the barbels, and intestine length. Certainly, KIRPICHNIKOV'S (1967) taxonomic conclusions have to be regarded with caution, but the assertion by BALON (1995a, b) that KIRPICHNIKOV (1943, 1967) did not present any evidence to support his views cannot be accepted. In the case of each subspecies, KIRPICHNIKOV (1967) reviewed the information on each of the nominal subspecific taxa that had been grouped together by BERG (1949), who had accepted only two of them as valid: *C. carpio* L. and *C. carpio haematopterus* (Temminck et Schlegel, 1842).

A basic problem that persists concerns the significance of the meristic and morphometric characters used by these authors for the intraspecific taxonomy of *C. carpio*. The data of MIŠÍK (1958a) on the wild Danube carp and GROMOV (1973) on the wild Amur carp populations still remain the only reliable sources on which an evaluation of the systematic value of individual characters can be based. According to the opinion of SVETOVIDOV (1933) and the much more precise conclusions by MIŠÍK (1958 a,c), from among the entire set of characters analyzed in the sections 'Variations and age' and 'Size variability', only the numbers of branched rays in *D*, scales in *l.l.*, and gill rakers on the interior and exterior edges of the first gill arch can be useful. None of the other characters used by KIRPICHNIKOV (1967) can be used to distinguish the subspecies of *C. carpio* because of their great variability or because enough data are still not available for a reliable analysis.

Also taken into account was the report of MIŠÍK (1958a) concerning the great adaptability of carps to different environmental conditions and their ability to develop somatic modifications. According to KIRPICHNIKOV (1943), wild carps can undergo somatic changes according to their physiological requirements. It should be remembered that both meristic and morphometric characters can change positively or negatively within a wide range.

BALON (1958a) stated that the morphological adaptability of wild carp in their variable environment is related to their great viability. He stated that populations of wild carp, in contrast to domestic forms, not only have a great ecological adaptability but also can respond in a variety of ways to changes in their external environment by rapid modification of their genotype.

Under conditions that might evoke changes in the morphometric and meristic characters, we analyzed selected meristic characters of wild carp populations only according to their pertinence to natural geographical distribution. The available data are too limited to permit statistical evaluation, but our method reveals the extent and direction of the individual changes in the characters or the degree of the differences encountered. According to the data on selected meristic characters that were reviewed (Table 24), the total range of variability, the range of mean values, and the cumulative values of these means in four geographical populations of wild carp were compiled (Table 27). For one character, the number of gill rakers on the outside of the first gill arch, we were able to use only the published data available for wild populations in the Danube (MIŠÍK 1958a, STEFFENS 1964a) and in East Asia, including those in the Amur Basin, China, and Japan (SVETOVIDOV 1933).

With regard to the dependence of the gill raker count on the size of the fish, KIRPICHNIKOV (1967) reported that this character increases in wild carps only in relation to the body mass while it is in the range from 40 to 50 g, that is, while the

Table 27: Selected meristic characters suitable for recognizing subspecies of the wild carp, according to different authors (see "Description" and Table 24).

Character	Area	Europe and Transcaucasia	Middle East	Far East, Northern China, Korea, Japan	Southern China and Vietnam
Db	total range	15-23	15-22	14-21	-
	range of means	18.10-20.30	17.60-19.40	16.50-19.40	19.77-20.33
	cumulative mean	19.20	18.80	18.09	19.99
	difference in %	-	-2.08	-5.78	+5.26
l.l.	total range	33-41	33-41	33-40	-
	range of means	36.28-39.08	35.60-39.00	37.30-39.60	31.45-32.60
	cumulative mean	38.14	36.98	38.03	32.06
	difference in %	-	-3.04	-0.29	-16.40
Sp.br. extern.	total range	18-29	21-28	17-28	-
	range of means	21.42-26.49	23.50-27.50	20.00-24.10	18.10-20.75
	cumulative mean	24.96	25.51	21.96	19.43
	difference in %	-	+2.20	-12.02	-20.24
Sp. br. intern.	total range	29-36	-	24-31	-
	range of means	30.20-31.87	-	-	-
	cumulative mean	31.04	-	27.32	-
	difference in %	-	-	-11.98	-

age group is not greater than 1+. However, STEFFENS (1964a) provided experimental evidence that the number of gill rakers on the external edge of the first gill arch increases with size and mass over a much longer period. The mean values were 26.95 for K_1, 28.60 for K_2, and 30.20 for K_3. Therefore, we used the value for K_3 in our calculations because the *Sl* of these fishes, 190 to 340 mm, seems to be closest to that of the fishes in MIŠÍK's (1958a) collection. By reviewing the variability and the mean values shown in Table 27, it is possible to find certain features that are of taxonomic importance for distinguishing the other subspecies from *C. carpio carpio*, which has been morphometrically defined in great detail by MIŠÍK (1958a) and STEFFENS (1964a) and also described in more complete detail by other authors (Tables 22–24).

After evaluating the number of branched rays in *D*, we were able to confirm the conclusion of MIŠÍK (1958a) that East Asian populations of wild carp tend to have fewer. In addition, it was found that in Southeast Asia, the carps tend to have more branched rays in *D*. The Central Asian populations do not differ markedly in this character from the nominotypic population.

The second character that was compared, the number of *l.l.* scales, exhibits a moderate, but apparently statistically insignificant, tendency to be lower in Central Asian populations. In contrast, the Southeast Asian populations can be easily distinguished from all others by their distinctly lower number of these scales. It can be concluded that, except of the case of the Southeast Asian populations, the number of *l.l.* scales is a character of minor value for distinguishing the subspecies of *C. carpio*.

Without doubt, the number of gill rakers on both the interior and exterior edges of the first gill arch had been considered to be the most important feature for determining the intraspecific systematics of *C. carpio* in the past (SVETOVIDOV 1933, MIŠÍK 1958a, BĂNĂRESCU 1964, STEFFENS 1964a, BALON 1974, 1995a, b). Our analysis showed that Central Asian populations differ from the type population in the number of *Sp.br.* on the external edge of the first gill arch only to a relatively minor degree. In contrast, there is a significant difference between the East and Southeast Asian populations in the number of gill rakers, which is lower among the Southeast Asian populations than among both the Central and East Asian carps. The statistical significance of the differences in the number of the *Sp.br.* on the exterior edge of the first gill arch was calculated by SVETOVIDOV (1933) and later by MIŠÍK (1958a), who considers this number sufficient for characterizing European and East Asian wild carp populations.

Almost all characters compared exhibit a considerable overlap within the range of their values. This was reported by MIŠÍK (1958a) and repeated by BALON (1974, 1995a, b). The systematic position of the Central Asian population of wild carp seems to present the most problems. Using a different approach, MIŠÍK (1958a) analyzed the features of the individual populations separately and concluded that the wild common carp from the Aral Sea and other Central Asian regions appears to be closely related to the European form in some characters but more closely related to the East Asian wild carp in others. However, it differs substantially from the East Asian carp in a number of features. The characteristics of the wild carp in western Central Asia were influenced during an earlier geological age. If western Central Asia was indeed the center of origin of the modern wild common carp, descendants of the original form had a longer time and possibility to evolve in isolation than did the wild carp that dispersed eastward and westward in the last postglacial period. Even though populations integrated later, more morphs may exist in the western part of Central Asia than elsewhere (BALON 1995a, b).

This situation has already been discussed by KIRPICHNIKOV (1943), who compared the Central Asian populations with the typical European subspecies and only later (KIRPICHNIKOV 1967) described them as an independent subspecies, *C. carpio aralensis*. After analyses of the meristic characters of Central Asian populations as well as evaluations of their systematic value and general differences within the species *C. carpio*, we do not consider *C. carpio aralensis* to be a valid subspecies. Independent systematic categories lower than subspecies, designated as 'races' by KIRPICHNIKOV (1967), probably exist in this area and seem to be manifested by various differences in the characters of different populations. However, after summarizing the results, the quality and value of these characters appear to be insignificant. We therefore have formed the opinion that Central Asian morphs belong to the typical European subspecies, *C. carpio carpio*, although they may belong to systematic groups that do not merit the status of subspecies.

Therefore we propose the hypothesis, supported by analyses of selected meristic characters, that only three subspecies of *C. carpio* can be recognized at the present time:

1. The European and Central Asian common wild carp, *Cyprinus carpio carpio*, formed as the ancestral population spread to the west;
2. The East Asian common wild carp, *Cyprinus carpio haematopterus*, originating from those that dispersed eastward;
3. The Southeast Asian wild carp, *C. carpio viridiviolaceus*, descended from the carps that spread southward.

It should be mentioned that the characters of the South Asian wild carp are most distinct from the other two subspecies. KIRPICHNIKOV (1967) also suggested the existence of a fourth wild carp subspecies in the Indonesian area. However, in this area, the carp is apparently not native but descended from individuals introduced from China, Europe, Taiwan, and Japan since the middle of the nineteenth century (SUMANTADINATA 1995).

We also took into account the recent opinions concerning certain fish species with disjunctive distribution ranges, such as the bitterling, *Rhodeus sericeus*. HOLČÍK & JEDLIČKA (1994) suggested that we dispense with such vague subspecific taxa altogether. There may be many altricial and precocial common carp forms within each population and geographical area of distribution (BALON 1989, BALON and GOTO 1989), which, however, should not produce any doubt that there is only one species of the wild carp, *Cyprinus carpio* L., in Europe, Transcaucasia, Central Asia, the Far East, Japan, Northern and Southern China, Korea, and Vietnam. Feral domestic carps, *Cyprinus carpio* morpha *domestica*, which descended from ancestors that escaped culture or were introduced (LELEK 1987a, b), are currently confusing the attempts to characterize this species. However, this should not be the cause for absolutely rejecting proposals to recognize subspecies, as suggested by BALON (1995a, b). We are not convinced that the same situation that exists in European rivers and lakes also exists in other large parts of the range of this species, especially those in Central and East Asia that are isolated and less influenced by man. Continued study of the intraspecific systematics of *C. carpio* is important to elucidate the evolutionary history of the species, especially in the separate geographical sites of their domestication. Thus, it seems important to define the precise taxonomy, not only of the wild carp but also of the domestic forms. We agree with BALON (1995a, b) that molecular biologists will have to confirm or revise the contemporary opinions and concepts concerning the intraspecific systematics of *C. carpio*.

Finally, the following key is presented to identify the subspecies of *C. carpio*. It can be applied effective only if mean values are employed, necessitating examinations of a sufficient number of specimens. The significant features described are characteristic of genetically pure wild carp with geographical origins that have been well established.

A. The range of the mean number of scales in *l.l.* is 31–33; that of *Sp.br.* on the exterior edge of the first gill arch is 18–21. Inhabits Southern China and Vietnam. . . . . *Cyprinus carpio viridiviolaceus*

B. The range of the mean number of scales in *l.l.* is 35–39(40) that of *Sp.br.* on the exterior edge of the first gill arch is 20–28. Inhabits Europe, Transcaucasia, the Middle East, Far East, northern China, Korea, and Japan. C

C. The range of the means of *Db* is 18–21, cumulative means* 18.80 – 19.20; the absolute range of *Sp.br.* on the interior edge of the first gill arch is 29–36, cumulative mean 31.04, and that on the exterior edge is 18–29, cumulative means 24.96 – 25.51. Distribution: Europe, Transcaucasia, and the Middle East. . . . . *Cyprinus carpio carpio*

D. The range of the mean of *Db* is 16–20, cumulative mean 18.09, the absolute range of *Sp.br.* on the interior edge of the first gill arch is 24–31, cumulative mean 27.32, and that on the exterior edge is 17–28, cumulative mean 21.96. Distribution: Far East, northern China, Korea, and Japan. . . . . *Cyprinus carpio haematopterus*

*"Cumulative mean" represents the average value of the mean characters found in analyzed populations of the wild common carp from particular geographical regions (derived from Table 24).

Hybrids

A natural hybrid of the wild common carp, *C. carpio carpio*, and the crucian carp, *Carassius carassius*, from Neusiedler See has been described as a separate species, *Cyprinus kollari* Heckel, 1836. HECKEL and KNER (1858) later placed it in a separate genus, *Carpio*. The same hybrid has also been found on the Crimean Peninsula (KESSLER 1859) and in Poland (DYBOWSKI 1862). This hybrid has two or sometimes only one pair of barbels, which are shorter and thinner than those of the wild common carp. In some cases, these barbels are completely lacking. The pharyngeal teeth are arranged in two rows with the formula 1.4–4.1.

The formula for the scales in *l.l.* is 34–39, *Squ. sup.* 6–8, *Squ. inf.* 6–7. The body length reaches only 300 mm, or in exceptional cases, it is longer (SABANEEV 1892, ANTIPA 1909, BERG 1949). BĂNĂRESCU (1964) diagnosed this hybrid from the Danube River in Romania as follows: *D* IV, 17–20 (21); *A* III, 4–5; *l.l.* 35–38 (39); *Vert.* 33–36 (37). He also provided an illustration of it that had been originally prepared by ANTIPA (1909).

According to NIKOLYUKIN (1972), experimental hybrids of female *C. carassius* and male *C. carpio* exhibit the following meristic characters: *D* III–IV, 17–20 (21), most frequently IV, 18–19, n = 167; *A* III, 5(6), n = 72; the number of scales in *l.l.* is 36–39, *Squ. sup.* 6–8, *Squ. inf.* 6–9 and most frequently 37–38, with a mean of 37.4, *Squ.sup.* 7, *Squ. inf.* 7, n = 60. The pharyngeal teeth formula for one-year-olds was (1).4–4.(1); the numbers in parentheses indicate a small projection, the future tooth, on the external row. Rarely other combinations of 1.4–4.(1) and (1).4–4.1 occur. The formula 1.4–4.1 is the most common among two-year-olds. However, other arrangements also appear, such as 4–4.1, 1.1.4–4.1, 1.1.4–4.1.1 and 4–4. Among 50 specimens, the number of gill rakers, *Sp.br.*, was (26) 27–30 (31, 32) with a mean of 28.3. A group of fishes very similar to the wild carp appears among the first filial F_1 generation, when another group similar to crucian carp also appears. However, specimens with intermediate characters also occur. Hybrids are seldom fertile, and among the rare fertile fishes, males considerably outnumber females. SABANEEV (1892) stated that this kind of hybrid usually occurs at sites where both common and crucian carps occur. Hybrids are apparently found in central Russia more frequently than to the South.

In most aquatic habitats where crucian carps occur together with domestic common carps, natural hybrids between them are often found. This hybrid is no longer found in Romania, where the crucian carp has practically become extinct. NIKOLYUKIN (1972) characterized the hybrids between female *C. carpio* morpha *domestica* and male *C. carassius* that he obtained experimentally as follows: *D* (III)IV–V, (16)17–21, most frequently IV, 18–19, n = 184; *A* III, 5–6, usually 5, n = 250. The number of scales in *l.l.* was 35–38, most frequently 35–36. The following combinations of pharyngeal teeth were found: 1.4–4.1, 1.1.4–4.1, 1.4–4.1.1, 1.1.4–4.1.1, 1.4–4.2, 2.4–4.2, 1.1.3–3.1.1, 1.4–4, 4–4.1.1, and others. The range of the total number of vertebrae was 33–36 (37) and most frequently, 35–36 with a mean of 35.1. These hybrids grow slower than domestic carps, but they resemble them in body form (SCHÄFERNA 1934). According to HANEL (1992), one eight-year-old hybrid from a backwater pool of the Labe River reached a *Tl* of 331 mm, and two six-year-old specimens from Klíčava Reservoir averaged 346 mm in *Tl*. They grew faster than crucian carps.

Many hybrids between different wild carp subspecies from various geographical regions and domestic races or landraces have been reared as part of a zootechnology selection program (STEFFENS 1975, 1976, BAKOS and GORDA 1995, FLAJŠHANS and GALL 1995, HULATA 1995). Crossbreeding of common domestic carp and species belonging to other genera of cyprinids, including *Carassius, Ctenopharyngodon, Hypophthalmichthys, Aristichthys, Labeo, Tinca,* and *Barbus,* have also been successful. Information on hybrids of the domestic common carp obtained prior to the end of the 19th century was compiled by LEONHARDT (1904). Much literature concerning carp hybrids is listed in the bibliographies of SCHWARTZ (1972, 1981).

Natural hybrids of the wild carp, *C. carpio,* and the silver crucian carp, *Carassius auratus gibelio*, are very rare. BERG (1949), reviewing information from NIKOLSKIY (1934, 1940), reported the discovery of one specimen of this hybrid, 200 mm in *Sl*, from the Aral Sea at Urga. The numbers of rays in the fins and scales in the lateral line (*l.l.*) were the same as those of the wild carp, and barbels were also present. However, the hybrids had a greater body depth, 49.4 % of *Sl*, and a characteristic silvery body colour. Another specimen of this natural hybrid, a female, was caught in Lake Balkhash in 1959 (SEROV 1966). It had 37 scales in the *l.l.* and a pharyngeal tooth formula of 1.4–4.1.

A natural hybrid (Fig. 29) of the silver crucian carp, *C. auratus gibelio*, with the domestic common carp, *C. carpio* morpha *domestica*, was caught by PROKEŠ and BARUŠ (1996) in Věstonice Reservoir at the Nové Mlýny waterworks in South Moravia in 1995. It was a male, 435 mm in *Sl* and weighing 2000 g, at an age of 7+. Its meristic characters were within the range of variability of the common carp: *D* IV, 17, the last ray being doubled; *A* III, 5; *P* I, 15; V II, 8; number of scales in the *l.l.*, 36, with 5 scale rows above and 5 below the *l.l.*; *Sp.br.* 25 on the exterior edge of the first gill arch and 31 on the interior; *Vert.* 35. The intensive silvery colouration of the whole body surface, including the dorsal part, is very characteristic of this hybrid. The abdominal region is silvery-white with the scales darkly margined. Two barbels, thinner and shorter than those usually found on the carp, were present on the right side of head, while only one dorsal one was present on the left side. The maximum body depth was 40.23 % *Sl*, and the head length was 27.58 % of *Sl*. The record of this hybrid is also remarkable in view of the fact that only gynogenetically reproducing female populations of the silver crucian carp were known from the Czech Republic (LUSK and BARUŠ 1978, BARUŠ and OLIVA 1995). If no bisexual population of *C. auratus gibelio* is encountered at this location, it would seem likely that the hybrids were released into this artificial lake together with

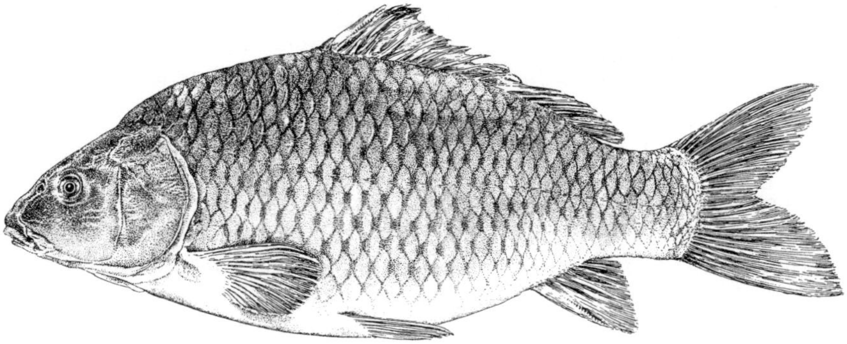

Fig. 29: Adult hybrid *Cyprinus carpio* x *Carassius auratus gibelio* captured in the Věstonická Reservoir, Czech Republic, on May 16, 1995. It is a male aged 7+ with an *Sl* of 435 mm and weight of 2000 g. (Drawing: PETR PELIKÁN, Brno).

carps of the Ropsha race, which have since become abundant members of the local fish fauna. Normally, bisexual populations of *C. auratus gibelio* are encountered in the eastern part of its range and also in the lower Danube (KUX 1982, FISTER and SOLDATOVIC 1991, BARUŠ and OLIVA 1995). Hybrids of wild or domestic carps with wild or domestic silver crucian carps have frequently been obtained experimentally. Many combinations of the different subspecies and colour varieties of both parental species are known. Increased attention has been given to the study of these hybrids for production as well as to determine their morphological and genetic characteristics (KHOKHLINA 1941, KEIZ and ADUMUA-BOSSMAN 1972, PULAN and SMITH 1987, CHERFAS *et al.* 1995). Comprehensive reviews on these extremely complex studies were provided by SCHWARTZ (1972, 1981) and HULATA (1995).

BĂNĂRESCU (1964) after COSTEA and POJOGA (1955) also described a hybrid of the wild carp, *C. carpio carpio,* with an apparently domestic form of the goldfish, *C. auratus auratus*, from Romania as follows: *D* III, 21; *A* III, 6; V II, 8(9); number of scales in the *l.l.* 38–39 with 5 scale rows above and 6 below the *l.l.*; *Sp.br.* 24 on the exterior edge of the first gill arch; *Vert.* 35; pharyngeal tooth formula 1.1.3–3.1.1.; maximum body depth 26.3–30.3 % of *Sl*; length of head 27.0–32.3 % of *Sl*; peritoneum light like that of the carp.

Distribution

The wild carp is a species that had been distributed over a vast contiguous area in temperate Eurasia prior to the last Ice Age (BERG 1909, 1949). NIKOLSKIY (1956) therefore included the wild carp in the archaic faunal complex of the upper Tertiary, which probably inhabited the entire northern part of both the Old and New World until the Pliocene. According to both of these authors, the wild carp disappeared from Siberia due to the cooling during the Ice Age, and its range was broken into a western and an eastern part. However, no evidence has yet been found to establish the occurrence of the wild carp in the New World. TURDAKOV (1963) expressed a different opinion, suggesting that the eastern and western areas of distribution had already been separated prior to the Ice Age by intensive orogenetic activity in Central Asia. However, this opinion has not been generally accepted. KIRPICHNIKOV (1967) explained the disjunct range of the wild carp in this part of Asia by stating that the entire range must have been in Siberia, which, of course, is not the case. Another important consideration in determining the origin of the wild European carp is that identification of the Ice Age remains of *Cyprinus carpio* in Western Europe (RUETIMEYER 1860, ZAUNICK 1925) is dubious (FOREL 1906, STEFFENS 1980), and its presence during that period has never been confirmed. The ancestral form, which possessed a greater number of pharyngeal teeth like *C. carpio anatolicus* Hankó, 1924, probably evolved in the area of the Caspian Sea at the end of the Pliocene. Some fossil forms from Hungary had numerous pharyngeal teeth, much like *C. c. anatolicus* (HANKÓ 1932).

It is generally assumed on the basis of documentary evidence that the carp's range shrank dramatically during various periods of glaciation. As stated by many authors, the modern wild common carp, *Cyprinus carpio,* evolved from an ancestor near the Caspian Sea and repopulated the basins of the Black and Aral Seas. Under the optimal temperature conditions during the postglacial period, some individuals spread as far as the Danube and possibly into the eastern and southeastern parts of continental Asia (BERG 1948, 1949, NIKOLSKIY 1956, BALON 1967, 1974, 1995a, b, BĂNĂRESCU 1960, 1964, 1992, KIRPICHNIKOV 1967).

BALON (1995a, b) did not believe that the eastern border of the wild carp's natural range is in Japan because the Japanese archipelago was last connected to mainland Asia during periods of glaciation in the Pleistocene (FUJITA 1973, IKEBE 1978), long before the modern wild carp could have arrived (ISHIKAWA *et al.* 1931). BĂNĂRESCU (1964) therefore did not include the Japanese archipelago within the natural range of the wild carp (Fig. 30).

Although KIRPICHNIKOV (1967) estimated that the wild carp has existed for at least 10 millions years, other authors have provided convincing arguments that the wild carp appeared in the western part of its present natural range, including the Danube River, only about 8,000 to 10,000 years ago (BORZENKO 1926, HANKÓ 1932, BALON 1967, 1974). It is therefore unlikely that the carp naturally occurred in Central and Western Europe beyond the Danube River System at the beginning of the Christian Era. Other evidence indicates that the Morava and Dyje Rivers (tributaries to Danube) formed the westernmost border of the wild carp's natural range (BALON 1967, 1969, 1974, 1995a, b). All subsequent records of the sporadic occurrence of the carp in Western European rivers may be explained by escapes from

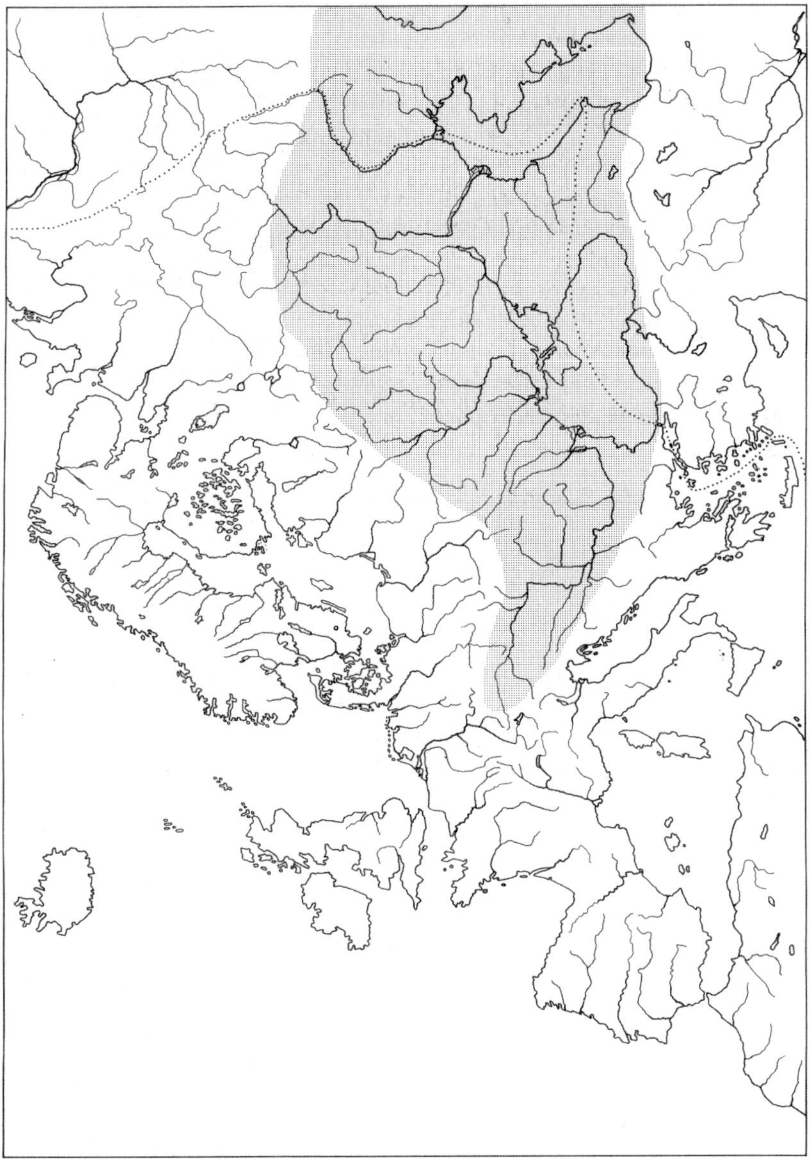

Fig. 30: The original natural range of the wild carp, *Cyprinus carpio*, after Bănărescu (1964).

ponds (HOFFMANN 1994a, b, 1995). The presence of the carp in the downstream sections of the Danube basin can be considered an extention of its natural range within the watersheds of the Black, Caspian, and Aral Seas (BERG 1949, BĂNĂRESCU 1964). Similarly, BERG (1949) reported that the wild carp is distributed mainly in the lower course of the Volga River downstream from Kamyshin. In contrast, it occurs very sporadically in the middle and upper parts of the Volga. BERG (1949) wrote, "they are nearly fully missing in some years, while they could be rather numerous during other years". In addition, the natural range of the wild carp encompasses the Kuban, Kuma, Terek, and Kura Rivers; the Aral Sea and its rivers, including the Amu-darya upstream to Piandzha and Zeravshan and the Syr-darya up to Kara-dari in Uzgen and Narina in Uch-kurgan. It also inhabits the lower and middle sections of the Sary-su River, the Chu River upstream to Tokmak, the Talas River at Dzhambul, and Lake Issyk-kul (BERG 1949). It also inhabits lakes as far as Uzboi, Iaskhan, Karategelek, and Topiatan. In contrast, the wild carp was absent

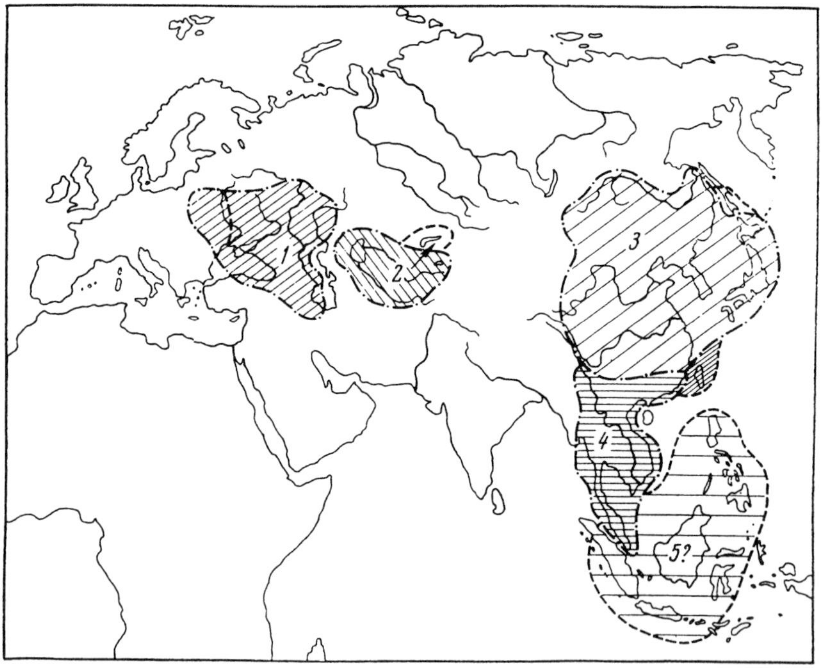

Fig. 31: The ranges of the individual subspecies of the wild carp, *Cyprinus carpio,* according to KIRPICHNIKOV (1967). **1** – The European-Transcaucasian *C. carpio carpio*; **2** – Central Asian *C. carpio aralensis*; **3** – Amurian-Chinese *C. c. haematopterus*; **4** – Vietnamese *C. c. viridiviolaceus*; 5 – presumed Indonesian subspecies for which only an approximate range is indicated.

from the southern part of Anatolia, the Tigris and Euphrates Rivers in Western Asia, Iran south of the Caspian Basin, Turkmenistan, India, the drainage area of Lake Balkhash prior to its introduction, and other water bodies of Central Asia.

BERG (1949) reported that the wild carp's range extends to the Dniepr River at Smolensk, where it is apparently rare; the Desna River as far upstream as Briansk; and the Volga River as far as its upstream section at Lake Seliger, as previously mentioned. It also occurs in the Sheksna, Oka, Sura, Vetluga, Sviiaga, and Kama Rivers as far as their upstream sections as well as in the Viatka and other rivers. It is rare in Lake Beloozero. In the Kura River, the wild carp can be found at relatively high elevations as far upstream as Lake Tarapavani. It also occurs in rivers flowing into the Caspian Sea from the South and the basin of the Atrek River. This part of the wild carp's natural range in Europe and Central Asia is inhabited by the subspecies, *C. carpio carpio*. In contrast, the rest of the Central Asian part of the range (Fig. 31) is inhabited by *C. carpio aralensis,* if indeed this is a valid subspecies.

The eastern part of the wild carp's range is inhabited by *C. carpio haematopterus*. According to BERG (1949), this includes the basins of the Amur and Onon Rivers, the Argun River upstream to its confluence with the Amur, the Sungari and Ussuri Rivers, Lake Khanka, and the Suifun River. Korea, Japan, and China southward to Fou Zhou are also included within the natural range of the subspecies (BERG 1949). Within Mongolia, the Onon, Kherulen, and Khalkhin-gol Rivers as well as Lake Buir-nur are considered to be inhabited by indigenous populations of wild carp (SHATUNOVSKIY 1983). That the wild carp is not native on the Japanese archipelago has already been mentioned (BĂNĂRESCU 1964, BALON 1995a, b).

According to SVETOVIDOV (1933), the Southeast Asian part of the range, which is inhabited by the subspecies *C. carpio viridiviolaceus*, encompasses Kwang-Tung in China, Vietnam, Laos, Cambodia, and Burma (Southeast Asian Peninsula). However, no reliable information is available concerning the native occurrence of wild carp in Indonesia, as suggested by KIRPICHNIKOV (1967). Unfortunately, the literature lacks more detailed information on the occurrence of the wild carp in East Asia.

Introductions: BERG (1949) reported that the wild carp has been introduced and cultured outside of its original area of distribution, in such countries as Poland, Germany, Denmark, southern Sweden, England, and Scotland. He mentioned that wild carps have been resistant enough to withstand the unfavorable climate in the Nieva River Basin, where they had obviously been introduced, as well. Introductions of wild carp populations to Central Asia include those in the Murab River of Turkmenistan in 1895 and in Lake Chelkar and the basin of the Ili River in 1905, from which they invaded Lake Balkhash, where they subsequently reproduced on a massive scale. From this locality, they were transported to Lake Büli-Kul and surrounding lakes in the basin of the Talas River. Later, in 1934, they were introduced into the Barabinsk lakes and Lake Alakul, east of Lake Balkhash. BERG (1949) remarked that wild carps were present elsewhere in Europe, Asia, and America without providing more precise localities or sources of information. He also stated that the wild carp occurs in France, Spain, Italy, and Greece, which may also be the result of introductions carried out in the past. Carp culture probably began in Italy during the time of the Roman Empire. In our opinion, the wild carp's present range must have been a result of such early introductions beyond its natural range (HOFFMAN 1995). SABANEEV (1892) mentioned the import of fishes alleged to be wild carp into England in 1504, Denmark and Sweden about 1560, and eastern Prussia around 1769. However, the report of the Prussian introduction remains somewhat doubtful.

We stated that all of the introductions mentioned above seem to have involved populations from European and Central Asian water bodies. These introduced carps therefore belong to the subspecies *C. carpio carpio* or *C. carpio aralensis*. Recent introduction of the wild carp from the Slovakian section of the Danube and from Lake Velence in Hungary to Germany, Poland, and the Czech Republic were undertaken for research purposes and to improve the stocks (STEFFENS 1959, 1964a, b, 1976, RUDZINSKI 1961, CHYTRA *et al.* 1962, 1963, WOJNO and GUZIUR 1972).

Only a few remarks in the literature concern the introduction of the East Asian subspecies, *C. carpio haematopterus*. SHATUNOVSKIY (1983) mentioned that this subspecies has been adapted to conditions in Lake Kenon. Introduction experiments conducted in Russia during the mid 1940s resulted in their invasion into the Selenga River. Recently, their range in Mongolia has increased to include the Tula River and Lake Ugii-nur.

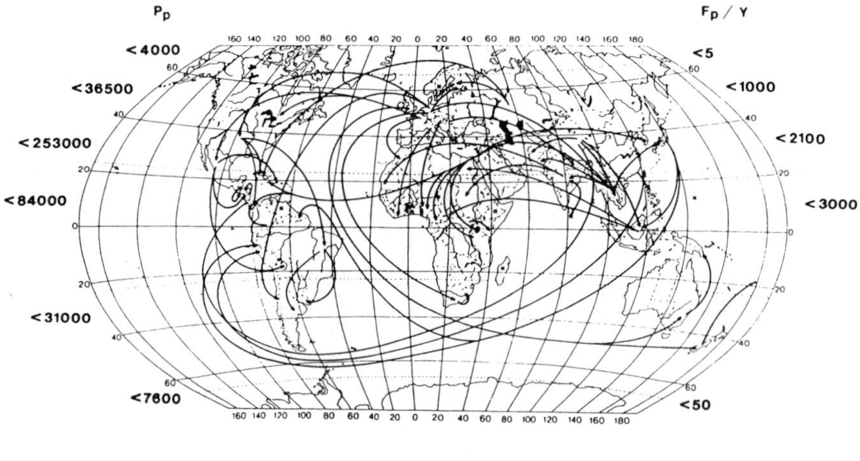

Fig. 32: Worldwide introductions and transfers of common carp, *Cyprinus carpio*, based on data from WELCOMME 1988 and HOLČÍK 1991. According to BIRÓ 1995.

There is no doubt that this subspecies of the wild carp was introduced into many water bodies of the former Soviet Union (SHIRKOVA 1955, POGODINOV and VOROBYOV 1954). It was hoped to use them for selection experiments and cross-breeding with local domestic races of carp aimed at increasing their resistance to low temperatures and during winter (BRIZINOVA and KIRPICHNIKOV 1952, KIRPICHNIKOV 1944, 1958, KIRPICHNIKOV and BERG 1952). *C. c. haematopterus* is considerably more resistant to low temperatures than *C. carpio carpio* of European origin.

Many races and landraces of the common domestic carp as well as the carp subspecies *C. carpio carpio* from the Tisza River; the East Asian subspecies, *C. carpio haematopterus* from the Amur River; and the Vietnamese scally carp, *C. carpio viridiviolaceus,* from northern Vietnam, are presently cultured in Hungary at the Fish Culture Research Institute at Szarvas (GORDA *et al.* 1995).

The intraspecific taxonomy of the domestic common carp, *C. carpio* morpha *domestica,* is complicated by the changes that occurred due to domestication and frequent transport of various races and landraces. The term "race" is used for a cultivated carp population inhabiting a limited geographical area, while the term "landrace" is used for individuals of a smaller population within the race. Historically, the carp has been the most widely cultivated warmwater fish in Europe, and its various cultured forms have been introduced into many other parts of the world, including at least 81 countries on all continents, except Antarctica (STEFFENS 1958, LELEK 1987b, WELCOMME 1988, HOLČÍK 1991, HOFFMAN 1995, BILLARD 1999). The purposes of these introductions have differed considerably and include aquaculture, sport, improvement of wild stocks, and the culture of ornamental fishes. Most risks due to transporting common domestic carp to other sites within its native range are considered negligible (BIRÓ 1995).

Correct zoological nomenclature should be applied to the different races as well as their landraces that have been introduced to other regions (KRATOCHVÍL 1966). For domestic races descended from wild Danube carp, we prefer *C. carpio carpio* morpha *domestica* as the name. In the case of the carp in China, which is another possible centre of domestication for wild carp races, the appropriate zoological name is *C. carpio haematopterus* morpha *domestica.* South Chinese and Vietnamese wild carps that might have been domesticated would be correctly called *C. carpio viridiviolaceus* morpha *domestica.*

Ecology

Within their natural range, wild carps are freshwater cyprinids that were, and in some places still are, very abundant. They inhabit lower courses of rivers in lowlands, usually with moderately flowing or standing water. In spite of the fact that the wild carps prefer lentic and slow lotic waters, they occasionally inhabit rapidly flowing waters with current speeds exceeding 2 m.s^{-1} (Nikolskiy and Evtukhov 1940). Other typical habitats for them are estuaries, back waters, and bays containing brackish water. Berg (1909) reported that wild carps inhabiting the Aral Sea occur within the entire range of salinity. This was also reported for carps in the Caspian Sea (Kazancheev 1981), where they usually inhabit waters along the western and southern coasts. Wild carp populations are also numerous in different sized lakes, mainly at lower elevations.

From this information, it is clear that wild carps occupy very diverse habitats facilitated by their great ability to adapt morphologically, physiologically, and ecologically. However, there are also some requirements for their survival, particularly those required for reproduction throughout their entire range (Kirpichnikov 1967). The ecological niche of the wild carp is considerably limited, according to the cited author, by the presence of competitors and predators. The species has therefore not been able to extend its range without setbacks. A suitable temperature is also necessary for their successful survival and reproduction, which gives some insights into the origin of this species (Sabaneev 1892). The populations of wild carp in the Far East are much more resistant to coolness and winter conditions than those in Europe and Central Asia. In spite of the fact that the wild carp is a very important fish economically, knowledge of its ecology, as already noted by Balon (1958a), is unfortunately insufficient.

Habitat: Because precise information is lacking, the biotope of the wild carp can be characterized only in a general way. For wild carps in the Danube, the ancestors of the domestic *C. carpio carpio* m. *domestica*, the biotope is found in lowland stretches of this large river and of its tributaries, which contain rich food sources and offer suitable conditions for reproduction. The carp prefers mainly lotic stretches of the rivers, where the sediment consists mainly of mud and sand but also may include smaller amounts of gravel. The stretches preferred are very sunny and support submerged vegetation in their backwaters. The carps also inhabit oxbows on flood plains and lentic water bodies in front of the dikes. Connections between the main channel and the oxbows on the flood plain also seem to be important for successful reproduction. In 1956, the wild carp was still distributed along nearly the whole Slovak and Hungarian sections of the Danube and some of its tributaries (Balon and Mišík 1956; Mišík 1958a, Jászfalusi 1960 and Žitňan 1985). A rapid decline in its abundance during subsequent years shows that other conditions are also important, such as water quality, the presence of waterways connecting the river with water bodies on its flood plain, rhythmic patterns of water discharge, and regular flooding. Recently, these conditions have changed considerably due to the damming, interruption of river continuum, construction of large hydroelectric plants and large navigation projects. Unfortunately, a serious decline has been observed among the wild carp populations even though they can be considered one of the most resistant fish species in the middle Danube. The former abundance of the carp in drainage area of the lower Danube was also determined by the existence of very large shallow lakes in the flood plain which offered excellent conditions for spawning and nursery of juveniles. The illogical decision to drain these shallow lakes during the late 1960-ties and early 1970-ties resulted in catastrophic decrease of the wild carp populations in Romania.

Another important habitat characteristic that facilitates the survival of the wild carp is the presence of deep waters along the stream, such as those in pools and backwaters, in which the fishes can spend the winter (SABANEEV 1892). Schools comprised of thousands of individuals congregate at such sites for dormancy in the Dniepr and Volga Rivers, as well as in the sea. The importance to the wild carp of being undisturbed during winter so that its rates of metabolism and activity can be considerably reduced, can be inferred from observations of analogous behaviour by the domestic carp (STEFFENS 1975).

Populations of wild carp inhabiting the sea usually prefer brackish, littoral habitats near estuaries and bays, which are often overgrown by reeds. This is the case in the northern part of the Caspian Sea, where the wild carps are distributed along the entire coast at depths reaching 4 to 6 m (BERG 1949). The same situation is encountered in the estuaries of large Ukrainian and Russian rivers.

Migrations: Within the range of the wild carp, two ecological forms are usually present together in the river basins: a resident, non-migratory form that permanently inhabits certain stretches of the river as far as the upstream limits of its estuary and a migratory or the semi-migratory form inhabiting parts of the estuaries and regularly migrating upstream for spawning (SABANEEV 1892, BERG 1949, KAZANCHEEV 1981). These fishes only undertake short trips along the courses of the rivers, related to patterns of dial and seasonal activity. When high water levels occur during spring, from April through June, spawning migrations are undertaken toward the flooded grassy areas near dikes, such as those in the Danube at Medved'ovo, or into backwaters connected to the main river, such as the Büdes and Ispanyos at Pal'kovičovo. The fishes also migrate into inundated meadows upstream from the canals flowing into Lake Lion at Čičovo. Wild carps in the Danube migrated from this lake into Čilizský potok Creek during the spawning season and spawned on freshly flooded meadows in the vicinity of these canals. The spawning sites are always located on flooded meadows, in backwaters, or near dikes on the flood plain of the Danube. Migrations away from the main channel may cover a distance from several hundred meters to 5 or 6 km, according to the depth of the flood. After spawning, the wild carps return from the flood plain to the river channel with the receding flood waters and continue in their resident mode of life. Spawning migrations occur only while the fishes are seeking suitable spawning sites, which are available only during the spring floods (BALON 1958a, MIŠÍK 1958a). Similar short migrations by resident wild carps were also observed by SABANEEV (1892). KAZANCHEEV (1981) observed that the body colouration of migratory and semi-migratory forms of the wild carp is usually brighter, especially dorsally. Resident, non-migratory populations of wild carp are also known from the basins of the Volga, Kura, Kuban', Dniepr, Don, and other rivers (Berg 1949). Non-migratory populations are distributed along the entire course of the Volga as far as the estuary. Sabaneev (1892), citing the observation by V. E. Jakovlev, reported that wild carps inhabiting seas near the mouths of estuaries migrate upstream as far as 200 versts, which equals 213.4 km, for spawning, and that they always return after spawning. He also mentioned that the populations of wild carp in rivers may undertake even longer migrations, depending upon the degree of gonad maturity and upon the accessibility of suitable spawning sites. After spawning, members of the populations in rivers migrate from the inundated areas downstream to their original habitats, and their behaviour again becomes fully resident within one month.

Berg (1949) reported morphological differences between members of the resident population in the Volga River and the migratory and semi-migratory wild carps that ascend that river from the Caspian Sea. The vertebrae in the wild carps that migrate in spring number 35 to 38, and their mean is 36.1, while the corresponding range and mean of the sedentary population are 35 to 39 and 37.1, respectively. There is also a difference in the growth rate, which is faster within the migratory populations. The onset of migration by wild carps was observed in the second half of March, and the run was completed by the second half of May. Spawning itself took place from the second half of May to mid-June. Berg (1949) also reported that the distance covered during migration in the Volga River is only about 15 km. Kazancheev (1981) remarked that the spring spawning migrations by the wild carps into the Volga River is no longer characterized by large numbers of fishes because the population has considerably declined. This decline is also the reason why the wild carps no longer migrate upstream in the Volga but rather remain in the sea near of estuary, where suitable spawning sites are also present. Berg (1949) also described a wild carp migration from the sea into the Volga that occurs during summer and autumn, from the beginning of August until November. However, the biological significance of this migration remains unexplained. The migrations of the populations in the Ural River are very similar to those of the Volga. The wild carps spend the winter in deep waters of the Ural Estuary and migrate to spawn during April and May, when they ascend the river for a distance of 40 to 60 km from the river mouth. After spawning during summer and autumn, they remain at shallow sites in the sea near the river mouth, where they also spend the winter.

Resident and semi-migratory populations of wild carp inhabit the lower course of the Terek River. According to Kazancheev (1981), the resident populations initiate a spawning migration into this river by mid-March. The semi-migratory ones begin somewhat later, usually before the beginning of April. Kazancheev (1981) also stated that migrations of wild carps into the Kura River were formerly very intensive and prolonged, and in addition to the spring run in April and May, a second spawning migration was observed during summer and autumn, from August through October. According to BORZENKO (1926) and BERG (1949), wild carps migrated upstream for spawning in the Kura as far as Lakes Alzhikabul and Sary-Su. The permanent spawning sites were located in vegetated waters along the coasts of Kizilagachskiy Bay. Only a few wild carps migrate into the Kura River at the present time because most of the fishes are captured in the sea near the river mouth (KAZANCHEEV 1981). The spring migration by wild carps into the Atrek River begins in March. It is particularly intensive during the second half of that month and at the beginning of April (KAZANCHEEV 1981).

The wild carp population in the Aral Sea could have been regarded as resident in the past because only short spawning migrations were undertaken similar to those of the populations in rivers. Wild carps were observed spawning very close to the coast, where they spawned in the salt water of the shallow littoral zone (BERG 1949) beside the entire shoreline of this sea except along the West Coast. Spawning occurred throughout a large range of salinities, reaching as much as 10 ‰ near Vozrozhdenia Island (NIKOLSKIY 1934). The spawning migration by adults toward the sea coast began in March. They prefer deep sites in the sea during summer, and even deeper sites were selected for wintering.

Wild carps younger than three years, including members of the resident populations, migrate to inundated areas on the flood plain during spring. Of course, this is not for spawning but rather for finding the food that is available in excess on flooded areas. In this way, the fishes also escape the strong currents in the rivers during the spring flood. On the other hand, the success of downstream migration by young wild carps from flooded areas into the river depends on the rate of decrease in the water level on the flood plain. Juveniles are more successful in returning to the river when the flood is prolonged and the water recedes slower. The longer the flood remains, the more successful is the recruitment of wild carps (KAZANCHEEV 1981). ŽITŇAN (1985) stated that larval and juvenile carps on the flooded meadows and areas between the dikes along the Slovakian section of the Danube often congregate in shallow depressions after a sudden decrease in the water level and cannot reach the river unless the water discharge rate in summer is high. Fry and juveniles of the wild carp are easily preyed upon by fish-eating birds, and they perish whenever the temporarily flooded habitats dry up.

SABANEEV (1892) was probably the first person to draw attention to a paradoxical phenomenon observed in large Russian rivers: spawning success during consecutive years is inversely related to the water level during the spring flood. Thus, when the discharge during the spring flood is relatively great and the inundation of the floodplains is correspondingly more extensive, spawners from the main rivers undertake longer migrations. The migration of the newly hatched fry and juveniles from the more remote spawning sites to the main river therefore becomes more difficult or even impossible as the water level decreases. This causes considerable losses, which might even reach 90 % of the year class or more. In contrast, when the level of the spring flood is low and only limited areas close to the rivers are inundated, most of the adult fishes, especially relatively large ones, spawn in backwaters and in shallow locations near the main channel. The fry can then reach the river more easily.

No semimigratory or migratory populations of carp living mainly in the sea and ascending the river only for spawning have not been distinguished in the lower Danube, as a consequence of the fact that the Black Sea is more salty along the Romanian coast than along the Ukrainian one. The carps normally live in the brackish lagoons near the Danube mouth, the populations from these lagoons being morphologically and biologically identical to those from fresh water. The specimens carried by high water into the salty parts of the Black Sea return soon into the Danube or die if they cannot (ANTIPA 1909).

The carp is listed in the Romanian ichthyological literature, alongside *Leuciscus idus, Rutilus rutilus, Abramis brama,* and *Stizostedion lucioperca* as a semi-migratory species, since they undergo migrations within the freshwater habitat. Most specimens from the lower reaches of the Danube, possibly also some from the lakes of the Danube Delta, migrated, until a few decades ago, into the very wide shallow lakes of the Danube floodplain which extend along the northern (left) side of the river, from Calafat, downstream from the Iron gate, almost to the Delta; the flood plain also included some shallow lakes between the arms of the Danube upstream from the confluence with the river Siret and several others on the right side of the Danube. Its total area includes 480,000 hectares of which only 90,000 were flooded permanently, while 390,000 hectares flooded only during the high water period.

The migration of the carp into the shallow lakes began early in the spring at a water temperature of 7–8 °C. Adult carp remained in the shallow lakes after having spawned and returned into the Danube between the end of June and beginning of August, when the water level began to become lower or the water temperature reached 24 to 26 °C. The enormous quantities of young produced in the shallow flood plain lakes migrated with the adults (or somewhat later) and spread through the entire area of the lower Danube, including the deeper lakes of the Danube Delta, which offer less favourable conditions for spawning and as a nursery for juveniles than do the shallow lakes (ANTIPA 1909). It is possible that the middle Danube, which has a much smaller flood plain in Hungary and Slovakia, was also populated by carps partially from the shallow lakes of the lower Danube. These migrations ceased due to drying of the flood plain, and the quantities of carp have strongly diminished throughout the Danube drainage area.

In the Far East, populations of adult wild carps undertake similar spring spawning migrations, and the fry migrate in the same manner from the flood plains to the rivers (KRYZHANOVSKIY *et al.* 1951). GROMOV (1973) studied the migrations of wild carps tagged in Lake Bolon' and recaptured in the basin of the Amur River, some of them up to four years later. He found that about 60 % of the tagged individuals undertake spawning migrations regularly and return to the same localities. The remaining individuals, constituting about 40 % of the population, migrate to spawning sites and feed in the nearest water bodies or flooded lowlands. Wild carps did not travel more than 50 to 80 km from their usual spawning and feeding sites. Longer migrations of up to 350 km were undertaken only in exceptional cases, obviously due to a lack of suitable conditions for spawning, feeding, or wintering. Only rarely do fishes, swimming alone, migrate through submountain stretches of tributaries in the Amur River System. Some adults ascend such rivers as the Sungari, which are located from 100 to 150 km upstream from the mouth of the Amur. GROMOV (1973), after completing his research, stated that local shoals of wild carp inhabit sections of the Amur River. Some of them are considerably isolated, as reflected by their morphometric differences. Both river and lacustrine forms of the wild carp are therefore thought to inhabit the Amur.

Hardiness: BALON (1967a, 1974) rightly concluded that wild carps are among the most resistant of all fish species in the Danube against oxygen deficiencies, inadequate food supplies, and even starvation. Relatively large spawning schools still occurred in westernmost parts of their natural ranges between the mouths of the Morava and Hron Rivers during the early fifties. River regulation and damming, such as that producing the Gabčíkovo disaster, and possibly also the complete draining of the large shallow lakes in the flood plain of the lower Danube, changed the entire area and destroyed convenient spawning sites. As a result, the wild carp was classified as a vulnerable species in the Slovakian stretch of the Danube until 1985, while it was listed as in danger of extinction in the Rhine (BARUŠ 1981, BARUŠ *et al.* 1989). The survival chances of this species deteriorated considerably, as recently documented by HOLČÍK (1995). Concerning the wild carp in the Slovakian section of the Danube, he stated, "at present, however, its distribution is restricted to the minor part of the main channel of the Danube. Only single specimens are occasionally caught. Among 1536 fish taken, tagged and released in the Danube at Gabčíkovo in 1993, only five (0.3 %) were carps with only one still displayed the typical characters of the wild form! ...The extinction of ...the wild carp is expected within a decade...". Thus, the wild carp in this section of the Danube will have to be rightfully considered in critical danger of extinction for several years to come (HOLČÍK 1983, 1989, 1996, ŽITŇAN 1985).

Wild carp populations have been seriously harmed during spawning when schools congregated on flooded meadows, where the fishes were easily captured in seines. When such fishing methods were employed, not only were the fishes prevented from spawning naturally due to the capture of most prior to the release of their eggs but even the eggs already sticking to the grass were destroyed (BALON 1958a). Such a harmful way of fishing for wild carps has been used not only along the Danube but also in other Eastern European rivers within the natural range of the species (BERG 1949, KAZANCHEEV 1981). In such situations, which probably still occur, not only the resident but also the semi-migratory and migratory wild carps are captured. The migratory populations in the Kura River are already decimated in the sea beyond the estuary, and that is the reason why the spring spawning migrations into this river are presently very sparce (KAZANCHEEV 1981). A similar situation also exists in the estuary of the Volga River. Similarly harmful methods have been used for capturing wild carps in the Amur River (KRYZHANOVSKIY *et al.* 1951). ŽITŇAN (1985) noted that losses were also caused during the migration of newly hatched fishes into the rivers when various stationary fishing devices were used at sites on the inundated areas between the dikes to catch the fishes returning to the river channel. This kind of fishing is considered poaching because the schools are fully protected by law during the spawning season.

The danger to larval and juvenile wild carps when they are prevented from escaping the temporary pools and depressions on the floodplain while it is rapidly drying was already pointed out. ŽITŇAN (1985) added that these losses increase due to the predation by young pikes, *Esox lucius*, and fish-eating birds as well as due to the increased occurrence of parasitic infections. It was also observed while seining the spawning schools of wild carp together with other fish species that the main predator on the wild carp eggs was *Chondrostoma nasus* (MIŠÍK 1957, BALON 1958a).

Feeding habits: Because its mouth protrudes from a position ventral to the longitudinal body axis, the carp is unable to prey upon rapidly moving animals. However, this mouth is well suited for ingesting benthic food items to a depth of 15 cm below the sediment surface, for consuming the leaves of submerged plants, and for capturing the minute planktonic organisms in the water column. The unsuitable food particles are expelled from the mouth by reversing the water flow (SCHÄPERCLAUS 1961). The ingested food is crushed by pharyngeal teeth and bound together by mucus as it passes into the long, tube-like intestine. The carp does not have a stomach. Trypsin is its main enzyme for digesting proteins.

The earliest published information on the food of the domestic common carp was provided by DUBRAVIUS (1547). Later, BLOCH (1783) and ŠUSTA (1884) added more details, particularly concerning the digestive tract. Extensive information is available in a great many publications on the domestic carp, which were listed by MUIR and ROBERTS (1982). However, only sketchy information has been provided concerning the feeding of the wild carp. BĂNĂRESCU (1964) reported some information on the diet of the wild carp in the Danube. He noted that the larvae fed on zooplankton after resorption of the yolk sac. Similar findings were reported by KAZANCHEEV (1981), who listed planktonic crustaceans as the main food of carp larvae. Both of these authors also found that the juveniles start to eat zoobenthos after they have reached a length of 18 mm. VOVK (1955), while rearing wild carp in experimental ponds in the Ukraine, found that chironomid larvae were included in the diet of individuals 7 mm long. Relatively large cladocerans and copepods only rarely occurred in ten-day-old larvae of wild carp. Even smaller quantities of plant remains and detritus were present in their intestines. A similar diet was observed for specimens 15.5 to 18 mm in *Sl*. The main items in the diet of carp fry 70 to 113 mm long observed in ponds during early autumn were cyclopids and chironomid larvae. Detritus is regarded as a second-rate food. Rarer food items include cladocerans of the families Bosminidae and Daphniidae, Monogononta of the genus *Keratella*, and body remains of insects and oligochaetes. The wild carp can generally be characterized as omnivorous, consuming both animal and plant food (SABANEEV 1892, BERG 1949, BĂNĂRESCU 1964, KAZANCHEEV 1981). The abundance of individual dietary components is strongly related to the food sources available in a given habitat.

BĂNĂRESCU (1964) stated that adult wild carps prefer lamellibranchs in the family Adacnidae in the Danube River, as well as dipterans, mainly in the family Chironomidae, and ostracods, amphipods, and mysids. They also consume crabs of the genus *Rhithropanopeus* and plants. The predominant food components in the complex of shallow lakes called Balta Brailei were found to be gastropods and lamellibranchs as well as adult and larval insects and plant detritus. Rarely, oligochaetes, filamentous algae, and diatoms were also found in the carp's diet. Planktonic crustaceans were consumed less intensely. In the shallow Calarasi Lake, the most abundant items in the diet of wild carp were larvae of chironomids and copepods of the family Cyclopidae, which were consumed together with plant detritus. The larvae of *Bezzia* sp. as well as cladocerans, oligochaetes, and molluscs were included in small quantities, while plants remains were rarely found. Two-year-old wild carps from the Danube Delta fed on copepods, ostracods, trichopterans, and diptera of the family Chironomidae, while molluscs and the remains of plants were rarely consumed.

In the Kura Estuary near the sea, plants, molluscs, and aquatic crustaceans were the most frequent components in the diet of adult carps, while in the Kura River, chironomid larvae were most abundant followed by oligochaetes, plants, and detritus (KAZANCHEEV 1981). The diet of adults from the Volga Estuary consists of gastropods, dipteran larvae of the families Chironomidae and Culicidae, and crustaceans of the family Gammaridae. BERG (1949) stated that molluscs, crustaceans, and oligochaetes are the main components in the diet of adult wild carps in the Caspian Basin. Ostracods and molluscs of the genera *Adacna* and *Dreissena* were the most abundant food component in the Aral Sea, while the larvae of dipterans of the family Culicidae and cladocerans are more important in its coastal zone. VOVK (1955) found that the predominant food item for adult wild carps from pond cultures in the Ukraine was chironomid larvae, followed in importance by crustaceans and plants. Much organic and inorganic detritus was always present in the intestines. Undoubtedly, the information on the food of the wild carp provided by SABANEEV (1892) is not only of historical value. This author stated that the wild carps, apparently the adults, prefer to consume plants rather than animals, such as worms, larvae, or adult aquatic insects. He considered the young shoots of the cattail, *Typha* sp., and other aquatic plants to be the main component of the diet. He also included the roe of other fish species in the diet of the carp during this period. Furthermore, he provided information that is quite interesting on the grains, such as oats, rye, wheat, and millet, that were the main food items of the wild carps in large, navigable Russian rivers with busy boat traffic during a 6 to 7 month period each year. He mentioned that thousands of puds (1 pud = 16.38 kg) of grain were often thrown into the river and that many boats sank with their freight in the Volga, Dniepr, and other rivers. This information, however, was limited to conditions at the end of 19th century. SABANEEV (1892) also mentioned cases of wild carps being "caught by glitter" and a perch being found in the intestine of a carp during late autumn. He deduced that small fishes could also be caught near the hibernation sites during winter.

Longevity: The wild carp is a fish with a long life span (BERG 1949, KAZANCHEEV 1981, KRUPKA 1982). BALON (1957) found 15-year-old specimens in a spawning shoal in the Lesser Danube River near Kolárovo, where males survive as adults for a period from 3 to 8 years, and females, for up to 9 years. At the age of 15, wild carps are senescent.

KAZANCHEEV (1981) mentioned that carps were captured at an age of 20 years. However, because of intensive fishing, old carps do not occur in the Caspian Sea. The oldest specimens in the estuary of the Ural River were 12 to 14 years old, while those in the Volga Delta were 9 to 10 years old. Wild carps of age of 7 to 10 years old constitute more than half of the total catch in the Volga Delta (KAZANCHEEV 1981). The maximum age of the wild carp recorded in the Aral Sea is 17 years (BERG 1949).

Data on the age of the domestic common carp are more abundant and reliable. KÁLAL (1995) reported that some carps live more than 40 years. For example, GIGER (1957) recorded a maximum age of 43 to 44 years, while BROWN (1957) mentioned an age of 47 years.

SABANEEV (1892) provided some historically interesting information on this subject. He stated "that a 30-year-old fish should have a weight of 50 Russian phunts (20.5 kg), a 40-year-old fish 2 Russian puds (i.e. 32.8 kg), a 50-year-old around 2.5 puds (40.9 kg), and finally a 100-year-old fish hardly reaches the weight of 3 puds (49.2 kg)." Furthermore, he stated that large specimens weighing 28.7 to 56.1 kg must have lived for many years, indicating that the carp must be a long-lived fish. Yet he doubted that the longevity of the wild carp is this great.

Information on carp of an "extreme age of 100 to 200 years" was reviewed by BLOCH (1783), whose paper contains a list of other authors. He included data on very old carps from the ponds at Charlottenburg palace, which he said were "being grown by moss". This incorrect information persisted in the literature and was the basis of a legend among fishermen, which persisted until the beginning of the 20th century. It is known that as many as 200 circuli form on the scale of a three-year-old carp. These were erroneously thought to be annuli or growth rings, each representing one year. In fact, each annual ring on the scale usually consists of several dozen circuli. Erroneous reports that individual carps were extremely old can be explained by a misinterpretation of the rings (OLIVA 1955).

Growth: Like that of other fish species, the growth rate of the wild and domestic carp is influenced by many factors, including the availability, variety, quality, and energy content of the food as well as the population density, length of the season suitable for macrophyte growth, water temperature, oxygen saturation, and the health condition and genotype of the carp. Hence, the growth potential of different wild carp populations under natural conditions differs from that of selected domestic races and landraces cultured under either extensive or intensive conditions in the temperate zone. It also differs from that of carp in the tropics and subtropics.

The mean standard length of the carp was back-calculated using data obtained mainly from the stock of one subspecies, *C. c. carpio,* in the Lesser Danube River of the Slovak Republic (BALON 1957, 1995a). It is clear from the values computed (Table 28) that the growth in length of males is slightly faster during the first year of life (l_1). However, females grow progressively faster during subsequent years of life. From the relatively large data set on the growth in length of the wild common carp within its natural range (Table 29), it is clear that the growth exhibits great variability.

Among the populations inhabiting the drainage area of the Danube, those from the Lesser Danube River and the Neusiedler See grew most rapidly, while the populations from the courses of the Virt and Asót in the Slovak Republic grew most slowly (UNTERÜBERBACHER 1963, BALON 1957, KRUPKA 1987, SEDLÁR et al. 1987). The wild carp population in the Kura River is considered to grow the most rapidly among those in the basin of the Caspian Sea (BERG 1949). It is also characterized by a delayed maturation. Some first reach maturity at an age of four, but most do not reach it until they are five years old. From the values of *Sl* observed among the populations in the basin of Caspian Sea by BERG *et al.* (1949) and NIKOLSKIY (1957), however, a more rapid growth, especially of individuals older than two years, can be inferred. The populations of wild carp from the Sea of Azov exhibit a growth rate similar to those of the Danube and Caspian populations. The fastest growth within the basin of the Aral Sea was recorded for populations in Lake Issyk-kul, while the slowest growth was observed among populations in the Amu-darya River. Slow-growing populations are also known from the Ili and Surkhan Rivers. The growth of the Eastern subspecies, *C. c. haematopterus,* is considered to be as fast as that of the particularly fast-growing populations of *C. c. carpio* (NIKOLSKIY 1956).

In addition to the information provided in the above review on the growth variability of the wild common carp in the subspecies *C. c. carpio* and *C. c. haematopterus* under natural conditions within their natural ranges, some experimentally calculated production indices are available to enable a comparison with

Table 28: Back calculated mean standard lengths (*Sl* in mm) for males and females of the wild common carp, *C. c. carpio*, from the spawning stock in the Lesser Danube above Kolárovo (Slovak Republic) after BALON (1957, 1995a).

Year of spawning	Age group	Mean *Sl* at time of capture	Back - calculated mean *Sl* at the end of each growth season														
			l_1	l_2	l_3	l_4	l_5	l_6	l_7	l_8	l_9	l_{10}	l_{11}	l_{12}	l_{13}	l_{14}	l_{15}
Males																	
1952	III	422	154	382	410												
1951	IV	400	158	298	360	391											
1950	V	428	149	298	354	391	414										
1948	VII	459	135	228	304	363	398	425	448								
1947	VIII	462	142	196	296	358	391	426	445								
1945	X	548	180	203	246	436	450	472	504	522	541						
1944	XI	478	150	230	283	302	341	375	410	440	460	470					
1940	XV	598	148	235	278	335	361	410	422	467	485	501	521	536	550	568	583
			152	259	316	368	392	422	446	476	495	501	521	536	550	568	583
Females																	
1950	V	418	120	321	366	390	411										
1949	VI	458	137	228	325	416	430	450									
1948	VII	491	136	276	343	385	422	453	475								
1947	VIII	483	150	271	379	404	456	450	468	478							
1946	IX	527	169	231	323	384	421	452	491	509	526						
			142	265	347	396	424	451	478	493	526						

Table 29: Selected data on mean growth of standard length (*Sl* in mm) in the wild common carp (*C. c. carpio* – localities 1–21 and *C. c. haematopterus*, locality 22) from various natural habitats; **1–5** Danube drainage system, Black Sea Basin; **6–14** Caspian Sea Basin; **15** Azov Sea; **16–19** Aral Sea Basin and **20–22** other localities. Authors: **A** – BALON (1957 b), **B** – SEDLÁR *et al.* (1987), **C** – BĂNĂRESCU (1974), **D** – UNTERÜBERBACHER (1963), **E** – BERG (1949), **F** – POSCHALUJEWA (1929), **G** – BERG *et al.* (1949), **H** – NIKOLSKIY (1957), **I** – NIKOLSKIY (1940), J – NIKOLSKIY (1956).

Area	Author	l_1	l_2	l_3	l_4	l_5	l_6	l_7	l_8	l_9	l_{10}	l_{11}	l_{12}
1. Lesser Danube	**A**	176	314	398	458	490	524	554	582	613	-	-	-
2. Asód channel	**B**	107	218	301	366	418	458	509	541	587	600	-	-
3. Virt channel	**B**	92	175	268	348	414	455	467	-	-	-	-	-
4. Greaca Lake	**C**	150	226	299	371	436	480	523	571	618	653	-	-
5. Lake Neusiedlersee	**D**	-	252	367	424	471	520	560	-	-	-	-	-
6. Don River Delta	**E**	130	245	325	375	425	470	-	-	-	-	-	-
7. Kuban River	**E**	130	255	340	410	455	490	-	-	-	-	-	-
8. Ural River	**E**	130	240	330	360-370	420	-	-	-	-	-	-	-
9 a. Volga River	**F**	125	245	325	350	420	495	-	-	-	-	-	-
9 b. Volga River	**E**	130	250	330	380	410-430	-	-	-	-	-	-	-
10. Kura River	**E**	145	280	385	455	500	-	-	-	-	-	-	-
11. Astara River	**E**	130	240	330	380	410	-	-	-	-	-	-	-
12. Gasan-Kuli	**E**	-	-	335	400	450	-	-	-	-	-	-	-
13. Argachanskiy bay	**E**	140	270	360	410	430-440	-	-	-	-	-	-	-
14 a. Caspian Sea	**G**	175	310	455	540	620	-	-	-	-	-	-	-
14 b. Caspian Sea, Southern part	**H**	147	200	359	437	504	561	-	-	-	-	-	-
15 a. Azov Sea	**G**	160	300	410	475	520	560	-	-	-	-	-	-
15 b. Azov Sea	**H**	130	255	349	410	455	490	-	-	-	-	-	-
16 a. Aral Sea, isle Vozrozhdeniya	**I**	126	215	293	360	416	465	-	-	-	-	-	-
16 b. Aral Sea, Karateren	**I**	128	215	287	343	384	-	-	-	-	-	-	-
17. Kamyshlabash Lake	**I**	121	195	253	300	340	-	-	-	-	-	-	-
18. Amu-darya River	**I**	79	128	172	233	302	362	400	-	-	-	-	-
19. Issyk-Kul Lake	**I**	135	260	350	417	471	525	590	-	-	-	-	-
20. Surchan	**H**	70	117	171	213	289	-	-	-	-	-	-	-
21. Ili River	**H**	91	148	190	223	249	265	-	-	-	-	-	-
22. Amur R., mean of 4 localities	**J**	132	244	318	380	436	482	530	588	637	665	686	698

similar indices for different races and landraces of domestic common carp and their hybrids (KIRPICHNIKOV 1943, STEFFENS 1959, 1964a, 1976, VOVK 1955, CHYTRA *et al.* 1962, 1963, RUDZINSKI 1961, BASTL 1962). Most of the available information on the growth in length and weight of wild carp reared in fishponds was obtained from studies of fishes in the 1+ to 3+ year classes. The results of all authors cited above generally confirm that wild carps in the subspecies *C. c. carpio* cultured in German, Czech, Slovak, Polish, Ukrainian, and European Russian fishponds containing both natural foods and a supplemental diet increase in both length and weight significantly more slowly than domestic races. Only during the first year of life is the significance of the differences in length and weight growth low. However, the decrease in the growth rate of the wild carp is quite remarkable during and after the second year of life.

From the first year of life, the wild carp is characterized by a body that is lower and narrower than that of the domestic races (CHYTRA *et al.* 1962), and this difference becomes more pronounced with increasing age. By their third year in pond cultures, wild carps have attained only approximately half the weight of domestic

Table 30: Mean standard length (*Sl* in mm) and weight (*w* in g) in the wild common (W) and domestic (D) carps in pond culture, according to STEFFENS 1964a. The origin of the wild carps, *C. c. carpio*, was Lake Velence, Hungary, of that of the domestic carps was Königswartha, Germany.

Age		*Sl* (mean)	*Sl* (range)	*v* (%)	*w* (mean)	*w* (range)	*v* (%)
W	K_1 n = 100	89.3	72 - 105	8.4	19.1	10.5 - 30.0	24.5
D	K_1 n = 100	111.3	84 - 137	9.9	48.3	19.0 - 83.0	30.1
W	K_2 unfed n = 60	206.7	160 - 250	10.5	231.3	105 - 357	26.5
W	K_2 fed n = 60	173.8	135 - 210	9.5	138.2	62 - 243	26.9
D	K_2 unfed n = 60	239.8	200 - 285	6.3	458.6	247 - 729	19.3
D	K_2 fed n = 60	219.8	185 - 260	7.4	345.6	197 - 587	24.7
W	K_3 unfed n = 97	261.1	190 - 340	14.1	448.5	183 - 995	40.8
W	K_3 fed n = 100	226.2	165 - 295	12.5	320.0	121 - 693	38.4
D	K_3 unfed n = 97	323.2	265 - 385	5.9	1120.5	644 - 1692	17.8
D	K_3 fed n = 100	303.1	255 - 355	6.4	1048.3	644 - 1449	18.6
W	K_4 fed n = 100	287.8	225 - 365	12.2	590.9	270 - 1170	34.7

common carps of the same age. These findings were well documented by STEFFENS (1964a; Table 30).

Additional data from the literature permitting comparisons of the standard length (*Sl*) and weight (*w*) of wild and domestic common carps reared under the same conditions were provided by BASTL (1962) and UNTERÜBERBACHER (1963). In Slovakian ponds, male wild carps attained a mean *Sl* of 385 mm and a mean weight of 1360 g while being cultured from K_1 to K_4. The females reached 402 mm and 1475 g, respectively. Corresponding values for male domestic K_4 carps were 480 mm in *Sl* and 3977 g in weight, while for the females, they were 484 mm and 4005 g, respectively. The mean *Sl* of four-year-old wild common carps from the Neusiedler See was 424 mm, and their mean weight was 985 g. The mean *Sl* of domestic common carps of equal age was 532 mm, and their mean weight was 2472 g. These averages were computed from combined data for both males and females.

In spite of their slower rate of growth in both length and weight, VOVK (1955) rated the wild carp favourably in comparison with the domestic races (VASILIU *et al.* 1947) because of their great ecological plasticity and resistance against diseases and unfavourable environmental conditions. These characteristics make them very suitable for pond culture, especially in southern Russia. Extensive experimental studies by KIRPICHNIKOV (1943, 1979) demonstrated that in the European part of the former USSR, wild *C. c. haematopterus* from the Amur River grow much faster in pond culture than Russian domestic landraces during the first year and, when there was competition for food, even during the second year of life. The growth potential and resistance of the wild Amur carp against cold and winter conditions are important advantages of this subspecies that should be effectively exploited in selection programmes for domestic common carp. These characteristics should also be considered when contemplating the introduction of this race into ponds in the northern parts of the region, where climatic conditions are less conducive to cultured races of carp descended from the subspecies *C. c. carpio* (KIRPICHNIKOV 1943, 1979).

Population dynamics: There is little information on the population dynamics of this species. BALON (1957) found that males were predominant among 78 specimens from a spawning shoal of *C. c. carpio* in the Lesser Danube River near Kolárovo. They accounted for 73 % of these specimens against only 27 % that were female. Some of these specimens were caught on May 7, 1955, when 69 % of the total were males and 31 % were females. The subsequent catch of carps spawning for the second time on June 7, 1955, included 76 % males and 24 % females. BALON (1957) reported the observations of local professional fishermen concerning year to year variations in the sex ratio and a frequent predominance of females in the catch. However, it has yet to be determined how the fishing method employed and the behaviour of fishes in the schools influence the sex ratio. MAKSUNOV (1955) found a slight predominance of males, which accounted for 53 % to 56 % of the wild

Table 31: Age composition (in %) of *C. carpio carpio* in samples from different areas. Data were provided by 1. BALON (1957) – Danube Basin in Slovakia, 2.–10. BERG (1949) – Caspian Sea Basin (both sexes); 11. NIKOLSKIY (1940) – Aral Sea.

Locality / Age class	1	2	3	4	5	6	7	8	9	10	11	12	13	14	15
1. Lesser Danube males	-	-	14	19	30	7	14	5	2	2	2	-	-	2	2
females	-	-	-	-	17	8	25	8	17	8	8	-	-	-	8
2. Ural R.	-	4	81	15	-	-	-	-	-	-	-	-	-	-	-
3. Volga - southern part of estuary	-	-	16	44	34	5	1	-	-	-	-	-	-	-	-
4. Volga - western part of estuary	-	-	56	27	16	1	-	-	-	-	-	-	-	-	-
5. Lagan'	-	-	41	41	16	2	-	-	-	-	-	-	-	-	-
6. Argachanskiy bay	-	0.2	8	36	45	10	1	0.2	0.1	-	-	-	-	-	-
7. Kura R.	-	-	1.5	5	41	27	16	7	1.5	0.2	-	-	-	-	-
8. Sara R.	6	21	31	40	2	-	-	-	-	-	-	-	-	-	-
9. Astara R.	4	63	32	1	-	-	-	-	-	-	-	-	-	-	-
10. Gasan - Kuli	-	-	38	47	12	2	0.5	0.1	-	-	-	-	-	-	-
11. Aral Sea males	-	-	9.8	25.8	31.0	24.0	8.4	0.5	0.5	-	-	-	-	-	-
females	-	-	1.2	9.2	29.2	28.8	18.6	8.4	2.6	1.6	0.4	-	-	-	-

carp spawning shoals in Lake Balkhash. Furthermore, a collection of 87 wild carps caught by MIŠÍK (1958a) in the Danube and Lesser Danube Rivers during spring, from March to June, in the years 1954 through 1956 consisted of 57 males, or 65.5 % of the individuals, and 30 females, or 34.5 %. The age distribution of some *C. c. carpio* populations is shown in Table 31. A comparison of the age structure of wild carp spawning shoals in the Aral Sea near Vozrozhdeniya Island (NIKOLSKIY 1940) and of the population in the Lesser Danube (BALON 1957) revealed that members of the Danube population attained sexual maturity later. In the Aral Sea, maturity was reached at an age of three or four, while the carps in the Danube matured at an age of five. However, the greatest number of females in the Danube population belong to age class 7.

Data on the age structure of wild carp populations from nine regions of the Caspian Sea, provided by BERG (1949), reveal that the largest numbers of specimens were usually in age classes 3 through 5. However, in the Kura River, age classes 5 through 7 were the largest, while in the Sara and Astara Rivers, the largest groups of specimens collected belonged to classes 2 and 3. It should be noted that the age structure determined has been influenced significantly by the inappropriate fishing methods employed during the spawning period, when the carps move into the shallow water on flooded meadows and become easy to capture using seines. The error is especially great for the oldest year classes. This kind of fishing interferes with natural spawning through the capture of the fishes before they can spawn, and the eggs sticking to plants are also destroyed. This fishing method, which is extremely destructive to wild carp stocks, has been practiced not only along the Danube River (BALON 1957) but also to a greater extent along such rivers as the Ural, Volga, Don, and Dniepr. Fishing with seines during the spawning period in the spring and at the beginning of summer is effective because wild carps do not display their usual fright reactions and lose their wariness. SABANEEV (1892) also described a fishing method, that is presently forbidden, which employs a two-pronged spear to stab the wild carp. This method of fishing during the spawning season of the wild carp was employed by many thousands of people throughout southern Russia south of 50° N latitude.

Reproduction Biology

Maturity: Male wild carps, *C. c. carpio*, first spawn at an age of about three years (BERG 1949, KAZANCHEEV 1981, KRYZHANOVSKIY **et al.** 1951). Spawners are usually longer than 360 mm in *Sl* and heavier than 1000 g. In the Aral Sea, individual carps can reach sexual maturity by an age of two years. However, mature carps are older in the Kura River, where they may reach maturity at an age of four but most frequently mature first during the fifth year of life (BERG 1949). The wild Amur carp also reaches maturity at this age (KRYZHANOVSKIY *et al.* 1951).

Gonads: BĂNĂRESCU (1964), on the basis of data by AURELIA NICOLAU, reported a maximum gonadosomatic index of 23.47 % for female wild carps. The absolute fecundity of a single female reaches a maximum of 1,900,000 eggs. The fecundity of the wild carp depends on the age of the spawners. For example, in the Danube, the following relationships between age and fecundity were observed (BANARESCU 1964): two to three year old fishes contain an average of 101,000 eggs, three to four-year-olds average 274,000; four to five-year-olds, 357,000; five to six-year-olds, 450,000 eggs; and six to seven-year-olds, 550,000 eggs. KAZANCHEEV (1981) examined three to ten year old wild carp from the Volga Delta, which ranged from 350 to 690 mm, with a mean of 445 mm in standard length. Their fecundity ranged from 179,000 to 1,130,000, with a mean of 530,000 eggs. It is also evident that the number of eggs is positively related with the standard length of the fish. This has been well documented for wild carps from the Volga Delta by BERG (1949) and KAZANCHEEV (1981), as shown in the following table:

Females, range of *Sl* in mm	Mean fecundity in thousands of eggs	
	BERG (1949)	KAZANCHEEV (1981)
310 – 350	162	–
360 – 400	222	298
410 – 450	278	–
460 – 500	365	–
510 – 550	508	–
560 – 600	810	963

In 80 wild carps from Kura River, BERG (1949) found that the fecundity ranged from 93,000 to 1,664,000 eggs. The maximum number of eggs was found in a female 920 mm in *Tl* and weighing 10.8 kg; the weight of the eggs was 1.7 kg. In the same river, KAZANCHEEV (1981) found that the mean fecundity of specimens ranging from 410 to 500 mm in *Sl* was 196,000 eggs. The average fecundity of specimens 510 to 600 mm in *Sl* was 363,000 eggs, and that of specimens 710 to 800 mm reached 650,000 eggs. Fecundity values lower than these was found by KAZANCHEEV (1981) for wild carps from the Atrek River, where the range was 16,000 to 543,000, and the mean was 125,000 eggs.

Spawn: Freshly spawned eggs of wild carps are yellowish. The mean size of the activated, swollen, and hardened eggs is usually 1.5 to 1.8 mm, and their yolk diameter is 1.2 mm (KRYZHANOVSKIY 1949, BĂNĂRESCU 1964, ANDRUSHCHENKO 1974, SMIRNOVA 1978, BALON 1995a, b). They seem to be somewhat larger than those of the domestic carp, according to information provided by PEŇÁZ et al. (1983). The external layer of the egg membrane is adhesive, enabling the eggs to stick to plants rather than falling to the bottom, where less oxygen is available (KRYZHANOVSKIY 1949).

Spawning period: The wild carp spawns by releasing only a portion of eggs at one time. A female normally releases two or three portions within 10 to 14 days. The carp has been placed in reproductive guild A.1.5, characterized as non-guarding, open substrate egg-scattering, obligatory phytophilous spawners (BALON 1975, 1981, 1990, 1995a, b). Spawning occurs from the end of April until the end of June or sometimes the beginning of July. It commences when the water temperature has reached 16 °C to 18 °C (BERG 1949), but it is interrupted if the temperature falls below 16 °C. Usually, only two portions of eggs are released, while some of the third portion are resorbed, and some are conserved for production of oocytes for spawning the following year. The first portion of eggs is always the largest, accounting for up to 86% of the total number of eggs released (KAZANCHEEV 1981). Wild carps from the Amur River, *C. c. haematopterus*, are known to spawn at lower temperatures, starting at 15 °C (KRYZHANOVSKYI *et al.* 1951). Wild carps seem to employ some ecological, reproductive, and behavioural mechanisms to retain their genetic distinction from the domestic forms (BALON 1995a, b).

The spawning habits seem to be very similar for wild carp populations in different river basins, including the Danube, Volga, and Amur. However, some indices reflecting certain morphological features of the various populations differed, just as they did for the wild and the domestic carps (BALON 1995a, b).

Spawning sites: In the basins of the Danube and other large central and eastern European rivers, the wild carp spawns during periods of high water on freshly flooded meadows at sites where the water depth is 25 to 50 cm (BERG 1949, BALON 1958a,c, 1974, 1995a, b; BĂNĂRESCU 1964). Numerous spawning schools congregate in flooded areas, backwaters, and in waters 25 to 50 cm deep in the littoral zone where aquatic or terrestrial plants are present. Wild carp populations in seas spawn along the coastline, where the water is shallow and aquatic plants are present. They may also spawn in big lakes at sites far away from estuaries of rivers and along the coasts of islands situated in the lakes (KAZANCHEEV 1981). According to BERG (1949), the wild carps in the Aral Sea spawn among the stands of *Phragmites* or in deeper water on *Potamogeton crispus* and *Myriophyllum* spp. As long as considerable changes in the hydrological and environmental conditions do not occur, the spawning sites are permanent and can be used year after year by the same wild carp populations.

Mating habits: At the beginning of the spawning season, wild carps undertake migrations toward the spawning sites. Mating groups usually consist of one female and several males. However, BALON (1958a, c) mentioned cases of one male being accompanied by a group of females. Young males at an age of three years predominate in spawning shoals when spawning begins, but females four to five years or older become more abundant as the spawning activity reaches its maximum. Finally, toward the end of the spawning season, the mean age of the fishes in the spawning shoals again decreases (BERG 1949).

Spawning habits: The wild carps usually spawn when the wind is almost calm during the early morning. Spawning is very violent. Mating groups, usually consisting of one female and several males, circle rapidly, often disturbing the surface of the water or with the dorsal parts of the body extending above the surface (BALON 1958c, 1995a, b). The released eggs stick to the vegetation and are often consumed in large numbers by other fishes, as observed by MIŠÍK (1957). During spawning, the wild carps lose their usual timidity and wariness.

Early ontogeny: The ontogeny of the wild carp is indirect with a metamorphosis of evolutive type. Fishes pass through a distinct larval period, and their development fits well into the general scheme described for cyprinid fishes (LANGE et al. 1974). Its developmental intervals are similar to those of the domestic carp, as suggested by PEŇÁZ *et al.* (1983).

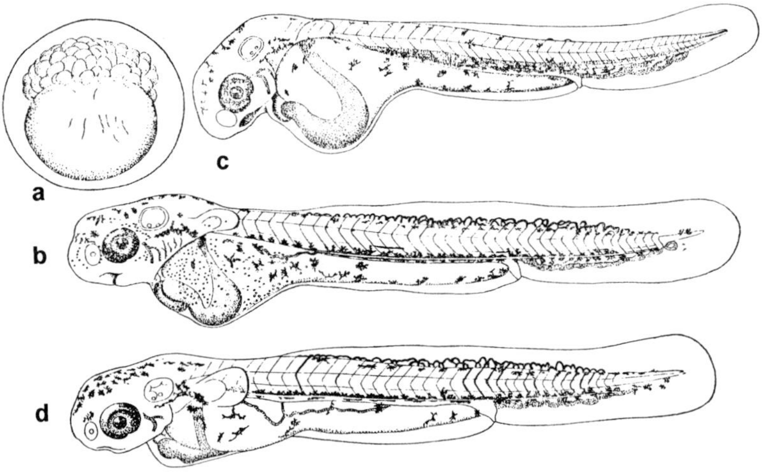

Fig. 33: Embryonic stages in the development of the wild carp, *C. carpio carpio*: **a)** an egg at the large-cell morula stage, age 3 hours and 10 minutes, egg diameter 1.69 mm, yolk sac diameter 1.3 mm; **b)** an embryo artificially removed from the egg envelopes before hatching, age 2 days and 8 hours, *Tl* 5.0 mm; **c)** a freshly hatched embryo at the beginning of the hatching process, age 4 days, *Tl* 6.7 mm; **d)** an embryo at the time most of the eggs are hatching, age 5 days, *Tl* 5 to 7 mm (after KRYZHANOVSKIY 1949).

The embryonic period, divided into nine embryonic steps, commences at the moment of fertilization and activation. At about 20 °C, it lasts for three days until the embryos commence external feeding, which occurs after they have reached a standard length of about 7 mm. The yolk-sac embryos hatch three to four days after activation, after they have attained an *Sl* of 4.5 to 5.5 mm, have relatively thick and short bodies, and are relatively richly pigmented on their external surfaces. The pigment is concentrated in two rows along the dorsal and ventral contours of the body. There are many asterisk-shaped melanophores on the head, yolk sac, and along the dorsum. Respiratory structures are well developed, including the ducts of Cuvier, anasthomosed interior caudal vein, segmented capillary loops in the dorsal finfold, and subintestinal vein. Sticky glands enable the embryos to

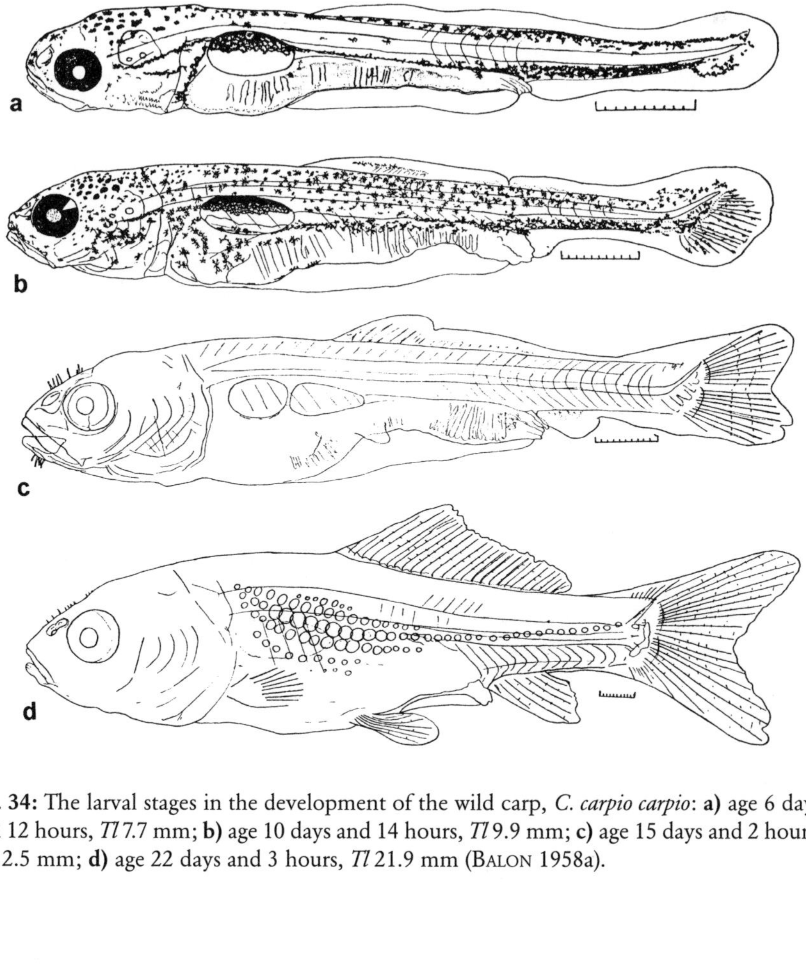

Fig. 34: The larval stages in the development of the wild carp, *C. carpio carpio*: **a)** age 6 days and 12 hours, *Tl* 7.7 mm; **b)** age 10 days and 14 hours, *Tl* 9.9 mm; **c)** age 15 days and 2 hours, *Tl* 12.5 mm; **d)** age 22 days and 3 hours, *Tl* 21.9 mm (BALON 1958a).

hang in a vertical position on submersed plants, where the oxygen supply is optimal (Fig. 33).

The larval period is usually divided into six steps. It commences at the transition to exogenous feeding and ends with morphological differentiation, characterized mainly by the completion of the fins and the replacement of provisional morphological structures by permanent ones. The initial steps of this period are characterized by yolk sac resorption, filling of the first swim bladder chamber, and start of digestive system functioning, including enzymatic transformation of exogenous nutrients. A typical pigment spot in the urostyle region is observed during this step. The subsequent steps in the larval period are characterized by the filling of the anterior chamber of the swim bladder, which rapidly becomes apparent by the improved maneuverability of the fish, both in the horizontal and vertical planes. Differentiation of the finfold into dorsal, caudal, and anal fins and the formation of paired fins also occur during these steps. The mouth shifts to a position lower on the head. The period ends about 23 days after activation (BALON 1958a), after an *Sl* of about 19 mm has been attained (Fig. 34).

The juvenile period commences at a standard length of about 17 mm, when the first scales appear. It lasts until the first oocytes mature. Usually, only the short initial part of this relatively long period is considered to be "early development". It is characterized by the rapid completion of all remaining organs and the scale cover of the entire body surface. On the wild carp, scales begin to develop on the anterolateral surface of the body after a *Tl* of about 18 mm is reached (BALON 1958a, b). According to KOBLICKAYA (1981), at approximately the same stage, after a *Tl* of 18 to 19 mm has been reached, both pairs of barbels appear (Fig. 35).

Early development is subject to great variation (CHERFAS 1940, DOLZHENKO 1953, VASNETSOV *et al.* 1957, EREMEYEVA 1960, SABLINA 1960, BAGIROVA 1963, 1964, KOKHANOVA 1969, SMIRNOVA 1978). Acceleration or retardation of the growth rate due to external factors, such as the food supply or temperature, is also reflected by

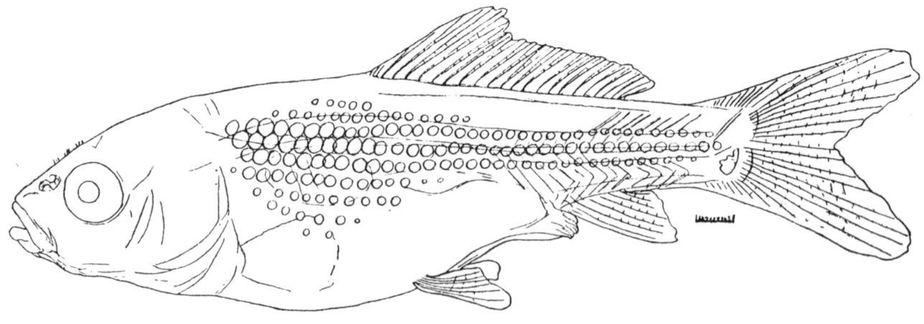

Fig 35: A juvenile wild carp, *C. carpio carpio*, 24 days and 3 hours old with a *Tl* of 24 mm, according to BALON (1958a).

corresponding changes in the rates of morphological differentiation of the organs. However, the rates of both growth and differentiation are independent from one another and not proportional. Older carps are usually more advanced morphologically than younger ones of the same size (GALKINA 1964, VERMA 1970, PEŇÁZ *et al.* 1983).

Hatchlings remain at the spawning site for only a short period. While the yolk sac is still not absorbed, they begin to disperse in the water body. During subsequent stages of their life history, they usually prefer benthic microhabitats, which they inhabit alone or in small groups. Zooplankton is the most important food during the larval period; however, a transition to omnivorous feeding begins during the next period of life.

Important Parasites and Diseases

The common carp, both wild and domestic, is apparently the most widely distributed fish species. Its cultured form, *C. carpio* f. *domestica,* is now found on almost all continents and under a wide range of geographical, climatic, and cultural conditions. As a result, there is a large group of pathogens that can parasitize the common carp or cause its diseases. As JENEY and JENEY (1995) stated, parasitic infestations are especially numerous.

Representatives of almost all major taxa of parasites have been found in the wild and domestic carp. With few exceptions, they have a world-wide distribution pattern. However, a few of them are characteristic of individual wild subspecies of the common carp, either *C. c. carpio* or *C. c. haematopterus*. Others provide examples of co-accumulation or co-adaptation of parasites on host species or on different races and landraces of the domestic common carp, including feral forms. These parasites were originally on various native fish species and adapted themselves to the carp after it was introduced into new areas. A list of parasites already found on the two Palearctic subspecies of the common wild carp within its native range and on the domestic common carp, *C. carpio* f. *domestica,* in regions to which it has been introduced is provided in Table 32.

The most complete checklist of parasites of the common carp previously compiled was published by LANDSBERG (1991). He recorded a total of 236 species that parasitize this host. The checklist presented in Table 32 includes a total of 310 parasite species of the common carp, 146 of them identified only on the subspecies *C. c. carpio*.

JENEY and JENEY (1995) also reviewed the fungal, bacterial, and viral diseases of the common carp based mainly on studies of the domestic carp. Fungal infections are a serious problem for both natural fish populations and those raised in intensive aquaculture. *Saprolegnia* spp., *Achlya* spp., and *Branchiomyces* spp. are probably the major fungi most pathogenic to the common domestic carp. Among the bac-

Table 32: Revised list of parasites found in two subspecies of wild common carp and of domestic common carp. **A** – *C. carpio carpio*; **B** – *C. carpio haematopterus*; **C** – *C. carpio* f. *domestica*. Compiled from papers and monographs by AMIN (1985), BAUER (1987), ERGENS and LOM (1970), GUSSEV (1985), HALDAR *et al.* (1996), HOFFMANN (1967), JENEY and JENEY (1995), KABATA (1985), LANDSBERG (1991), LOM and DYKOVÁ (1989, 1992), LOM and MORAVEC (1995), MARGOLIS and ARTHUR (1979), MOLNÁR and MORAVEC (1997), MORAVEC (1987, 1994, 1998), OSMANOV (1976), PAPERNA (1964), PROTASOVA *et al.* (1990), SCHOLZ (1989), SCHOLZ and HANZELOVÁ (1998), SCHULMAN (1984), WANG (1981) and ŽITŇAN (1985).

Species of parasite	Presence in the hosts		
	A	B	C
PROTOZOA			
1. *Acanthamoeba ameiuri* (Pushkarev, 1913)	o	o	+
2. *Ambiphrya ameiuri* Kirkegard et Jahn, 1947	o	o	+
3. *Amphileptus branchiarum* Wenrich, 1924	+	+	+
4. *Apiosoma carpelli* Scheubel, 1973	+	o	+
5. *A. minutum* Chen, 1961	o	+	o
6. *A. oviforme* Chen et Hsieh in Chen, 1973	o	o	+
7. *A. piscicolum* Blanchard, 1885	+	o	+
8. *A. tenerum* Kashkowski, 1975	o	o	+
9. *Chilodonella hexasticha* Kiernik, 1909	+	+	+
10. *Ch. piscicola* Zacharias, 1894	+	+	+
11. *Chloromyxum cristatum* Léger, 1906	+	+	+
12. *Ch. fluviatile* Thélohan, 1892	+	+	+
13. *Ch. koi* Fujita, 1913	o	+	+
14. *Ch. legeri* Touraine, 1931	o	o	+
15. *Cryptobia branchialis* Nie in Chen, 1956	+	+	+
16. *Cr. cyprini* (Plehn, 1903)	+	o	+
17. *Cr. indistincta* Khaibulaev, 1984	+	o	+
18. *Dermocystidium cyprini* Červinka et Lom, 1974*	o	o	+
19. *D. erschowi* Garkavi, Denisov et Afanasjev, 1980	o	o	+
20. *D. kamilovi* Allamuratov, 1965	+	o	o
21. *D. kobiakovae* Allamuratov, 1965	+	o	o
22. *D. koi* Hoshina et Sahara, 1950	o	o	+
23. *Epistylis lwoffi* Fauré-Frémiet, 1943	+	+	+
24. *Goussia carpelli* (Léger et Stankovitch, 1921)	+	+	+
25. *G. haichengensis* Chen, 1984	o	o	+
26. *G. subepithelialis* (Moroff et Fiebiger, 1905)	+	o	+
27. *Hartmanella vermiformis* Page, 1967	o	o	+
28. *Henneguya cutanea* Dogel et Petrushevski, 1933	+	+	+
29. *Hofferelus cyprini* (Doflein, 1898)	+	+	+
30. *Ichthyobodo necator* (Henneguy, 1883)	+	+	+

* Miscellaneous microorganisms traditionally classified together with protozoans: *Dermocystidium, Ichthyochytrium, Sporozoon,* and *Mucophilus*.

Table 32: (cont.)

Species of parasite	Presence in the hosts		
	A	B	C
31. *Ichthyophthirius multifiliis* Fouquet, 1876	+	+	+
32. *Ichthyochytrium vulgare* Plehn, 1920*	o	o	+
33. *Mucophilus cyprini* Plehn, 1920*	o	o	+
34. *Myxidium cuneiforme* Fujita, 1924	o	+	+
35. *M. pfeifferi* Auerbach, 1908	+	+	+
36. *M. rhodei* Léger, 1905	+	+	+
37. *Myxobolus acinosus* Nie et Li in Chen, 1973	o	o	+
38. *M. acutus* Wu et Chen, 1987	o	o	+
39. *M. achmerovi* Schulman, 1966	o	+	+
40. *M. alacaudus* Yukhimenko, 1986	o	+	o
41. *M. albovae* Krasilnikova in Shulman, 1966	+	o	o
42. *M. amurensis* Akhmerov, 1960	o	+	o
43. *M. anisocapsularis* Shulman, 1962	o	o	+
44. *M. artus* Akhmerov, 1960	+	+	o
45. *M. basilamellaris* Lom et Molnár, 1983	o	o	+
46. *M. bellus* Hoffman, 1967	o	o	+
47. *M. branchialis* Markevitch, 1932	+	o	+
48. *M. carassii* Klokacheva, 1914	o	o	+
49. *M. circulus* (Akhmerov, 1960)	o	o	+
50. *M. cordis* Keysselitz, 1908	o	o	+
51. *M. cristatus* Shulman, 1962	+	o	+
52. *M. cuttacki* Haldar, Samal et Mukhopadhyay, 1996	o	o	+
53. *M. cyprini* Doflein, 1898	+	o	+
54. *M. cyprinicola* Reuss, 1906	+	o	+
55. *M. dermatobius* Hoshina, 1953	o	o	+
56. *M. dispar* Thélohan, 1892	+	+	+
57. *M. disparoides* Shulman, 1962	+	o	o
58. *M. diversicapsularis* Slukhai et Shulman, 1966	+	o	+
59. *M. dogieli* Bykhovskaya et Bykhovski, 1940	+	+	+
60. *M. dujardini* (Thélohan, 1892)	+	o	+
61. *M. eliptoides* Wu et Chen, 1987	o	o	+
62. *M. ellipsoides* Thélohan, 1892	+	+	+
63. *M. encephalicus* (Mulsow, 1911)	+	o	+
64. *M. exiguus* Thélohan, 1895	+	+	+
65. *M. haematopterus* Yukhimenko, 1986	o	+	o
66. *M. infundibulatus* Donec et Kulakovskaya in Shulman, 1962	+	o	+
67. *M. koi* Kudo, 1920	o	+	+
68. *M. kubanicus* Bykhovskaya et Bykhovski, 1940	+	+	+
69. *M. macrocapsularis* Reuss, 1906	+	o	+

* Miscellaneous microorganisms traditionally classified together with protozoans: *Dermocystidium, Ichthyochytrium, Sporozoon,* and *Mucophilus.*

Table 32: (cont.)

Species of parasite	Presence in the hosts		
	A	B	C
70. *M. microlatus* Li et Nie in Chen, 1973	o	o	+
71. *M. minutus* Nemeczek, 1911	+	o	+
72. *M. muelleri* Buetschli, 1882	+	+	+
73. *M. musculi* Keysselitz, 1908	+	+	+
74. *M. musseliussae* Yakovchuk, 1979	o	o	+
75. *M. nemachili* Weiser, 1949	+	o	+
76. *M. obliquus* Kudo, 1934	o	o	+
77. *M. obovoides* Li et Nie in Chen, 1973	o	o	+
78. *M. oviformis* Thélohan, 1892	+	+	+
79. *M. pfrille* Fanthaus, Porter et Richardson, 1939	o	o	+
80. *M. rotundatus* Akhmerov, 1956	o	+	o
81. *M. rotundus* Nemeczek, 1911	+	o	+
82. *M. squamae* Keysselitz, 1908	+	+	+
83. *M. toyamai* Kudo, 1915	o	+	+
84. *M. vastus* Kudo, 1934	o	+	+
85. *M. yunchisi* Yukhimenko, 1986	o	+	o
86. *Scyphidia globularis* Solomatova, 1977	o	o	+
87. *S. doliaris* Chernova, 1977	+	o	o
88. *Sphaerospora angulata* Fujita, 1912	+	+	+
89. *S. carassii* Kudo, 1919	+	+	+
90. *S. chinensis* Lom, Dyková, Pavlásková et Grupcheva, 1983	o	+	+
91. *S. cyprini* (Fujita, 1912)	o	+	+
92. *S. hangzhouensis* Li et Wu, 1983	o	+	o
93. *S. molnari* Lom, Dyková, Pavlásková et Grupcheva, 1983	+	o	+
94. *S. renicola* Dyková et Lom, 1982	o	o	+
95. *Spironucleus elegans* Lavier, 1936	o	o	+
96. *Sporozoon tincae* Volf et Dvořák, 1928*	o	o	+
97. *Tetrahymena pyriformis* (Ehrenberg, 1830)	+	+	+
98. *Thelohanellus acuminatus* Ky, 1971	o	o	+
99. *T. callisporis* Ky, 1971	o	o	+
100. *T. cyprini* Hoshina et Hosoda, 1957	o	o	+
101. *T. dogieli* Akhmerov, 1955	o	+	+
102. *T. fuhrmanni* Auerbach, 1909	o	+	+
103. *T. hovorkai* Akhmerov, 1960	o	o	+
104. *T. kitauei* Egusa et Nakajima, 1981	o	+	+
105. *T. oculileucisci* (Trojan, 1909)	o	o	+
106. *T. pyriformis* Kabata, 1985	o	o	+
107. *T. rohitae* Southwell et Prashad, 1918	o	o	+
108. *T. sagittarius* Li et Nie in Chen, 1973	o	o	+
109. *Trichodina acuta* Lom, 1961	+	+	+
110. *T. centrostrigata* Basson, van As et Paperna, 1983	o	o	+

Table 32: (cont.)

Species of parasite	Presence in the hosts		
	A	B	C
111. *T. compacta* van As et Basson, 1989	o	o	+
112. *T. dallii* Zhukov, 1964	o	+	+
113. *T. heterodentata* Duncan, 1977	o	o	+
114. *T. modesta* Lom, 1970	+	o	+
115. *T. mutabilis* Kazubski et Migala, 1968	+	o	+
116. *T. nigra* Lom, 1960	+	o	+
117. *T. nobilis* Chen, 1963	+	o	+
118. *T. pediculus* (O.F. Müller, 1786)	+	+	+
119. *T. perforata* Lom, Golemanski et Grupcheva, 1976	o	o	+
120. *T. rectangli* Chen et Hsieh, 1964	o	+	+
121. *T. reticulata* Hirschmann et Parsch, 1955	+	+	+
122. *T. rostrata* Kulemina, 1968	+	o	+
123. *Trichodinella carpi* Duncan, 1977	o	o	+
124. *T.epizootica* (Raabe, 1950)	+	+	+
125. *T. subtilis* Lom, 1959	+	+	+
126. *Tripartiella bulbosa* (Davis, 1947)	+	o	+
127. *T. copiosa* Lom, 1959	+	+	+
128. *T. lechridens* Basson et van As, 1987	o	o	+
129. *Trichophrya piscium* Bütschli, 1889	+	+	+
130. *Trypanoplasma borreli* Laveran et Mesnil, 1901	+	+	+
131. *Trypanosoma carassii* Mitrofanov, 1884	+	+	+
132. *T. gracilis* Khaibulaev, 1984	o	+	o
133. *T. pseudobagri* Dogel et Akhmerov, 1959	o	+	+
134. *Zschokella carassii* Nie et Li in Chen, 1973	o	o	+
135. *Z. cyprini* Quadri, 1962	o	o	+
136. *Z. linghnensis* Chen et Hsieh, 1984	o	o	+
137. *Z. minuta* Nie et Li in Chen, 1973	o	o	+
138. *Z. nova* Klokacheva, 1914	o	o	+
MONOGENEA			
139. *Dactylogyrus achmerowi* Gussev, 1955	+	+	+
140. *D. anchoratus* (Dujardin, 1845)	+	+	+
141. *D. biwaensis* Ogawa et Egusa, 1982	o	+	+
142. *D. crassus* Kulwiec, 1927	+	o	+
142. *D. extensus* Mueller et van Cleave, 1932	+	+	+
144. *D. falciformis* Akhmerov, 1952	o	+	o
145. *D. minutus* Kulwiec, 1927	+	+	+
146. *D. mollis* (Wedl, 1857)**	o	o	+

** A species described from the domestic common carp found near Vienna has been evaluated taxonomically as *species inquirenda*.

Table 32: (cont.)

Species of parasite	Presence in the hosts		
	A	B	C
147. *D. molnari* Ergens et Dulmaa, 1969	o	+	o
148. *D. mrazeki* Ergens et Dulmaa, 1969	o	+	+
149. *D. lopuchinae* Yukhimenko, 1981	o	+	o
150. *D. osmanovi* Urasbaiev, 1966	+	o	o
151. *D. sahuensis* Ling, 1965	o	+	+
152. *D. sphyrna* Linstow, 1878	o	o	+
153. *D. takahashii* Ogawa et Egusa, 1982	o	+	+
154. *D. vastator* Nybelin, 1924	+	+	+
155. *D. yinwenyingae* Gussev, 1962	+	+	+
156. *Eudiplozoon nipponicum* (Goto, 1891)	+	+	+
157. *Gyrodactylus cyprini* Diarova, 1964	+	o	+
158. *G. derjavini* Mikailov, 1975	+	o	o
159. *G. katharineri* Malmberg, 1964	+	o	+
160. *G. kherulensis* Ergens, 1974	+	+	+
161. *G. longoacuminatus* Žitňan, 1964	o	+	+
162. *G. medius* Kathariner, 1893	+	o	+
163. *G. nagibinae* Gussev, 1962	+	o	+
164. *G. ophiocephali* Gussev, 1955	o	+	+
165. *G. shulmani* Ling, 1962	+	+	+
166. *G. sprostonae* Ling, 1962	+	+	+
167. *G. stankovici* Ergens, 1970	+	o	+
168. *Paradiplozoon alburni* Khotenovsky, 1982	+	o	+
169. *P. cyprini* Khotenovsky, 1982	o	+	o
170. *P. homoion homoion* (Bykhovski et Nagibina, 1959)	+	o	+
171. *P. zeller* (Gyntovt, 1967)	+	o	o
172. *Pseudacolpenteron pavlovskii* Bykhovski et Gussev, 1955	+	o	+
DIGENEA – adults			
173. *Acrolichanus auriculatum* (Wedl, 1857)	+	o	o
174. *Allocreadinum carparum* Odening, 1959	+	o	+
175. *A. isoporum* (Looss, 1894)	+	o	+
176. *A. montanum* Sidorov et Butenko, 1960	+	o	o
177. *A. transversale* (Rudolphi, 1802)	o	o	+
178. *Aponurus tshugunovi* Issaitschikoff, 1927	+	o	o
179. *Aspidogaster conchicola* Baer, 1927	o	+	o
180. *A. enneatis* Ekman, 1932	o	o	+
181. *A. limacoides* Diesing, 1835	+	o	o
182. *Asymphylodora kubanica* Issaitschikoff, 1923	+	o	o
183. *A. tincae* (Modeer, 1790)	o	o	+
184. *Budonera luciopercae* (Mueller, 1776)	o	o	+
185. *Crepidostomum cooperi* Hopkins, 1931	o	o	+

Table 32: (cont.)

Species of parasite	Presence in the hosts		
	A	B	C
186. *Nicolla skrjabini* (Iwanitzky, 1928)	+	o	+
187. *Orientotrema cyprini* Wang, 1982	o	o	+
188. *O. payangensis* Wang, 1982	o	o	+
189. *Paraasymphylodora markewitschi* (Kulakovskaya, 1947)	+	o	+
190. *Phyllodistomum elongatum* Nybelin, 1926	o	o	+
191. *Sanguinicolla armata* Plehn, 1905	o	o	+
192. *S. inermis* Plehn, 1905	+	+	+
193. *Sphaerostomum bramae* (Mueller, 1776)	+	o	+
DIGENEA – larvae			
194. *Apharyngostrigea cornu* (Zeder, 1800)	+	+	+
195. *A. sogdiana* (Pavlovsky et Anitschkov, 1923)	+	o	o
196. *Apophalus venustus* Ransom, 1920	o	o	+
197. *Ascocotyle coelostoma* Looss, 1896	+	o	o
198. *Bolbophorus confusus* Krause, 1914	+	o	o
199. *Bucephalus polymorphus* Baer, 1827	+	+	o
200. *Conodiplostomum perlatum* (Ciurea, 1911)	+	o	o
201. *Diplostomum spathaceum* (Rudolphi, 1819)	+	+	+
202. *Echinochasmus perfoliatus* (Ratz, 1908)	o	o	+
203. *Hysteromorpha triloba* (Rudolphi, 1819)	+	+	+
204. *Ichthyocotylurus pileatus* (Rudolphi, 1802)	o	o	+
205. *Metagonimus yokogawai* Katsurada, 1912	+	+	+
206. *Opistorchis felineus* (Rivolta, 1884)	+	+	+
207. *Ornithodiplostomum ptychocheilus* (Faus, 1917)	o	o	+
208. *Posthodiplostomum brevicaudatum* (Nordmann, 1832)	+	o	+
209. *P. cuticola* (Nordmann, 1832)	+	o	+
210. *P. minimum* (MacCallum, 1921)	o	o	+
211. *Rhipidocotyle campanula* (Dujardin, 1845)	+	o	+
212. *Tylodelphys clavata* (Nordmann, 1832)	+	+	+
213. *Tetracotyle echinata* Diesing, 1858	o	o	+
CESTOIDEA – adults			
214. *Archigetes iowensis* Calentine, 1962	o	o	+
215. *A. sieboldi* Leuckart, 1878	o	o	+
216. *Atractoloytocestus huronensis* Anthony, 1958	o	o	+
217. *Biacetabulum appendiculatum* (Szidat, 1937)	o	o	+
218. *Bothriocephalus acheilognathi* Yamaguti, 1934	+	+	+
219. *Caryophyllaeides fennica* (Schneider, 1902)	+	o	+
220. *Caryophyllaeus brachycollis* Janiszewska, 1951	o	o	+
221. *C. fimbriceps* Annenkova-Chlopina, 1919	+	o	+
222. *C. laticeps* (Pallas, 1781)	o	+	+

Table 32: (cont.)

Species of parasite	Presence in the hosts		
	A	B	C
223. *Khawia iowensis* Calentine et Ulmer, 1961	o	o	+
224. *K. japonensis* (Yamaguti, 1934)	+	+	o
225. *K. sinensis* Hsü, 1935	+	+	+
226. *Markewitschia sagittata* Kulakovskaya et Akhmerov, 1956	o	+	o
227. *Proteocephalus percae* (O. F. Müller, 1780)	o	o	+
228. *P. torulosus* (Batsch, 1786)	+	o	o
CESTOIDEA – larvae			
229. *Diagramma interrupta* (Rudolphi, 1810)	+	o	+
230. *Ligula intestinalis* (Linnaeus, 1758)	o	o	+
231. *Neogryporhynchus cheilancristrotus* (Wedl, 1855)	+	+	+
232. *Paradilepis scolecina* (Rudolphi, 1819)	+	o	+
233. *Triaenophorus nodulosus* (Pallas, 1781)	o	o	+
234. *Valipora campylancristrota* (Wedl, 1855)	+	o	+
NEMATODA – adults			
235. *Camallanus ancylodirus* Ward et Magath, 1916	o	o	+
236. *C. lacustris* (Zoega, 1776)	o	o	+
237. *Capillaria catenata* Van Cleave et Mueller, 1932	o	o	+
238. *C. patzcuarensis* Osorio-Sarabia, Pérez-Ponce de Léon et Salgado-Maldonado, 1986	o	o	+
239. *Cucullanus cyprini* Yamaguti, 1941	o	+	+
240. *C. dogieli* Krotas, 1959	+	o	+
241. *Philometroides cyprini* (Ishii, 1931)	o	+	+
242. *Pseudocapillaria tomentosa* (Dujardin, 1843)	+	+	+
243. *Rhabdochona cascadilla* Wigdor, 1918	o	o	+
244. *R. gnedini* Skrjabin, 1946	+	o	o
245. *Schulmanela petruschewskii* (Shulman, 1948)	o	o	+
246. *Skrjabillanus cyprini* Molnár et Moravec, 1997	o	o	+
247. *Spinitectus gracilis* Ward et Magath, 1917	o	o	+
NEMATODA – larvae			
248. *Agamospirura* sp.	+	o	o
249. *Anguillicola crassus* Kuwahara, Niimi et Itagaki, 1974	o	o	+
250. *Anisakis* sp.	+	o	o
251. *Contracaecum microcephalum* (Rudolphi, 1809)	o	o	+
252. *C. micropapillatum* (Stossich, 1890)	o	o	+
253. *C. multipapillatum* (Drasche, 1882)	o	o	+
254. *C. rudolphii* Hartwich, 1964	o	o	+
255. *Contracaecum* sp.	o	o	+

Table 32: (cont.)

Species of parasite	Presence in the hosts		
	A	B	C
256. *Desportesius brevicaudatus* (Dujardin, 1845)	o	o	+
257. *Eustrongylides excisus* Jägerskiöld, 1909	o	o	+
258. *Gnathostoma hispidum* Fedtschenko, 1872	o	o	+
259. *G. spinigerum* Owen, 1836	o	o	+
260. *Porrocaecum reticulatum* (Linstow, 1899)	+	o	+
261. *Porrocaecum* sp.	o	+	o
262. *Raphidascaris acus* (Bloch, 1799)	o	o	+
263. *Spiroxys contortus* (Rudolphi, 1819)	+	o	+
264. *Syncuaria squamata* (Linstow, 1885)	o	o	+
ACANTHOCEPHALA – adults			
265. *Acanthocephalus alabamensis* Amin et Williams, 1983	o	o	+
266. *A. anguillae* (O. F. Müller, 1780)	+	o	+
267. *A. clavula* (Dujardin, 1845)	o	o	+
268. *A. dirus* (Van Cleave, 1931)	o	o	+
269. *A. lucii* (O. F. Müller, 1776)	+	+	+
270. *Leptorhynchoides thecatus* Linton, 1891	o	o	+
271. *Neoechinorhynchus australe* Van Cleave, 1931	o	o	+
272. *N. rutili* (O. F. Müller, 1780)	+	+	+
273. *Paracanthocephalus curtus* Akhmerov et Dombrowskaja-Akhmerova, 1941	o	+	o
274. *P. tenuirostris* Akhmerov et Dombrowskaja-Akhmerova, 1941	o	+	o
275. *Pomphorhynchus bulbocolli* (Linkins, 1911)	o	o	+
276. *P. laevis* (O. F. Müller, 1776)	+	o	+
ACANTHOCEPHALA – larvae			
277. *Corynosoma strumosum* (Rudolphi, 1802)	o	o	+
HIRUDINEA			
278. *Caspiobdella fadejewi* (Epstein, 1961)	+	o	o
279. *Cystobranchus respirans* Troschel, 1850	o	o	+
280. *Hemiclepsis hubeiensis* Yang Tong, 1981	o	o	+
281. *H. marginata* (O. F. Müller, 1774)	+	+	+
282. *Limnotracheobdella sinensis* (Blanchard, 1896)	o	+	o
283. *L. turkestanica* (Stschegolew, 1912)	+	o	o
284. *Piscicola geometra* (Linnaeus, 1761)	+	+	+
285. *Placobdella montifera* Moore, 1912	o	o	+
MOLLUSCA – larvae			
286. *Anodonta cygnea* (Linnaeus, 1758)	o	o	+

Table 32: (cont.)

Species of parasite	Presence in the hosts		
	A	B	C
CRUSTACEA			
287. *Argulus appendiculosus* Wilson, 1907	o	o	+
288. *A. catastomi* Dana et Herrick, 1837	o	o	+
289. *A. coregoni* Thorell, 1864	o	o	+
290. *A. ellipticaudatus* Wang, 1960	o	+	o
291. *A. flavescens* Wilson, 1916	o	o	+
292. *A. foliaceus* (Linnaeus, 1758)	+	o	+
293. *A. indicus* Weber, 1892	o	o	+
294. *A. japonicus* Thiele, 1900	+	+	+
295. *A. yui* Wang, 1958	o	+	o
296. *Caligus lacustris* Steenstrup et Lütken, 1861	o	o	+
297. *C. macarovi* Gussev, 1951	o	o	+
298. *Ergasilus caeruleus* Wilson, 1911	o	o	+
299. *E. sieboldi* Nordmann, 1832	+	+	+
300. *Lamproglena pulchella* Nordmann, 1832	+	o	+
301. *Lernaea esocina* Burmeister, 1833	o	o	+
302. *L. ctenopharyngodontis* Yin, 1960	o	o	+
303. *L. cyprinacea* Linnaeus, 1758	+	o	+
304. *Paraergasilus medius* Yin, 1956	o	+	o
305. *P. rylovi* Markewitsch, 1937	+	+	+
306. *Tracheliastes polycolpus* Nordmann, 1832	o	o	+
307. *Sinergasilus undulatus* Markewitsch, 1940	o	+	o
ARACHNIDA – larvae			
308. *Elylais* sp.	+	o	o
309. *Hydrachna* sp.	+	o	o
310. *Hydryphantes* sp.	+	o	o

terial diseases, motile *Aeromonas* septicaemia (MAS) is an infection produced by *Aeromonas hydrophila.* Carp erythrodermatitis is caused by the atypical *A. salmonicida* (CE bacterium), and columnaris disease is an infection produced by *Flexibacter columnaris,* which is distributed worldwide.

Among the viral diseases, two are important for the common domestic carp in Europe. The first is spring viremia of carp (SVC), which is caused by *Rhabdovirus carpio,* and the second is carp pox, caused by *Herpesvirus cyprini.* Carp iridovirus is one of the viral infections of the carp with indeterminate pathogenicity. Of 17 viruses that are suspected of causing diseases but have not yet been isolated, one is the coronavirus of the common domestic carp (Wolf 1988).

Most pathogens of the common domestic carp are considered to be facultative (JENEY and JENEY 1995). This means that they are present almost everywhere in the water inhabited by the fishes, but they can cause disease only under certain conditions, namely, when the resistance of a fish is reduced by adverse changes in the environment, especially in aquaculture. SNIESZKO (1974) developed a theory that the role of environmental stress in the outbreak of fish diseases is a primary one and is especially important in warm water aquaculture (LANDSBERG 1991, JENEY and JENEY 1995), such as the culture of the common domestic carp, for which mainly surface waters are used and in which the basic water quality parameters are subject to a wide range of fluctuations. Furthermore, management of such cultures is generally lax.

Because of stress in the culture system, it is not surprising that several diseases of common domestic carps are typically "stress-mediated." These are various parasite infestations, such as costiosis and trichodinosis, as well as bacterial gill disease, flexibacteriosis, and spring viremia of carp. Environmental stress factors that produce a predisposition to diseases in the domestic common carp are listed in Table 33.

Table 33: Environmental stress factors that predispose common domestic carps to diseases (WEDEMEYER and MCLEAY, 1981, and JENEY and JENEY, 1995).

Disease	Environmental stress factors predisposing to disease
Parasite infestation (*Costia, Trichodina*)	Overcrowding of fry and fingerlings; low oxygen concentrations; excessive size variations among fishes in ponds
Bacterial gill disease	Overcrowding of fry and fingerlings; low oxygen concentrations; excessive size variations among fishes in ponds
Columnaris disease	Crowding or handling during warm (above 15 °C) water periods if carrier fish are present in the water supply; temperature increase to about 30 °C if the pathogen is present, even if the fishes are neither crowded nor handled
Haemorrhagic septicaemia (*Aeromonas* and *Pseudomonas* spp.)	Pre-existing protozoan infestations such as *Costia* or *Trichodina*; increased bacterial load in the water; particulate matter in the water; handling (especially after overwintering at low temperatures); crowding; low oxygen concentrations; chronic sublethal exposure to heavy metals or PCBs
Spring viremia of carp	Handling after overwintering at low temperatures, rapid warming of the water temperature

Economic Importance

Undoubtedly, the wild common carp has been an important food item caught by humans since ancient times. This is attested to by numerous local names that are used along the Danube River downstream from the westernmost boundary of the natural range of the carp. The remains of this fish have been found in archaeological excavations of early human settlements (Balon 1995a, b; Hoffmann 1994a, b, 1995) along the Danube and in the drainage area of the Black Sea. The wild carp must have been well-known to the ancient Romans living in Carnutum, Gerulata, Ad Flexum, Quadrata, Arrabona, Ad Mures, Ad Statuas, Brigetio, and Celamantina, which were all major Roman colonies or settlements in the western part of the empire along the Amber road. This road was built along large flood plains, on which schools of wild carps congregated every spring for spawning (Balon 1964, 1974, 1995a, b).

Archaeological excavations of Roman forts were intensified during the last 15 years because of their expected submersion after the construction of the Gabčíkovo-Nagymaros dams (Rajtár 1990, 1992). In all earlier excavations, such as those at Carnutum and Brigetio, the focus of the interest had been on Roman artefacts. However, the more recent excavations near the Slovak village of Iža included the collection of fish remains, perhaps to test the hypotheses of Balon (1969, 1974). These samples were given to the prominent Slovak ichthyologist, Professor Karol Hensel, for examination (Balon 1995a, b). The samples included abundant bones from the wild common carp, which were much more numerous than those of any other fish species, such as sturgeons and wels. This discovery has provided the first direct evidence that the Romans knew about the wild common carp in the Danube, which is very important for elucidating the origin of cultured carp races and their domestication.

The earliest quantitative data on the catches of wild common carps throughout a considerable part of their range were provided by Sabaneev (1892). He estimated that the catches of this species in the downstream sections of rivers in the southern part of Imperial Russia amounted to more than one million puds annually, which totals at least 16,380.000 kg of total mass. He also mentioned that almost all of this catch had been consumed at locations very near the capture site, so the significance of the wild common carp on the market was slight. In Moscow, for example, the demand for this fish was quite small during the last century, and the wild common carp was sold for a price similar to that demanded for relatively large chub and other cyprinid species.

Berg *et al.* (1949) emphasized the fact that the catch of the wild common carp constituted a significant proportion of the total harvest of freshwater fishes in Russia. For example, in the northern part of the Caspian Sea, an average of 12,000 metric tons of wild common carp were captured annually until 1914. The catches increased to 60,600 metric tons in 1917 but decreased again to 15,200 in 1926 and further to 8,500 in 1939 (BERG 1949). The following data are available on catches of the wild common carp in the region of the Caspian Sea and Kura River during the period from 1921 through 1925 (BERG 1949):

Year	Catches of carp (in metric tons)	Proportion of the carp of the total catch	Proportion of the carp of the catch of cyprinids
1921	4,474	41.3	54.0
1922	5,168	25.9	56.0
1923	4,071	20.4	27.9
1924	3,533	15.2	22.2
1925	3,295	14.0	35.7

Concerning this summary of data, the author remarked that more than 300,000 wild common carps were captured during April 1923 alone in the lower stretches of the Kura River and its tributary, the Akusha River. The total catch of wild common carps in the estuary of the Kuban River was 934 metric tons in 1926. The data in this table also seem to suggest that both the overall weight of the wild common carps caught and their proportion in the total catch exhibited a decreasing tendency during the period.

NIKOLSKIY (1940) provided information concerning the catch of this species in the Aral Sea. Very large catches, which are of historical interest and do not reflect the present situation, were recorded during eight consecutive years (1930 – 1937). In thousands of metric tons, they amounted to 8.44, 9.65, 6.53, 6.84, 7.33, 7.69, 6.08, and 10.66. The increase in 1937, the last year in this sequence, was extraordinary.

The percentage of wild common carps in the total production of freshwater fishes in the former Soviet Union remained quite large during subsequent years. From a review by BERG et al. (1949) for the individual regions during the period from 1936 through 1939, the following information was obtained:

Region or watershed	Catch in metric tons (range)
Sea of Azov	650 – 1,400
Black Sea	75 – 225
Caspian Sea	5,500 – 12,400
Aral Sea	4,650 – 6,300
Lake Balkhash	4,850 – 6,000
Amur River Basin*	600 – 1,050

* Catches of the eastern subspecies, *C. c. haematopterus*. It should be noted that in 1936, about 55,000 metric tons of wild common carp were caught in China.

Data from the former Soviet Union are also available for later periods, such as 1974 through 1978 (KAZANCHEEV 1981). These data are provided for each of the following regions (in thousands of metric tons):

Region	Years				
	1974	1975	1976	1977	1978
Volga-Caspian	3.29	5.60	10.05	6.04	5.24
Ural River	–	–	0.03	–	
Dagestan	0.09	0.46	0.28	0.30	0.07
Caspian-Kura	0.97	0.68	0.18	0.15	0.03
Turkmen	0.20	0.16	0.06	0.08	0.17
Total	4.54	6.90	10.60	6.57	5.51

These data show that the total catches of wild carps during the period from 1974 through 1978 did not come close to matching those during the years 1930 through 1937 in the Aral Sea alone (NIKOLSKIY 1940).

Despite great year to year variations in the catches, a decreasing tendency is very obvious, especially in the Caspian-Kura Region. At the present time, the largest population of the wild common carp inhabits the lower section of the Volga River and its estuary (LUKIN 1948, KAZANCHEEV 1981). In contrast, this species has become least abundant in the region of the Caspian Sea and Kura River, where this species presently has commercial value only because of an intensive artificial culture and release of numerous fry. Almost 652 millions fry were released into the Kura River in 1971, while about a thousand million fry were set free in the watershed of the Volga River.

There are presently very few wild common carps left in other rivers, as well, such as the Terek and the Ural (SHAPOSHNIKOVA 1954). The stock of this species is also rather low in the Atrek River, but there has been considerable fluctuation in the size of the year classes, as reflected by the variations in the catches from year to year (KAZANCHEEV 1981).

IVANNIKOV (1997) brings the most recent data on the exploitation of the wild common carp from the Volga-Caspian region. The annual fishing yield ranged there from 4,100 metric tons in 1982 to 7,600 in 1987. The catches decreased during the following period, 1989 to 1993, from 6,700 to 7,200 metric tons, and since then have stabilized at about 2,200 to 2,500 metric tons from 1994 to 1996. This trend in the annual catches of the wild common carp reflects mainly the changes in fishing intensity and in the hydrological conditions in the Volga-Caspian region. The author stated that the distribution and abundance of the wild common carp doubled during the last 20 years in this region; however, the abundance decreased in the traditional fishing areas. In contrast, the abundance increased in the open Volga Delta and in brackish waters where there was no fishing. The harvestable stock of the wild common carp in this region was estimated to be 12 to 15 million fishes with a biomass of 40,000 to 50,000 metric tons. The recommended size of the annual catch was 4,500 to 5,000 metric tons (IVANNIKOV 1997).

The wild carp had been an important object of commercial fisheries in natural waters of Romania prior to the drying of the flood plain of the Lower Danube. The annual catches of the wild common carp in the Danube Delta alone amounted to about 1,000 metric tons. In addition, it totaled hundreds of metric tons in each major shallow lake on the flood plain during the first years of the 19th century (ANTIPA 1909). The annual catches in all of Romania during the period from 1950 to 1959 ranged from 1,100 to 13,100 metric tons of carps each weighing 2 kg, 48 to 400 metric tons of carps between 2 and 4 kg, and 16 to 103 metric tons of carps exceeding a body weight of 4 kg. About 254 to 1,800 kg of fishes under the legal size were also caught. The greatest quantities were caught in shallow lakes on the flood plain, followed by the Danube Delta and the Razelm lagoon complex south of the Delta. The catches of the wild common carp underwent a catastrophic decline after the drying of the shallow lakes on the Danube flood plain

Similarly, the situation in the Europe is also serious. According to LELEK (1987b), "it may seem unbelievable, but the wild common carp is in great danger". Stable, self-sustaining stocks can still be found in the lower section of the Danube River and its tributaries. In the Slovakian and Hungarian sections of the Danube, the wild common carp is a critically endangered species (HOLČÍK 1983, 1996, KRUPKA 1992, KRUPKA and MESZÁROS 1989, MESZÁROS 1990), and in the lower sections of the Morava and Dyje (Thaya) River, its wild population has become extinct (LUSK 1996). In the delta of the Rhone River, the last wild common carps were observed in 1967. It has not yet been determined whether all the populations in the Rhine River are feral (BALON *et al.* 1986, LELEK 1987a, b). In contrast, PAVLOV *et al.* (1994) do not classify the wild common carp as a threatened fish. This is explained by the fact that the decrease in the abundance of this species in the European and Asian parts of the former Soviet Union has still not been sufficient to endanger its existence. It continues to be an economically important species that is frequently caught. The most important reasons for the widely observed decline in the wild common carp populations are obviously anthropogenically induced changes along the lower reaches of water courses. The most detrimental of these include the straightening of rivers, water flow regulation and damming, fragmentation of rivers into isolated sections unpassable for migrating fishes, isolation of the main river channel from backwaters along flood plains, and control of the flood plain vegetation by drainage. Because of the decline, stocking with domestic common carps has been increasing. It is often falsely believed that high-bodied, scaly carps are fast-growing wild ones. The scarcity or absence of the indigenous form is not noticed because feral and domestic forms are not distinguished by managers.

LELEK (1987b) proposed several conservation measures. Protection of the wild common carp must start with the conservation of habitats suitable for natural spawning. The connections between main streams and adjacent spawning places must be maintained. The slope in the flooded area should be directed toward the main channel or a permanent water body, which is, at least seasonally, connected to the whole river system. The size limit for the catch must be at least 45 cm in *Tl* wherever wild carps occur, as the females may reach maturity of this size. The spawning places must be protected from human visitors and possible poachers because the spawning carps do not react to possible danger. Monitoring the pH seems necessary to detect high alkality, which may be lethal during early stages of juvenile development (HOCHMAN and KIRKA 1990). It should be noted that in European countries situated in the western part of the wild common carp's range, the species is usually protected by the law, and its capture is not permitted.

It is necessary to emphasize that the wild common carp has assumed a world-wide importance as the ancestor of the domestic common carp, *C. carpio* f. *domestica*. The economic importance of the domestic forms is extraordinarily great. According to the FAO statistics for 1992 (TACON 1994), 1,022,887 metric tons of domestic common carps were produced, which equals 10.9 % of the total fish production by aquaculture of 9,417,153 metric tons worldwide. Included in the total were 125,893 metric tons produced in European ponds, 706,119 produced in China, and 87,000 in Indonesia (KESTEMONT 1995). At the end of this chapter, problems in the domestication of the wild common carp will be briefly mentioned.

Domestication: The common carp is probably the oldest cultured and most frequently domesticated fish in the world. They have been raised for about 4000 years in China and several hundred years in Europe (WOHLFAHRT 1984, HULATA 1995, BILLARD 1999). According to SUZUKI (1979), the common carp has been cultured for at least 2400 years in China and more than 1900 years in Japan, where, of course, the wild common carp is not native. The probable sites at which the Amur-Chinese subspecies of the carp, *C. carpio haematopterus,* was domesticated are located in China, Korea, and Japan. In China and neighbouring countries of Southeast Asia, "big-belly" is the common name of the major race of domestic common carp, which is probably descended from wild *C. c. haematopterus* (MOAV *et al.* 1979). It has been raised and bred for over 2000 years on Chinese farms, where predominant practices, such as polyculture, high stocking rates, and manure feeding, combine to create a very harsh environment. Under these conditions, the big-belly carp has had to survive and adapt. The presence of several local varieties of common carp in Vietnam (THIEN and TRONG 1995) suggests the possibility that this region may have been an additional site of domestication, which may have involved the Southeast Asian subspecies, *C. carpio viridiviolaceus.*

The domestic European carp, which originated from the subspecies *C. carpio carpio*, has had quite a different evolutionary history from that of its Chinese and

Vietnamese relatives. The history of the domestication process of the European common carp has been analyzed in detail by RUDZINSKI (1962), BALON (1966a, b, 1967, 1969, 1974, 1995a, b), STEFFENS (1967), HARSÁNYI (1986), LASSLEBEN (1990), HOFFMANN (1994a, b, 1995), and others. BALON (1966a, 1967, 1969, 1974, 1995a, b) presented palaeological, morphological, ecological, physiological, linguistic, archaeological, and historical information concerning the domestication of the wild common carp, *C. carpio carpio,* on the European subcontinent. There are reasons to believe that the Romans were the first to culture carps collected from the Danube River, and that the tradition of "piscinae dulces" was continued in monasteries throughout the Middle Ages. Much better documentation is available on the culture of common carp in ponds of lay and clerical landowners in western Europe after the 11th century. The spread of the common carp west of the Danube's piedmont zone was clearly brought about by humans, who were also responsible for introductions throughout the other continents. Concerning the domestication of the wild common carp, BALON (1995a, b) and HOFFMANN (1995) agree that at the present time, there are two possible explanations: (a) a Roman and monastic domestication, and (b) a 12th or 13th century secular pond culture in France and Burgundy. Additional evidence will be required before the actual origin of carp culture can be identified with certainty.

By the 1500's, complete treatises as well as reports of simple studies about the rearing of the common carp and pond culture of the species were published (DUBRAVIUS 1547, STRUMIENSKI 1573). Undoubtedly, the increase in the numbers of publications after the 17th century reflects more the frequency of writing reports and the greater number of copies preserved than an actual beginning of common carp farming. BALON (1995a, b) stated that no record from Europe exists of a domestic common carp with distinctive scale cover, body shape, or colour prior to the 16th century.

The progress in domestication of the wild common carp during last four centuries has led to the production of standardized breeds maintained by constant artificial selection (KIRPICHNIKOV 1961, 1981, WOLNY 1974, JHINGRAN & PULIN 1985, HULATA 1995). At the present time, numerous varieties, races, and landraces of domestic common carps are available in Russia, Ukraina, Central Europe, Southeast Asia, and Japan. These are cultured for food or as ornamental fishes. In the course of domestication, numerous varieties, races, and landraces of the common carp developed through a combination of factors, including geographical isolation, adaptation, accumulation of mutations, and both natural and artificial selection (HULATA 1995). Cross-breeding and introgression also played a role in the development of different varieties, mainly through introduction of domestic European carp varieties into other countries. A partial list of famous varieties (KIRPICHNIKOV 1981, JHINGRAN & PULIN 1985, POKORNÝ et al. 1995, HULATA 1995) includes the Chinese big belly carp and long-bodied carp; Indonesian Punten carp; Malayan and other colour varieties (SUMANTADINATA 1995); Japanese Asagi and

Yamato races; Ukrainian carp; Russian Ropsha, Para, and Krasnodar carps; European Galician and Aischgrunder (Fig. 36); German Lausitzer; Czech Bohemian (Fig. 37); French Royale; Hungarian Dinnyes; Croatian Nasice; and Israeli Dor-70. Many other less famous strains, races, and landraces of edible domestic common carp occur throughout the world. It should be noted, however, that some of these races and landraces are no longer distinct varieties due to intermixing with other local stocks. Live gene banks to preserve genetic material from some of the varieties were established at the Fish Culture Research Institute, Szarvas, Hungary (GORDA *et al.* 1995) and the Research Institute of Fish Culture and Hydrobiology, Vodňany, Czech Republic (POKORNÝ *et al.* 1995). In addition, a cryopreserved sperm bank was established at the Research Institute of Pond Fisheries, Rybnoe, Moscow Province, Russia (HULATA 1995).

The following four basic forms of domestic common carps being raised in ordinary culture facilities are recognized (BRYLIŃSKA 1986). Their scales are controlled by a pair of alleles: scalelessness by N and n (nudus) and scalation by S and s (squammatus):

1. *Scaled carp*, homozygous SSnn or heterozygous Ssnn, with regular imbricating scales over the entire body, much like the ancestral wild common carp;
2. *Line carp*, heterozygous SSnN or SsNn, with a regular single mid-lateral row of scales and elsewhere none or only a few scales;
3. *Mirror carp*, recessive homozygous ssnn with rows of large scales along the dorsum and bordering the gill openings and some scales at the base of the caudal fin. Three varieties are recognized:
   - *a. Scaled mirror carp* with three lateral rows of scales;
   - *b. Line mirror carp* with interrupted rows of scales along the lateral line and isolated or small groups of scales elsewhere;
   - *c. Scattered mirror carp* with a few scales along the body contours;
4. *Leather* or *naked carp*, heterozygous ssnN, with no scales and sometimes coloured blue.

The Nishikigoi ornamental carp strains, sometimes called "koi," have been developed in Japan during the last two centuries (MCDOWALL 1990), probably through colour mutations occurring among wild common Magoi carps from Central Asia (HULATA 1995). The earliest mutants were mainly red or occasionally white (KAMIHATA 1988). Several varieties, recognized by their colour patterns, now exist. Intensive selection and production of coloured carps was begun after the Second World War, probably in the 1950's. It developed along with the improving living standards of the people in Japan and the West. Genetically engineered forms of coloured carps kept in garden ponds only for ornamental purposes as "swimming flowers" have become very successful commercially (BALON 1995a, b).

Because the domestic common carp has become the leading species produced for freshwater fisheries and in aquaculture in many countries, the multifaceted

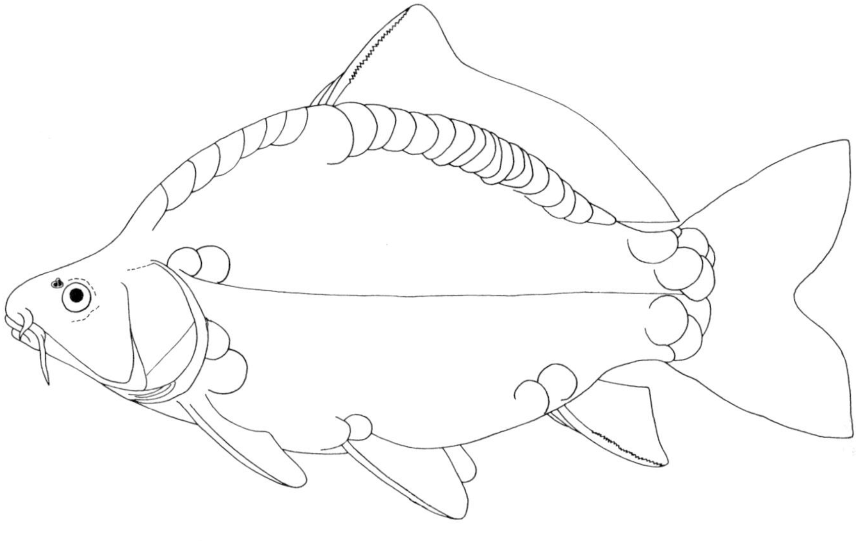

Fig. 36: The "Aischgrunder" race of the domestic common carp, *C. carpio carpio* f. *domestica*, according to GROTE *et al.* (1909).

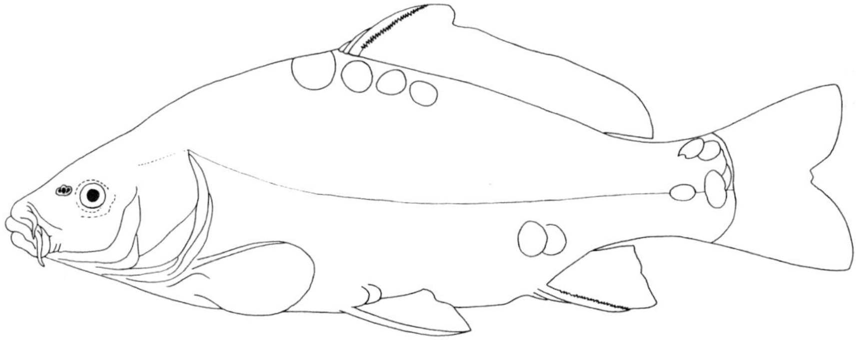

Fig. 37: The "Bohemian" race of the domestic common carp, *C. carpio carpio* f. *domestica*, according to GROTE *et al.* (1909).

problems encountered in the culture of this species have been studied worldwide, and the results have been published in thousands of scientific papers and dozens of monographs, as surveyed in *Aquaculture* 129: 1–468 (BILLARD 1999).

Finally, it should also be emphasized that the common domestic carp is the most important species for the European freshwater sport fishery. Cultured carps are regularly released into water bodies utilized by fishermen in many European countries. They are stocked in both running waters and reservoirs, including those that are less conducive for them. Management usually involves a "put and take" system. More carps are caught than most other sport fishes, both in numbers and weight (BARUŠ & OLIVA 1995).

As the ancestor of many domestic forms and because of its enormous potential for production, the wild common carp has a prominent position among the European ichthyofauna.

Literature

AMIN, O. M., 1981: Hosts and geographic distribution of *Acanthocephalus* (Acanthocephala: Echinorhynchidae) from North American freshwater fishes, with a discussion of species relationships. – Proc. Helminthol. Soc. Wash. **52** (2): 210–220.

ANDRUSHCHENKO, A. I., 1974: (Some morphological peculiarities in the embryogenesis of the common carp in commercial culture). – Vopr. rybokhoz. osvoeniya vodokhranilishch **95:** 168–169. (in Russian).

ANTIPA, G., 1909: Fauna ihtiologica a Rominiei. Bucuresti, Acad. Rom., Publ. fond. Adamachi. 294 pp.

BAGIROVA, Sh. M., 1963: (Steps in the development of the common carp in the Ust-Kurinsk spawning and culture facility). – Izv. AN Azerb. SSR, ser. biol. i med. nauk **5:** 53–64. (in Russian).

BAGIROVA, Sh. M., 1964: (Ecological and morphological peculiarities in the development of young bream, common carp, and Caspian roach in the Ust-Kurinsk spawning and culture facility). – Dokl. AN Azerb. SSR 7: 53–56. (in Russian).

BAKOS, J., and S. GORDA, 1995: Genetic improvement of common carp strains using intraspecific hybridization. – Aquaculture **129:** 183–186.

BALON, E., 1957: Vek a rast neresového stáda dunajského kapra – sazana (*Cyprinus carpio* morpha *hungaricus* Heck.) z Malého Dunaja pod Kolárovom (Age and growth of the spawning school of the Danubian wild carp). – Pol'nohospodárstvo (Bratislava), **4** (5): 961–986. (in Slovak).

BALON, E., 1958a: Vývoj dunajského kapra (*Cyprinus carpio carpio* L.) v priebehu predlarválnej fázy a larválnej periódy (Development of the Danubian carp during the prelarval phase and larval period). – Biol. práce SAV, Bratislava **4** (6): 5–54. (in Slovak).

BALON, E., 1958b: Die Entwicklung der Beschuppung des Donau-Wildkarpfens. – Zool. Anz., Leipzig **160** (3/4): 68–73.

BALON, E. K., 1958c: Rozmnožovanie, vývoj a rast dunajského sazana (Reproduction, development and growth of Danubian wild carp). – Pol'ovníctvo a rybárstvo **10** (7): 162–163. (in Slovak).

BALON, E., 1964: Predbežný zoznam kruhoústých a rýb Slovenska (Petromyzones a Teleostei) (The preliminary list of lampreys and fishes of Slovakia). – Biológia (Bratislava) **19** (5): 342–358. (in Slovak).

BALON, E., 1966a: Ryby Slovenska (Fishes of Slovakia). Vyd. Obzor, Bratislava. 413 pp. (in Slovak).

BALON, E. K., 1966b: Od dunajského kapra k vianočnému (From the Danubian wild carp to the Christmas carp). Pol'ovníctvo a rybárstvo **18** (12): 26–27. (in Slovak).

BALON, E., 1967a: O pôvode kapra (On the origin of carp). – Vesmír (Prague) **46** (11): 344–347. (in Slovak).

BALON, E. K., 1967b: Vývoj ichtyofauny Dunaja, jej súčasný stav a pokus o prognózu d'alších zmien po výstavbe vodných diel (Evolution of the Danube ichthyofauna, its recent state and an attempt to predict further changes after the construction of the planned water diversion schemes). – Biologické práce SAV, Bratislava, **13** (3): 3–121. (in Slovak).

BALON, E. K., 1969: Studies on the wild carp *Cyprinus carpio carpio* Linnaeus, 1758. I. New opinions concerning the origin of the carp. – Práce Laboratória rybárstva (Bratislava) **2:** 9 - 120.

BALON, E. K., 1974: Domestication of the carp *Cyprinus carpio* L. - Royal Ont. Mus. Life Sci. Misc. Publ., Toronto 1974: 1–37.

BALON, E. K., 1975: Reproductive guilds of fishes: a proposal and definition. – J. Fish. Res. Board Can. **32:** 821–864.

BALON, E. K., 1981: Additions and amendments to the classification of reproductive styles in fishes. - Env. Biol. Fish. **6:** 377–390.

BALON, E. K., 1989: The epigenetic mechanisms of bifurcation and alternative life-history styles. In: M. N. BRUTON (Ed.), Alternative Life-History Styles of Animals. Perspectives in Vertebrate Science 6. Kluwer Academic Publishers, Dordrecht. pp. 467–501.

BALON, E.K., 1990: Epigenesis of an epigeneticist: the development of some alternative concepts on the early ontogeny and evolution of fishes. – Guelph Ichthyol. Rev. **1:** 1–42.

BALON, E. K., 1995a: Origin and domestication of the wild carp, *Cyprinus carpio*: from Roman gourmets to the swimming flowers. – Aquaculture **129:** 3–48.

BALON, E. K., 1995b: The common carp, *Cyprinus carpio*: its wild origin, domestication in aquaculture, and selection as colored nishikigoi. – Guelph Ichthyol. Rev. **3:** 1–55.

BALON, E. K., S. S. CRAWFORD, and A. LELEK, 1986: Fish communities of the upper Danube River (Germany, Austria), prior to the new Rhein – Main – Donau Connection. – Env. Biol. Fish. **15** (4): 243–271.

BALON, E. K., and A. GOTO, 1989: Styles in reproduction and early ontogeny. In: A. GOTO and K. MAEKAWA (Eds.), Reproductive Behavior in Fishes: Styles and Strategies. Tokai University Press, Tokyo. pp. 1 – 47.

BALON, E., and V. MIŠÍK, 1956: Zoznam nových dokladov o výskyte niektorých málo známych alebo nových druhov rýb na Slovensku (New data on the occurrence of some less known or new species in Slovakia). – Biológia (Bratislava) **11** (3): 168–176. (in Slovak).

BĂNĂRESCU, P., 1946: Les poissons des environs de Timisoara. – Notationes biol. (Bucuresti) **4:** 135–165.

BĂNĂRESCU, P., 1960: Einige Fragen zur Herkunft und Verbreitung des Süsswasserfischfauna der europäisch-mediterranen Unterregion. – Arch. Hydrobiol. **57:** 16–134.

BĂNĂRESCU, P., 1964: Fauna Republicii Populare Romine. Pisces – Osteichthyes. Vol. XIII. Bucarest. 962 pp.

BĂNĂRESCU, P., 1992: Zoogeography of fresh waters. Vol. 2. Distribution and dispersal of freshwater animals in North America and Eurasia. AULA-Verlag, Wiesbaden. pp. 519–1091.

BARUŠ, V., 1981: Návrh seznamu ohrožených taxonů obratlovců (Vertebrata) fauny ČSSR (Proposal of the list of endangered vertebrate taxa in Czechoslovakia). – Vertebrat. Zprávy, 1981, Brno. pp. 35–42. (in Czech).

BARUŠ, V., Z. BAUEROVÁ, J. KOKEŠ, B. KRÁL, S. LUSK, J. PELIKÁN, J. SLÁDEK, J. ZEJDA, and J. ZIMA, 1989: Červená kniha ohrožených a vzácných druhů rostlin a živočichů ČSSR. II. Kruhoústí, ryby, obojživelníci, plazi a savci (Red data book of the endangered and rare species of plants and animals of the Czechoslovakia, II. Lampreys, fishes, amphibians, reptiles and mammals). Stát. zem. nakl. Prague. 136 pp. (in Czech).

BARUŠ, V., and O. OLIVA (Eds.), 1995: Fauna ČR a SR, svazek 28/2. Mihulovci Petromyzontes a Ryby – Osteichthyes (Fauna of the Czech and Slovak Republics, Vol. 28/2. Lampreys and fishes). Academia Prague. 698 pp. (in Czech).

BASTL, I., 1961: Zmena morfometrických znakov dunajského kapra počas rastu v rybničnom chove (Change of morphometric features of the wild common carp during pond culture). Final Report, Laboratory of Fisheries, Bratislava. 159 pp. (unpublished, in Slovak).

BASTL, I., 1962: Rast divého dunajského kapra (*Cyprinus carpio* L.) v rybníkoch na Kamennom Mlyne (Growth of the wild Danubian carps in ponds). – Biológia (Bratislava) **12** (10): 757–770. (in Slovak).

BAUER, O. N. (Ed.), 1987: [Parasites identified on freshwater fishes in the fauna of the U.S.S.R. III. Parasitic metazoans (Part 2).]. Izd. Nauka, Leningrad. 584 pp. (in Russian).

BERG, L. S., 1909: (Fishes of the Amur Basin). – Zap. Akad. nauk po Fiz.-mat. otd. **24** (9): 1–270. (in Russian).

BERG, L. S., 1948: (Freshwater Fishes in the U.S.S.R. and neighbouring countries). Vol. 1. Izd. AN SSSR, Moscow – Leningrad. pp. 1–466. (in Russian).

BERG, L. S., 1949: (Freshwater Fishes in the U.S.S.R. and neighbouring countries). Vol. 2. Izd. AN SSSR, Moscow – Leningrad. pp. 469–925. (in Russian).

BERG, L. S., A. S. BOGDANOV, N. I. KOZHIN, and T. S. RASS (Eds.), 1949: (Commercial Fishes of the U.S.S.R.). Pishchepromizdat, Moscow. 787 pp. (in Russian).

BILLARD, R. (Ed.), 1999: Carp: biology and culture. (Springer/Praxis series in aquaculture & fisheries). Praxis Publishing Ltd., Chichester, UK. 342 pp.

BIRÓ, P., 1995: Management of pond ecosystems and trophic webs. – Aquaculture **129:** 373–386.

BLAHÁK, P., 1994: Kapitální kapři z našich vod (Trophy carps from our waters). Rybářské noviny (květen 1994). pp. 2–3. (in Czech).

BLOCH, M. E., 1783: Ökonomische Naturgeschichte der Fische Deutschlands. Hesse, Berlin, Vol. 1. 332 pp.

BORZENKO, M. P., 1926: [Data on the biology of the common carp (*Cyprinus carpio* Linnaeus)]. – Izv. Bakinskoi Ikhtiol. Lab., Baku **1** (2): 1–131. (in Russian).

BRIZINOVA, P. N., and V. S. KIRPICHNIKOV, 1952: (Problems of increasing the winter hardiness of yearling carps, Amur carps, and their hybrids). – Zool.zh., AN SSSR Moscow **31** (4): 595–604; **31**(6): 897–905. (in Russian).

BRODY, T., R. MOAV, Z. W. ABRAMSON, G. HULATA, and G. WOHLFAHRT, 1979: Application of electrophoretic genetic markers to fish breeding. II. Genetic variation within maternal half-sibs in carp. – Aquaculture **9:** 351–365.

BROWN, M. E., 1957: Experimental studies on growth. In: M. E. BROWN (Ed.), Physiology of Fishes. I. Academic Press, New York. 447 pp.

BRYLIŃSKA, M. (Ed.), 1986: Ryby slodkowodne Polski. Publ. House PWN, Warsaw. 429 pp.

BURMAKIN, E. V., 1956: (On the morphological changes of common carps becoming adapted to the basin of Lake Balkhash). – Zool.zh. **35** (12): 1887–1891. (in Russian).

CEND-AIUUSH, Ya., 1967: (The biology of the crucian carp and the common carp in Lake Buir-Nur). – Trudy Inst. Biol. AN Mong. Narod. Resp. **2:** 71–76. (in Russian).

CHERFAS, B.I., 1940: (Survival of young of the wild common carp under different ecological conditions). Rybnoe khozyaystvo, 8 (cit. after VOVK 1955, in Russian).

CHERFAS, N. B., O. V. EMELYANOVA, V. A. RECOUBRATSKY, and B. I. GOMELSKY, 1995: Cytogenetic and fishery properties of hybrids between the crucian carp and the common carp: An application of genetic methods in treatment of distant hybrids. – Aquaculture **129:** 217–218.

CHYTRA, P., V. KRUPAUER, and O. PÍCHA, 1962: Růst dunajských kaprů v rybnících v prvých třech letech života (Growth of the Danube wild carp during first three years of life). – Práce VÚRH Vodňany **1:** 73–91. (in Czech).

CHYTRA, P., V. KRUPAUER, O. PÍCHA, and J. TESARČÍK, 1963: Růst a odolnost kříženců mezi dunajským a rybničním kaprem (Growth and resistance of the hybrids between the Danube wild and pond carps). Závěrečná zpráva, VÚRH Vodňany. 53 pp. (unpublished, in Czech).

COSTEA, E., and POJOGA, I., 1955: Un hibrid natural intre crapul salbatio de greaca si carasul rosu (*Carassius auratus*) gasit in bazinul din gradina Cismiglu. – Bul. Inst. Cerc. (Bucuresti), **14:** 85–88.

CZHAN, DIN CHZHONG, 1967: (Information on the variations in inland waters, biology, and spread of carps in northern Vietnam). – Genetika (3) 2 (cit. after KIRPICHNIKOV 1967, in Russian).

DE LIGNY, W., 1969: Serological and biochemical studies on fish populations. – Oceanogr. Mar. Biol. Ann. Rev. **7:** 411–513.

DEMENTEV, A. F., and A. Ya. SHCHERBUKHA, 1974: (Morphological variations in the common carps in Kakhovsk Reservoir). – Rybnoe khozyaystvo, Kiev **18:** 88–94. (in Russian).

DOLZHENKO, K. F., 1953: (The biology of common carp becoming adapted to the lakes of western Siberia). – Zool. zh. **32** (6): 1217–1221. (in Russian).

DRYAGIN, P. A., 1936: [Fishes and fish management of the Tchu river]. – Trudy kirg. Kompleksnoy ekspedicii, AN SSSR 1932–1933, Vol. III/1: 49–83 (in Russian).

DUBRAVIUS, J., 1547: De piscinis ad Antonium Fuggerum. Rev. ed. In: A. SCHMIDTOVÁ (Ed.), Sborník filologický ČSAV, Prague 1953 **1:** 11–75.

DYBOWSKI, B., 1862: Versuch einer Monographie der Cyprinoiden. Livlands, Dorpat. 215 pp.

EREMEYEVA, E. F., 1960: (Comparative steps in the development of the common carp, *Cyprinus carpio* L., in the Kuban Liman and the Volga). – Trudy Inst. Morph. Zhiv., AN SSSR **25:** 37–46. (in Russian).

ERGENS, R., and J. LOM, 1970: Původci parazitárních nemocí ryb (Agents of parasitic diseases in fishes). Academia Prague. 384 pp. (in Czech).

FANG, P.W., 1936: Chinese freshwater fishes referring to Cyprininae. – Sinensia **7** (6): 686–712.

FIEBIGER, J., 1918–1919: Über den Körperbau des Karpfens. – Österr. Fischerei Ztg. **15:** 17–20, 26–27, 34–36, 49–50, 77–79, 110–11, 118–119, 125–127, 134–135, 177–178, 190–191; **16:** 2–5, 22–25, 53–55.

FISTER, S., and B. SOLDATOVIC, 1991: Karyotype analysis of male and diploid female *Carassius auratus gibelio* Bloch (Pisces, Cyprinidae) caught in the Danube at Belgrade. Evidence for the existence of a bisexual population. – Acta Vet. Belgrade **41** (1): 80–90.

FLAJŠHANS, M., and B. A. E. GALL, 1995: (1) Selection and hybridization. (Genetics: Chairperson's Summary). – Aquaculture **129:** 221–222.

FOREL, F. A., 1906: La peche du Leman. – Arch. Sci. Phys. Nat., Geneve **22:** 188–189.

FUJITA, Y., 1973: (Formation of the Japanese island arcs, Green Tuff movement). Tsukii - Shokan, Tokyo. 257 pp. (in Japanese).

GALKINA, Z. I., 1964: [Development of juvenile cyprinids and the Amur common carp (*Cyprinus carpio* L.) with different rates of growth]. – Izv. GosNIORKH **58:** 167–179. (in Russian).

GIGER, F., 1957: Die Karpfen von Schlossteich Anif. – Österr. Fischerei **10:** 137–138.

GORDA, S., J. BAKOS, J. LISKA, and C. KAKUK, 1995: Live gene bank of common carp strains at the Fish Culture Research Institute, Szarvas. – Aquaculture **129:** 199–202.

GRODZIŃSKI, Z., 1961: Anatomia i embriologia ryb. PWRiL, Warsaw, 319 pp.

GROMOV, I. A., 1973: (Morphological characteristics and local populations of the Amur common carp, *Cyprinus carpio haematopterus*). – Vopr. Ikhtiol. **13** (3): 408–422. (in Russian).

GROTE, W., C. VOGT, and B. HOFER, 1909: Die Süsswasserfische von Mittel-Europa. Verl. Schlütter u. Mass, Halle a. S. 558 pp.

GUSSEV, A. V., 1985: [Parasites identified on freshwater fishes in the fauna of the U.S.S.R. II. Parasitic metazoans (First Part)]. Izd. Nauka, Leningrad. 426 pp.

HAFEZ, R., R. LABAT, and R. QUILLER, 1978: Etude cytogenétique chez quelques espéces de cyprinides de la region Midi-Pyrenées. – Bull. Soc. Hist. Nat. Toulouse **114** (1–2): 1–38.

HALDAR, D. P., K. K. SAMAL, and D. MUKHOPADHYAY, 1996: Studies on the protozoan parasites of fishes in Orissa: eight species of *Myxobolus* Bütschli (Myxozoa: Bivulvulidae). – Journ. Bengal. Nat. Soc. (new series) **16:** 3–24.

HANEL, L., 1992: Poznáváme naše ryby (We identify our fishes). Zem. nakl. Brázda, Prague. 288 pp. (in Czech).

HANKÓ, B., 1932: Ursprung und Verbreitung der Fischfauna Ungarns. – Arch. Hydrobiol. **23:** 520–556.

HARSÁNYI, A., 1986: Der Karpfen – *Cyprinus carpio* L. – Verbreitungsgeschichte, Domestikation, seine Bedeutung in Seen und Flüssen. – Fischer und Teichwirt **37** (7): 201–203.

HECKEL, J. J., 1836: Über einige neue oder nicht gehörige unterschiedene Cyprinen, nebst einer systematischen Darstellung der europäischen Gattungen dieser Gruppe. – Ann. Wien. Mus. Nat. **1**(2): 219–223, 2 pl.

HECKEL, J. J., and R. KNER, 1858: Die Süsswasserfische der österreichischen Monarchie mit Rücksicht an die angrenzenden Ländern. W. Engelmann, Leipzig. 388 pp.

HERMANN, O., 1887: A Magyar halászat könyve. I., II. K.M. Természettudományi Társulat, Budapest. 860 pp.

HOCHMAN, L., and A. KIRKA, 1990: Usmernovanie chovu rýb v špecifických prirodzených a experimentálnych podmienkach (Management of fish culture in specific natural and experimental conditions). Final Report, Ústav rybárstva a hydrobiológie Bratislava. 23 pp. (in Slovak, unpublished).

HOFFMANN, G. L., 1967: Parasites of North American fishes. University of California Press, Berkeley and Los Angeles. 486 pp.

HOFFMANN, R. C., 1994a: Remains and verbal evidence of carp (*Cyprinus carpio*) in medieval Europe. In: W. VAN NEER (Ed.), Fish Exploitation in the Past. Proc. of the 7th Meeting of the ICAZ Fish Remains Working Group. – Ann. Mus. Royal l'Afrique Centrale, Sci. Zool., Tervuren **274:** 139–150.

HOFFMANN, R. C., 1994b: Medieval Cistercian fisheries natural and artificial. In: L. PRESSOUYRE (Ed.), L'espace cistercien. Comité des travaux historiques et scientifiques, Paris. pp. 401–414.

HOFFMANN, R. C., 1995: Environmental change and the culture of common carp in medieval Europe. – Guelph Ichthyol. Rev. **3:** 57–85.

HOLČÍK, J., 1970: K rekordnému úlovku dunajského kapra z Med'vedovského vel'kého ramena (On the trophy Danube carp caught in the Large Medvedovo Arm). – Pol'ovníctvo a rybárstvo **22** (5): 21 (in Slovak).

HOLČÍK, J., 1983: Ichtyofauna Československa a problematika jej ochrany (Ichthyofauna of Czechoslovakia and problems of its protection). In: Aktuální problémy ochrany fauny v ČSSR, Vyd. ÚVO ČSAV Brno. pp. 72–80 (in Slovak).

HOLČÍK, J., 1989: Návrh červeného zoznamu ohrozených kruhoústych a rýb Slovenska (Proposal for red list of threatened cyclostomes and fishes of Slovakia). – Pamiatky a príroda **20:** 26–28 (in Slovak).

HOLČÍK, J., 1991: Fish introductions in Europe with particular reference to its central and eastern part. – Can. J. Fish. Aquat. Sci. **48** (Suppl. 1): 13–23.

HOLČÍK, J., 1995: Rybárstvo pred a po prehradení slovenského úseku Dunaja (Fisheries before and after the damming of the Slovak section of the Danube). In: A. SVOBODOVÁ and M. J. LISICKÝ (Eds.), Výsledky a skúsenosti z monitorovania bioty územia ovplyvneného vodným dielom Gabčíkovo. Ústav zoológie a ekosozológie SAV Bratislava. pp. 144–152 (in Slovak).

HOLČÍK, J., 1996: Vanishing freshwater fish species of Slovakia. In: A. KIRCHHOFER and D. HEFTI (Eds.), Conservation of Endangered Freshwater Fish in Europe. Birkhäuser Verlag, Basel, Switzerland. pp. 79–88.

HOLČÍK, J., and L. JEDLIČKA, 1994: Geographical variation of some taxonomically important characters in fishes: the case of the bitterling *Rhodeus sericeus*. – Env. Biol. Fish **41:** 147–170.

HULATA, G., 1995: A review of genetic improvement of the common carp (*Cyprinus carpio* L.) and other cyprinids by crossbreeding, hybridization and selection. – Aquaculture **129:** 143–155.

IKEBE, N., 1978: (Bio- and chronostratigraphy of Japanese Neogene, with remarks on paleogeography). Prof. Nobuo Okebe Mem. Vol., pp. 13–34 (in Japanese).

ISHIKAWA, C., K. OKAMURA, S. TANAKA, A. TERAO, H. MARUKAWA, T. HIGURASHI, and H. SENÔ (Eds.), 1931: Illustrations of Japanese aquatic plants and animals. Vol. 1. Fisheries Soc. of Japan, Tokyo. 50 plates. (in Japanese).

IVANNIKOV, V. P., 1997: [Status of stocks and commercial exploitation of the wild carp (*Cyprinus carpio* L.) in the Volga-Caspian region]. First Congress of ichthyologists of Russia (Theses of Reports). VNIRO Publ. House Moscow, pp. 420–421 (in Russian).

JÁSZFALUSI, L., 1960: Vadponty és nemesponty tenyésztési kérdései. Halászat **7** (3): 46–47.

JENEY, Zs., and G. JENEY, 1995: Recent achievements in studies on diseases of common carp (*Cyprinus carpio* L.). – Aquaculture **129:** 397–420.

JHINGRAN, V. G., and R. S. V. PULLIN, 1985: A hatchery manual for the common, Chinese and Indian major carps. ICLARM Stud. Rev. 11. 191 pp.

KABATA, Z., 1985: Parasites and diseases of fish cultured in the tropics. Taylor and Francis, London. 318 pp.

KÁLAL, L., 1995: 1. druh *Cyprinus carpio* Linnaeus, 1758 (First species *Cyprinus carpio* Linnaeus, 1758). In: V. BARUŠ and O. OLIVA (Eds.), Fauna ČR a SR, Mihulovci – Petromyzontes a Ryby – Osteichthyes, Vol. 28/2. Academia, Prague. pp. 234–262. (in Czech).

KAMIHATA, S., 1988: Nishikigoi. – Trop. Fish Hobbyist **36** (8): 53–68.
KAZANCHEEV, E. N., 1981: (Fishes of the Caspian Sea). Izd. Legk. i Pishch. promyshl., Moscow. 168 pp. (in Russian).
KEITZ, G., and J. A. ADUMUA-BOSSMAN, 1972: Über biometrische Untersuchungen an Karpfen (*Cyprinus carpio* L.), Giebel (*Carassius auratus gibelio* Bloch) und deren Bastarden. – Verh. int. Ver. theor. angew. Limnol. **18:** 1155–1165.
KESSLER, K. T., 1856: Zur Ichthyologie des südwestlichen Russlands. – Bull. Soc. Nat., Moscow **29:** 335–393.
KESSLER, K. T., 1859: (A journey, with zoological goals, to the northern shore of the Black Sea and the Crimea Peninsula in the year 1858). Univ. izv., Kiev. pp. 1–248. (in Russian).
KESTEMONT, P., 1995: Different systems of carp production and their impacts on the environment. – Aquaculture **129:** 347–372.
KHOKHLINA, E. A., 1941: (Hybridization experiments with goldfish and common carp). – Doklady AN SSSR **30** (cit. after NIKOLYUKIN 1972).
KIRKOVÁ, Ž., 1990: Základní hematologické ukazovatele kapra dunajského a kapra kultúrneho (Basic haematological indices in the wild Danube and cultured common carps). – Živočišná výroba **35** (10): 889– 894. (in Slovak).
KIRPICHNIKOV, V. S., 1943: [Experimentation for the systematics of the common carp (*Cyprinus carpio* L.). I. Growth and the morphological characteristics of the Tarapavan, Volga-Caspian, and Amur common carp under conditions of pond culture]. – Izv. AN, ser. biol., Izd. AN SSSR, Moscow **4:** 189–220. (in Russian).
KIRPICHNIKOV, V. S., 1944: [Winter hardiness of the Amur common carp, *Cyprinus carpio viridiviolaceus,* (under conditions in the central and northern zones of the European part of the U.S.S.R.)]. – Doklady AN SSSR **43** (1): 37–41. (in Russian).
KIRPICHNIKOV, V. S., 1958: (Resistance to low temperatures and winter conditions by juvenile cyprinids, common carp, and their hybrids). – Tr. soveshch. ikhtiol. komisii AN SSSR **8:** 261–270. (in Russian).
KIRPICHNIKOV, V. S., 1967: [Homologous hereditary variation and evolution of the common carp (*Cyprinus carpio* L.)]. – Genetika (2) **3:** 34–47. (in Russian).
KIRPICHNIKOV, V. S., 1979: (Genetic basis of fish selection). Izd. Nauka, Leningrad. 391 pp. (in Russian).
KIRPICHNIKOV, V. S., 1981: Genetic bases of fish selection. Springer Verlag, Berlin. 410 pp.
KIRPICHNIKOV, V. S., 1987: (Genetics and selection of fishes). Izd. Nauka, Leningrad. 520 pp. (in Russian).
KIRPICHNIKOV, V. S., and R. L. BERG, 1952: (Problems in increasing the winter hardiness of first year cyprinids, common carps, and their hybrids). – Zool. zh., Izd. AN SSSR, Moscow **31** (6): 897–905. (in Russian).
KIRPITSCHNIKOW, V. S., 1961: Die genetische Methoden der Selektion in der Karpfenzucht. – Z. f. Fischerei, Vol. **10**, No. 7(1–3): 137–163.
KLUST, G., 1939/1940: Über Entwicklung, Bau und Funktion des Darmes beim Karpfen (*Cyprinus carpio* L.). – Int. rev. Hydrobiol. **39:** 498–536; **40:** 88–173.
KOBLICKAYA, A. F., 1981: (Identifying juvenile freshwater fishes). Izd. Legkaya i pishchevaya promyshlennost', Moscow. 208 pp. (in Russian).
KOKHANOVA, G. D., 1969: (Certain questions on the biology and fishery of the common carp in the Kremenchug Reservoir). – Rybn. khoz., Kiev **8:** 133–136. (in Russian).

KOTTELAT, M., 1997: European Freshwater Fishes. An heuristic checklist of the freshwater fishes of Europe (exclusive of former USSR), with an introduction for non-systematists and comments on nomenclature and conservation. – Biológia (Bratislava) **52** (Suppl. 5): 1–271.

KRATOCHVÍL, J., 1966: (Die wissenschaftlichen Namen der Haustiere). – Acta Univ. Agriculturae, Brno **3:** 417–420. (in Czech).

KRUPKA, I., 1982: Vývoj, vek a rast kapra (Development, age and growth of wild carp). – Pol'ovníctvo a rybárstvo **34** (5): 32. (in Slovak).

KRUPKA, I., 1987: O raste kapra (On the growth of the wild carp). – Pol'ovníctvo a rybárstvo **39** (11): 483 (in Slovak).

KRUPKA, I., 1992: Zvel'adenie genofondu dunajského kapra (Improvement of gene pool of the Danube wild carp). Záverečná výsk. zpráva, Ústav rybárstva a akvakultúry, Bratislava. 20 pp. (in Slovak, unpublished).

KRUPKA, I., and J. MESZÁROS, 1989: K problematike súčasného stavu genofondu dunajského kapra (On the problem of the actual state of gene pool of the Danube wild carp). In: Zmiešaná komisia medzinárod. dohody o rybolove vo vodách Dunaja – materiály 29. Zasedania. (in Slovak, unpublished).

KRUPKA, I., J. MESZÁROS, P. RÁB, and V. ŠLECHTOVÁ, 1989: Towards the preservation of endangered fish gene pools – example of wild Danube carp (*Cyprinus carpio carpio*). – Práce VÚRH Vodňany **18:** 27–33.

KRYZHANOVSKIY, S. G., 1949: [Regular eco-morphological development of cyprinoid and siluroid fishes (Cyprinoidei and Siluroidei)]. – Trudy Inst. Morphol. Zhivot. **1:** 5–332. (in Russian).

KRYZHANOVSKIY, S. G., SMIRNOV, A. I., and S. G. SOIN, 1951: (Information on the development of fishes in the Amur River). – Trudy Amursk. Ikhtiol. Eksped. 1945 – 1949, Vol. **2:** 5–222. (in Russian).

KUX, Z., 1982: Příspevek k problematice křízenců rodu *Carassius* (Contribution to the hybrids of the genus *Carassius*). – Čas. Mor. muzea, Brno **67:** 181–188. (in Czech).

LANDSBERG, J. H., 1991: Parasites and associated diseases of fish in warm water aquaculture, with special emphasis on intensification. In: M. SHILO and S. SARIG (Eds.), Fish Culture in Warm Water Systems. Problems and Trends. CRC Press, Cleveland. pp. 195–252.

LANGE, N. O., E. N. DMITRIEVA, E. N. SMIRNOVA, and M. PEŇÁZ, 1974: (Methods for the study of morphological and ecological peculiarities of fish development during the embryonic, larval and juvenile periods). In: R. S. VOLSKIS (Ed.), (Standard methods for research on the productivity of fish species within their areas). Publ. House Mintis, Vilnius. pp. 56–57. (in Russian).

LASSLEBEN, P., 1990: Wildkarpfen. – Fischer und Teichwirt **41** (8): 214–217.

LELEK, A., 1987a: Notes on the reproduction ecology of the feral form of the common carp, *Cyprinus carpio carpio*, in the Rhine River. – Proc. V. Congr. Europ. Ichthyol., Stockholm. pp. 169–173.

LELEK, A., 1987b: Threatened fishes of Europe. The Freshwater Fishes of Europe. Vol. 9. AULA-Verlag, Wiesbaden. 343 pp.

LEONHARD, E., 1906: Der Karpfen. Geschichte, Naturgeschichte und wirtschaftliche Bedeutung unseres wichtigsten Zuchtfisches. J. Neumann, Neudamm. 104 pp.

LEONHARDT, Ch., 1904: Die Bastarde der deutschen karpfenähnlichen Fische. – Fisch. Ztg., Neudamm **6:** 1–58.

LIN, S.Y., 1933: Contribution to a study of Cyprinidae of Kwantung and adjacent provinces. – Lingnan Sci. Journ., Canton, **12:** 75–91, 197–215, 237–448, 489–505.

LINNAEUS, C., 1758: Systema naturae per regna tria naturae, secundum classes, ordines, genera, species, cum characteribus differentiis, synonymis, locis. Tomus I, Ed. Decima, reformata. L. Salvii, Holmiae, 823 pp.; Ed. XII, 532 pp.

LOM, J. and I. DYKOVÁ, 1989: Protozoární paraziti užitkových ryb (Protozoan parasites of commercial fishes). Vyd. ČRS v SZN Prague. 103 pp. (in Czech).

LOM, J. and I. DYKOVÁ, 1992: Protozoan parasites of fishes. Developments in Aquaculture and Fisheries Science, Vol. 26. Elsevier, Amsterdam - London - New York - Tokyo. 315 pp.

LOM, J., and F. MORAVEC, 1995: 1. druh *Cyprinus carpio* Linnaeus, 1758 (First species *Cyprinus carpio* Linnaeus, 1758. Parasites). In: V. BARUŠ and O. OLIVA (Eds.), Fauna ČR a SR, sv. 28/2, Mihulovci - Petromyzontes a ryby - Osteichthyes. Academia, Prague. pp. 261–262. (in Czech).

LUKIN, A. V., 1948: (Common carp in the middle Volga and the fishery for its commercial exploitation). – Tr. Tatarsk. Otd. VNIORCH **4:** 103–122. (in Russian).

LUSK, S., 1996: The status of the fish fauna in the Czech Republic. In: A. KIRCHHOFER and D. HEFTI (Eds.), Conservation of Endangered Freshwater Fish in Europe. Birkhäuser Verlag, Basel, Switzerland. pp. 89–98.

LUSK, S., and V. BARUŠ, 1978: Morphometric features of *Carassius auratus* from the drainage area of the Morava River. - Folia Zool. **27** (2): 177–190.

MAI DING, Y., 1978: (Classification of freshwater fishes from the north provinces of Vietnam). Hanoi. 340 pp. (in Vietnamese).

MAKSUNOV, V. A., 1955: (On the biology of the common carp in Lake Balkhash). - Vopr. Ikhtiol. **5:** 71–77. (in Russian).

MARGOLIS, L., and J. R. ARTHUR, 1979: Synopsis of the parasites of fishes of Canada. - Bull. Fish. Res. Board Can. **199:** 1–269.

MCDOWALL, R.M., 1990: New Zealand Freshwater Fishes. Heynemann Reed, Auckland, 553 pp.

MESZÁROS, J., 1990: Overenie biotechnológií odchovu násad ohrozených druhov rýb (Checking of methods for culture of endangered stockfish species). Záverečná zpráva výsk. úkolu, Ústav rybárstva a hydrobiológie Bratislava. 81 pp. (in Slovak, unpublished).

MIŠÍK, V., 1957: Technika lovu rýb zátahovou sietou na Malom Dunaji pri Kolárove (Fishing techniques with seine at the Lesser Danube near Kolárovo). - Pol'nohospodárstvo **1** (4): 135–156. (in Slovak).

MIŠÍK, V., 1958a: Biometrika dunajského kapra (*Cyprinus carpio* L.) z dunajského systému na Slovensku (Biometry of the Danubian carp of Slovakia). - Biol. práce SAV (Bratislava) **4** (6): 55–128. (in Slovak).

MIŠÍK, V., 1958b: Význam dunajského kapra a jeho chov ve volných vodách (Importance of the Danubian carp and their culture in free waters). - Pol'ovníctvo a rybárstvo **10** (2): 40–41. (in Slovak).

MIŠÍK, V., 1958c: Posudzovanie exteriéru kapra dunajského (Recognition of the shape of Danubian carp). - Pol'ovníctvo a rybárstvo **10** (3): 61–63. (in Slovak).

MIŠÍK, V., and V. TUČA, 1965: Posudzovanie exteriéru dunajského kapra so zretelom na výber plemenného materiálu (Evaluation of the form of Danubian carp with respect to the selection of breeders). - Pol'nohospodárstvo **11** (1): 35–44. (in Slovak).

MOAV, R., T. BRODY, G. WOHLFAHRTH, and G. HULATA, 1979: A proposal for the continuous production of F_1 hybrids between the European and Chinese races of the common carp in traditional fish farms of Southeast Asia. In: T. V. R. PILLAY and W. A. DILL (Eds.), Advances in Aquaculture. Papers presented at the FAO Technical Conference on Aquaculture, Kyoto, Japan, 26 May – 2 June 1976. Fishing News Books Ltd., Farnham – Surrey, England. pp. 635–638.

MOLNÁR, K. and F. MORAVEC, 1997: *Skrjabillanus cyprini* n.sp. (Nematoda: Dracunculoidea) from the scales of common carp *Cyprinus carpio* (Pisces) from Hungary. – Syst. Parasitol. **37:** 1–5.

MORAVEC, F., 1987: Revision of capillariid nematodes (subfamily Capillariinae) parasitic in fishes. – Studie ČSAV, Academia Prague **3:** 1–141.

MORAVEC, F., 1994: Parasitic nematodes of freshwater fishes of Europe. Academia, Prague. 476 pp.

MORAVEC, F., 1998: Nematodes of freshwater fishes of the Neotropical Region. Academia, Prague, 456 pp.

MUIR, J. F., and R. J. ROBERTS (Eds.), 1982: Recent advances in aquaculture. Croom Helm, London. 458 pp.

NEKRASHEVICH, N. G., 1934: (The Balkhash common carp in Lake Chany). – Tr. Zap.- Sib. otd. VNIORCH **1:** 44–69. (in Russian).

NICHOLS, J. T., 1943: The fresh-water fishes of China. The American Mus. of Nat. History (Natural history of Central Asia), Vol. 9, pp. 1–322.

NIE, P., 1995: Communities of intestinal helminths of carp, *Cyprinus carpio* in highland lakes in Yunnan province of Southwest China. – Acta Parasitologica **40** (3): 148–151.

NIKOLSKIY, G. V., 1934: (The Aral carp. Information on its systematics, biology, and fishery). – Trudy Aralskogo otdel. Vsesoy. inst. morsk. rybn. khoz., Aralsk **3:** 5–75. (in Russian).

NIKOLSKIY, G. V., 1940: (The fishes of the Aral Sea. Information on the fauna and flora of the U.S.S.R.). – Izd. MOIP, Moscow. 216 pp. (in Russian).

NIKOLSKIY, G. V., 1949: (The common carp – *Cyprinus carpio* Linnaeus). In: (Commercial Fishes of the U.S.S.R.). Pishchepromizdat, Moscow. pp. 449–452. (in Russian).

NIKOLSKIY, G. V., 1956: (Fishes of the Amur Basin). Itogi Amurskoi ikhtiologicheskoi ekspedicii 1945 – 1956]. Izd. AN SSSR Moscow. 551 pp. (in Russian).

NIKOLSKIY, G. V., 1957: Spezielle Fischkunde. VEB Deutscher Verlag der Wissenschaften, Berlin. 632 pp.

NIKOLSKIY, G. V., and N. A. EVTUKHOV, 1940: (Fishes in the flood plain of the Ili River. Part I. Cyprinids). – Byul. MOIP, otd. biol. **49** (3–4): 187–189. (in Russian).

NIKOLYUKIN, N. I., 1972: [Distinguishing hybridized acipenserids and bony fishes (theory and practice)]. Izd. Pishchevaya promyshlennost', Moscow. 336 pp. (in Russian).

OHNO, S., J. MURAMOTO J., and L. CHRISTIAN, 1967: Diploid-tetraploid relationship among old-world members of the fish family Cyprinidae. – Chromosoma **23:** 1–9.

OLIVA, O., 1955: Příspěvek k biologii a rychlosti růstu kapra (*Cyprinus carpio*) v Polabí [Contribution to the biology and growth rate of the common carp (*Cyprinus carpio*) from the Elbe flood plain]. – Univ. Carolina, Biologica **1** (3): 225–273. (in Czech).

OSMANOV, S. O., 1976: [Questions on parasitology of the Aral Sea (on changes in the parasitological situation during the past 40 years)]. Izd. FAN Uzbek. SSR, Tashkent. 200 pp. (in Russian).

PAAVER, T., 1983: (Biochemical genetics of carp, *Cyprinus carpio* L.). Valgus Publ., Tallin. 122 pp. (in Russian).

PAPANDOPOL, M., and I. CAUTIS, 1965: Contributii la studiul variatili morfologice a crapului – *Cyprinus carpio* L. – din bazinul Dunarii inferioare. – Soc. de stiinte nat. si geografie din R.P.R., Comunicari de soologie **3** (cited after DEMENTEEV A.F. & SHCHERBUKHA A.Ya. 1974).

PAPERNA, I., 1964: The metazoan parasite fauna of Israel inland water fishes. – Bamidgeh **16** (1/2): 3–66.

PAVLOV, D. S., K. A. SAVVAITOVA, L. I. SOKOLOV, and S. S. ALEKSEEV, 1994: (Rare and disappearing animals. Fishes). Izd. Vyssh. shkola, Moscow. 334 pp. (in Russian).

PAVLOV, P. I., 1964: [Adaptational changes in the common carp (*Cyprinus carpio* L.) of the Danube region]. – Vopr. Ikhtiol., **4** (cited after Dementeev A.F. & Shcherbukha A.Ya. 1974). (in Russian).

PEŇÁZ, M., M. PROKEŠ, J. KOUŘIL, and J. HAMÁČKOVÁ, 1983: Early development of the carp, *Cyprinus carpio*. – Acta Sc. Nat. Brno **17** (2): 1–39.

POGODINOV, F. O., and V. A. VOROBYOV, 1954: (Acclimatization of the Amur common carp in Siberian water bodies). – Ryb. khoz. (11): 33–34. (in Russian).

POKORNÝ, J., M. FLAJŠHANS, P. HARTVICH, P. KVASNIČKA, and I. PRUŽINA, 1995: Atlas kaprů chovaných v České republice (The atlas of carps bred in the Czech Republic). Victoria Publishing, Prague. 71 pp. (in Czech).

POSCHALUJEWA, K., 1929: Beiträge zur Kenntnis des Alters und Wachstumtempos des Karpfen aus dem Aralsee (Referat). – Fischerei Ztg. **32:** 659.

PROKEŠ, M., and V. BARUŠ, 1996: On the natural hybrid between common carp (*Cyprinus carpio*) and Prussian carp (*Carassius auratus gibelio*) in the Czech Republic. – Folia Zool. **45** (3): 277–282.

PROTASOVA, E.N., B. I. KUPERMAN, V. A. ROYTMAN, and L. G. PODDUBNAYA, 1990: (Caryophyllaeid fauna of the U.S.S.R.). Izd. Nauka Moscow. 240 pp. (in Russian).

PULLAN, S., and P. J. SMITH, 1987: Identification of hybrids between koi (*Cyprinus carpio*) and goldfish (*Carassius auratus*). – N. Z. J. Mar. Freshw. Res. **21:** 41–46.

RÁB, P., 1994: Cytogenetická studie vybraných skupin holarktických sladkovodních ryb (A cytogenetic study of selected groups of Holarctic freshwater fishes). Doktorská disertační práce. Ústav živočišné fyziologie a genetiky AV ČR, Libechov. 98 pp. (in Czech, unpublished).

RÁB, P., 1995: 1. druh *Cyprinus carpio* Linnaeus, 1758. Karyotyp (First species *Cyprinus carpio* Linnaeus, 1758. Karyotype). In: V. BARUŠ and O. OLIVA (Eds.), Fauna ČR a SR, Mihulovci – Petromyzontes a Ryby – Osteichthyes, Vol. 28/2. Academia, Prague. p. 246 (in Czech).

RÁB, P., and M. J. COLLARES-PEREIRA, 1995: Chromosomes of European cyprinid fishes (Cyprinidae, Cypriniformes): a review. – Folia Zool. **44** (3): 193–214.

RAICU, P., E. TAISESCU, and A. CRISTIAN, 1972: Diploid complement of the carp. – Cytologia **37:** 355–358.

RAJTÁR, J., 1990: Anfänge der römischen Bautätigkeit im Vorfeld von Brigetio. In: H. VETTERS and M. JANDLER (Eds.), Akten des 14. Int. Limeskongresses 1986 in Carnutum. Der Römische Limes in Österreich, Wien. pp. 771–778.

RAJTÁR, J., 1992: Das Holz-Erde-Lager aus der Zeit der Markomannenkriege in Iža. In: Probleme der relativen und absoluten Chronologie ab Laténzeit bis zum Frühmittelalter. Kraków. pp. 149 – 170.

RUDZINSKI, E., 1961: Vergleichende Untersuchungen über den Wildkarpfen der Donau und den Teichkarpfen. – Z. f. Fischerei **10** (1/2): 105–135.

RUDZINSKI, E., 1962: Fragmenty historii europejskiego kapra hodowlanego. – Gospod. rybna **14** (4): 3–5.

RUETIMEYER, L., 1860: Untersuchung der Thierreste aus den Pfahlbauten der Schweiz. Mitteil. Antiquarische Gesellschaft, Zurich (cit. after BALON 1974).

SABANEEV, L. P., 1892: [Fishes of Russia]. – Zhizn' i lovlya (uzhenie) nashikh presnovodnykh ryb. **2**: 7–576. (Reprint by Izd. Fizkult. i sport, Moscow, 1982). (in Russian).

SABLINA, T.B., 1960: [Late stages in the development of the roach, *Rutilus rutilus heckeli* (Nordmann), bream, *Abramis brama* (L.), and common carp, *Cyprinus carpio* (L.) in the fore-delta of the Volga and the northern Caspian Sea.]. – Trudy Inst. Morph. Zhiv. AN SSSR **25:** 137–160. (in Russian).

SAPOSHNIKOVA, G. K., 1954: (The common carp in the Ural River and possibilities for its utilization in culture in artificial water bodies in the Ural region). – Tr. Zool. Inst. AN SSSR **16:** 576–609. (in Russian).

SCHÄFERNA, K., 1934: Křiženci některých našich ryb (Hybrids of some of our fishes). 48. ročenka 1. ryb. Klubu v Praze za rok 1934. pp. 6–10. (in Czech).

SCHÄPERCLAUS, W., 1961: Lehrbuch der Teichwirtschaft, 2nd Ed. P. Parey, Berlin. 582 pp.

SCHOLZ, T., 1989: Amphilinida and Cestoda, parasites of fish in Czechoslovakia. – Acta Sc. Nat. Brno **23** (4): 1–56.

SCHOLZ, T., and V. HANZELOVÁ, 1998: Tapeworms of the genus *Proteocephalus* Weinland, 1858 (Cestoda: Proteocephalidae), parasites of fishes in Europe. Studie AV ČR, 2/98, Academia Prague, 120 pp.

SCHULMAN, S. S. (Ed.), 1984: (Parasites identified on fishes in the fauna of the U.S.S.R. I. Parasitic protozoans.). Izd. Nauka, Leningrad. 429 pp. (in Russian).

SCHWARTZ, F. J., 1972: World literature to fish hybrids with an analysis by family, species and hybrid. – Gulf Coast Res. Lab. Mus. 3, Ocean Springs, Mississippi. 328 pp.

SCHWARTZ, F. J., 1981: World literature to fish hybrids with an analysis by family, species and hybrid. Suppl. 1. U.S. Dept. Commerce ND/AA Techn. Rep. 750, Seattle, Wash. 507 pp.

SEDLÁR, J., I. STRÁŇAI, and A. MAKARA, 1987: Kapor (The carp). Vyd. Príroda, Bratislava. 192 pp. (in Slovak).

SEROV, N. P., 1966: (The hybrid between the silver crucian carp x wild carp). – Trudy In-ta Ikhtiol. i Rybn. Khoz., Alma-Ata **5:** 352–353. (in Russian).

SHATUNOVSKIY, M. I. (Ed.), 1983: [Fishes of the Mongolian People's Rebublic (Living conditions, systematics, morphology, and zoogeography)]. Izd. Nauka, Moscow. 278 pp. (in Russian).

SHIRKOVA, A. P., 1955: [Successful acclimatization of Amur common carp in Pskov-Chudsk water bodies]. – Ryb. khoz. **10:** 32. (in Russian).

SIEBOLD, C. T. C., 1863: Die Süsswasserfische von Mitteleuropa. W. Engelmann, Leipzig, 430 pp.

SMIRNOVA, E. N., 1978: [The development of cultured common carps during the embryonic period]. In: B. V. KOSHELEV (Ed.), Ekologo-morfologicheskie issledovaniya razvitiya ryb. Izd. Nauka, Moscow, pp. 56–71. (in Russian).

SNIESZKO, S. F., 1974: The effect of environmental stress on outbreaks of infectious diseases of fishes. – J. Fish Biol. **6:** 197–208.

STEFFENS, W., 1958: Der Karpfen. Die Neue Brehm-Bücherei. A. Ziemsen Verlag, Wittenberg Lutherstadt. 90 pp.

STEFFENS, W., 1959: Wachstum und Körperform ungarischer Wildkarpfen in deutschen Teichen. – Deutsche Fischerei Ztg. **6** (7): 213–218.

STEFFENS, W., 1962: Der Karpfen. 2nd ed., A. Ziemsen Verlag, Wittenberg Lutherstadt.

STEFFENS, W., 1964a: Vergleichende anatomisch-physiologische Untersuchungen an Wild- und Teichkarpfen (*Cyprinus carpio* L.). Ein Beitrag zur Beurteilung der Zuchtleistungen beim deutschen Teichkarpfen. – Z. f. Fischerei, N.F. **12** (8, 9, 10): 725–800.

STEFFENS, W., 1964b: Der heutige Stand der Zuchtleistungen beim Karpfen in unseren Teichwirtschaften. – Deutsche Fischerei Ztg. **11** (11): 328–335.

STEFFENS, W., 1967: Das Domestikationsproblem beim Karpfen (*Cyprinus carpio* L.). – Verh. Int. Ver. theor. angew. Limnol. **16** (2): 1441–1448.

STEFFENS, W., 1975: Der Karpfen (4th ed.)., A. Ziemsen Verlag, Wittenberg Lutherstadt., 215 pp.

STEFFENS, W., 1976: Wachstum und Körperproportionen von Rückkreuzungshybriden (Teichkarpfen x Wildkarpfen) x Teichkarpfen (*Cyprinus carpio* L.). – Biol. Zbl. **95:** 675–682.

STEFFENS, W., 1980: Der Karpfen, *Cyprinus carpio*. (5th ed.). A. Ziemsen Verlag, Wittenberg. 215 pp.

STRUMIENSKI, O., 1573: O sprawie, sypaniu, wymieszaniu i rybieniu stawow. Publ. F. Kucharewski, Kraków. 87 pp.

SUMANTADINATA, K., 1995: Present status of common carp (*Cyprinus carpio* L.) stocks in Indonesia. – Aquaculture **192:** 205–209.

ŠUSTA, J., 1884: Výživa kapra a jeho družiny rybniční (The diet of the carp and some other pond fishes). Nezměněný otisk vydání z r. 1884, vyd. Čs. akademie zemědělská v r. 1937, Prague. 224 pp.

SUZUKI, R., 1979: The culture of common carp in Japan. In: T. V. R. PILLAY and W. A. DILL (Eds.), Advances in Aquaculture. Papers presented at the FAO Technical Conference on Aquaculture, Kyoto, Japan, 26 May – 2 June 1976. Fishing New Books Ltd., Farnham – Surrey, England. pp. 161 – 166.

SVETOVIDOV, A. N., 1933: Über die europäischen und ostasiatischen Karpfen (*Cyprinus carpio*). – Zool. Anz., **104:** 257–261.

TACON, A. G. J., 1994: Feed ingredients for carnivorous fish species. Alternatives to fishmeals and other fishery resources. FAO Fisheries Circular 881, FAO, Rome. 35 pp.

TARONOVA, L. V., 1959: [Morphological characteristics of the common carp in the delta of the Ili River]. –Sbornik rabot po ikhtiol., i gidrobiol., Alma-Ata **2:** 87–90. (in Russian).

THANG, T.L., 1933: The study of Chinese cyprinoid fishes. Part I. – Zool. Sinica, Peiping, Ser. B., **2** (1): 1–247.

THIEN, T.M., and T. D. TRONG, 1995: Genetic resources of the common carp in Vietnam. – Aquaculture **129:** 216.

TUČA, V., 1958: Štúdium genetickej (exteriérovej) čistoty dunajského kapra (The research on genetic purity of the Danube wild carp). Final Research Report, Laboratory of Fishery Research, Bratislava. 109 pp. (in Slovak, unpublished).

TUČA, V., and I. BASTL, 1961: Zmena morfometrických znakov dunajského kapra počas rastu v rybničnom chove (The change of morphometric characters of the Danube wild carp during the pond culture). Final Research Report, Laboratory of Fishery Research, Bratislava. 159 pp. (in Slovak, unpublished).

TURDAKOV, F. A., 1956: (Outline of the biology of the commercial fishes in Lake Issyk-Kul.). Ikhtiol. sbornik, AN Kirghiz. SSR, Frunze. pp. 3–66. (in Russian).

TURDAKOV, F. A., 1963: (The fishes of Kyrgyzstan). Izd. AN Kirghiz. SSR, Frunze. 283 pp. (in Russian).

UNTERÜBERBACHER, H., 1963: Über Wachstum und Lebensweise der Karpfen im Neusiedlersee. – Z. f. Fischerei, N.F. **11:** 481–532.

URBANOWICZ, K., 1956: Osteologia karpia (Osteology of carp). PWN, Warsaw. 96 pp.

VASILIU, G. D., M. N. DUVAZ, and A. NICOLAU, 1947: Contributions a la connaissance du rythme de la croissance chez quelques races de carpe des étanges systematiques de la

Station de Pisciculture de Nucet. – Annalele Insitutului de Cercetari Piscicole al Romaniei, Bucarest **3**–6: 80–114.

VASNECOV, V. V., E. F. EREMEYEVA, and N. O. LANGE, 1957: (Steps in the development of semi-migratory commercial fishes in the Volga and Don – the bream, common carp, Caspian roach, Black Sea roach, and pike-perch). – Trudy Inst. morph. zhivot. AN SSSR **16:** 7–76. (in Russian).

VERMA, P., 1970: Normal stages in the development of *Cyprinus carpio* var. *communis* L. – Acta biol. Acad. Sci. Hung. **21**(2): 207–218.

VOVK, P. S., 1955: [Ecological conditions for raising common carp (*Cyprinus carpio* L.) in ponds of the U.S.S.R.]. – Vopr. prudovogo rybn. chozyaystva USSR, Trudy Inst. gidrobiol., AN USSR, Kiev **32:** 125–148. (in Russian).

WANG, P. G., 1981: [Notes on some species of Acanthocephala from fishes of China]. – Acta Zootax. Sinica **6** (2): 121–129. (in Chinese).

WANG, Y., 1979: [On classification, distribution, origin and evolution of the fishes referred to the subfamily Cyprinidae of China, with description of a new species]. – Acta Hydrobiol. Sinica, **6:** 419–438. (in Chinese).

WEDEMEYER, G. A., and D. J. MCLEAY, 1981: Methods for determining the tolerance of fishes to environmental stressors. In: A. D. PICKERING (Ed.), Stress and fish. Academic Press, London – New York – Toronto – Sydney – San Francisco. pp. 247–276.

WELCOMME, R. L., 1988: International introductions of inland aquatic species. FAO Fish. Techn. Paper 294. 318 pp.

WOHLFARTH, G. W., 1984: Common carp. In: I. L. MASON (Ed.), Evolution of domesticated animals. Longman, London. pp. 375–380.

WOJNO, T., and J. GUZIUR, 1972: Rast a preživanie divého dunajského kapra (*Cyprinus carpio carpio* L.) v rybníkoch severného Polska (Growth and survival of the Danube wild carp in ponds of northern Poland). –Pol'ovníctvo a rybárstvo **24**(4): 20–21. (in Slovak).

WOLF, K., 1988: Fish viruses and fish viral diseases. Cornell University Press, Ithaca, NY.

WOLNY, P., 1974: Karp. PWRiL, Warsaw. 235 pp.

WU, H. W., 1977: (Cyprinid fishes of China). Vol. 2. Sci. Tech. Press, Shanghai. pp. 231–598. (in Chinese).

WUNDER, W., 1934: Beobachtungen über Knochenerweichung und nachfolgende Wirbelsäuleverkümmerung beim Karpfen (*Cyprinus carpio* L.). – Z. f. Fischerei **32:** 37–67.

WUNDER, W., 1949: Fortschrittliche Karpfenteichwirtschaft. Stuttgart. 391 pp.

ZAN, R. and Z. SONG, 1980: Studies of the karyotypes of eight species in *Cyprinus* and *Anabarilius*. – Zool. Res., Kumming, **1:** 141–150.

ZAUNICK, R., 1925: Tritt der Karpfen schon in Diluvium Norddeutschlands auf? – Mitteil. Fischereivereine Brandenburg-Pommern **17:** 80–83.

ZHOU, W., 1989: [Phylogeny of the subfamily Cyprininae (Pisces, Cyprinidae)]. – Acta Zootax. Sinica, **14:** 247–256. (in Chinese).

ZHOU, W., and X. CHU, 1986: Systematic study of the genus *Cyprinus* (Pisces – Cyprinidae) in Yunnan, China. – Zool. Research **7** (3): 297–310.

ŽITŇAN, R., 1985: Naše chránené a ohrozené ryby. Neresové húfy divého kapra dunajského (Our protected and endangered fishes. Spawning shoals of the Danubian wild carp). Pol'ovníctvo a rybárstvo **37** (10): 28. (in Slovak).

ZUSSER, S., 1938: [Biology and fishery of the common carp in the northern Caspian Sea]. – Rybn. khoz. **3:** 6–13. (in Russian).

Catla Valenciennes, 1844

E = Catla carp; F = -; G = -; R = ?

Petru Bănărescu

Synonyms

Gibelion Heckel, 1843: 1014; type species: *Cyprinus catla* Hamilton, by subsequent designation (Jordan, 1912: 211).
Catla Valenciennes, in Cuvier et Valenciennes, 1844: 410 : type species : *Catla buchanani* Valenciennes = *Cyprinus catla* Hamilton, by monotypy.
Tambra (as subenus of *Hypselobarbus*) Bleeker, 1860: 275; type species: *Cyprinus abramoides* Sykes = *Cyprinus catla* Hamilton, by monotypy.

Etymology

The generic name *Catla* is derived from the specific name catla, proposed by Hamilton, 17, 1822, in the combination *Cyprinus catla.* This is a latinization of the vernacular Hindi, Bengali, and Punjabi names of the species.

Description

Large-sized cyprinid fishes with deep, strongly compressed bodies. The body profile is convex dorsally and flattened ventrally. There is neither a dorsal nor a ventral keel. The caudal peduncle is short, deep, and compressed. The head is very large; the mouth is terminal and directed slightly upward. There is no upper lip; the lower lip is well developed, smooth and continuous, being separated from the mentum by a narrow groove. It has a movable articulation at the symphysis but no prominent tubercles. There are no barbels. The gill opening is large. The numerous gill rakers are large and closely set. The pharyngeal teeth are arranged in three rows. There are almost always five teeth in the main row. The teeth are plough-shaped.

The scales are moderate in size and number 40 to 43 along the lateral line. They cover the entire body, including the isthmus, and they also extend onto the base of the caudal fin. The lateral line is straight and complete. The scales bear numerous parallel striae.

The dorsal fin is long with 14 to 16 branched rays; its outer edge is deeply notched. The anal fin has five branched rays. The last simple rays of both fins are slender.

Morphologically, this genus closely resembles *Catlocarpio* Boulenger from the Chao Phraya and Mekong drainage areas (KOTTELAT 1989). It differs from *Catlocarpio* in the number of dorsal rays and the formula of the pharyngeal teeth. *Catlocarpio* has only five branched dorsal rays, and its teeth are arranged in a single row (SMITH 1945). Since the ranges of both genera are vicariant, they may be closely related sister taxa. *Catla* includes only a single species.

Catla catla (Hamilton, 1822)

E = catla carp; G = –; F = –; R = –

Petru M. Bănărescu

Synonyms

Cyprinus catla Hamilton, 1822: 287 (Ganges River)
Leuciscus catla, Valenciennes, 1843: 379 (Ganges River)
Cyprinus abramoides Sykes, 1840: 35, pl. 62 (India)
Gibelion catla, Heckel, 1843: 1014 (India)
Cyprinus catla, Valenciennes, in Cuvier et Valenciennes, 1843: 433 (India)
Catla buchananini Valenciennes, in Cuvier et Valenciennes, 1844: 411, pl. 515 (Ganges River)
Hypselobarbus abramoides, Bleeker, 1860: 430 (northern India)
Catla buchanani, Günther, 1868: 34 (Bengal, Assam, Ganges River)
Catla buchanani, Day, 1878 (new reprint, 1958: XXX pl., fig. (northern Indian subcontinent, south to Kistna River).
Catla catla, Jayaram, 1981: 131, fig. 65 (Indian subcontinent from northern India south to the Kistna River)
Catla catla, Talwar & Jhingran, 1992: 153 (northern and central areas of India, Pakistan, Nepal, and Bangladesh)

Holotype

Day's specimens, on which the original description of the species was based, have not been preserved. Specimens from the lower Ganges River can be considered typical.

Etymology

The specific name *catla* is the vernacular name of the species in the Hindi, Bengali, and Punjabi languages (Day 1878).

Description

Du 3–4; *Db* (15) 16–18; *Au* 2–3; *Ab* 5; *Pu* 1; *Pb* 19–21; *Vu* 2; *Vb* 7; *l.l.* 40–43 ; *Squ.sup.* 8–10 ; *Squ. inf.* 7–8 ; *D.phar.* 5–3, 2–2, 3–5.

The main morphological characteristics of the species are mentioned in the diagnosis of the genus. The eyes are located in the anterior half of the head; their diameter equals about 14 % to 16 % of the head length and about 30 % to 40 % of the interorbital width. The insertion of the dorsal fin is anterior to that of the ventrals. The pectoral fins reach as far as the insertion of the ventral and anal fins, and those of males reach to beyond the anterior margin of the caudal fin. The caudal fin is deeply forked.

Colour pattern: The body is greyish above, silvery on the lateral surface, and whitish ventrally. The fins are dark, and in some specimens, even blackish.

Karyotype: Not studied

Sexual dimorphism: There is little difference between the sexes, but the fins, especially the anal, are larger in males.

Variability: Not studied; apparently slight.

Subspecies: No subspecies have been described.

Hybrids: Not mentioned.

Distribution

The catla carp is native to the northern and central areas of the Indian subcontinent, including the basins of the Indus, Ganges, and lower Brahmaputra Rivers in Pakistan, India, Nepal, and Bangladesh. It inhabits rivers in both watersheds on the Indian Peninsula south to the Krishna or Kistna River (JAYARAM 1981). Its occurrence in rivers of Burma, from the Irrawaddy to the Salween, is uncertain. Reports from Thailand by earlier authors refer to misidentified specimens of *Catlocarpio siamensis* (SMITH 1945).

Introductions

The catla carp has been introduced into the drainage system of the Cauvery or Kaveri River in the western watershed of the Indian Peninsula (JAYARAM 1981) and possibly into some parts of Southeast Asian. It has also been introduced into European Russia, but it did not successfully colonize the water bodies (N. BOGUTSKAYA, pers. comm.)

Ecology

Habitat. The normal habitats of the catla carp are large lowland rivers and standing water bodies.

Migrations, Hardiness, Feeding habits, Longevity, Growth, Reproductive biology, Important parasites and **diseases:** No information is available.

Economic importance

The catla carp is highly esteemed as food in India, although the flesh of fishes exceeding two feet is coarse (DA 1876). It is an important fishery product from natural waters and is also reared in tanks and on fish farms.

Literature

BLEEKER, P., 1860: Ichthyologiae Archipelagi Indici Prodromus, Vol. III, Cyprini. – Acta Soc. Sci. Indo-neerl. **7**; 492 + XIII pp.

CUVIER, G., and A. VALENCIENNES, 1843: Histoire Naturelle des Poissons. Vol. 16, Strasbourg. 472 pp.

CUVIER, G., and A. VALENCIENNES, 1844: Histoire Naturelle des Poissons. Vol. 17, Strasbourg. XXII + 497 pp.

DAY, F., 1876 : The fishes of India, being a natural history of the fishes known to inhabit the seas and fresh waters of India, Burma and Ceylon. Part. 2: p. 169–368, pls. 79–138, London.

GÜNTHER, A., 1868: Catalogue of the fishes in the British Museum. Vol. 7, Taylor and Francis, London. XX + 512 pp.

HAMILTON, F., 1922: An account of the fishes found in the river Ganges and its branches. Edinburgh & London. VIII + 405 pp.

HECKEL, J., 1843: Abbildungen und Beschreibung der Fische Syriens, nebst einer neuen Classification und characteristic sämmtlicher Gattungen der Cyprinen. In: J. VON RUSSEGGER (Ed.), Reisen in Europa, Asien und Afrika, 1 (Part 2): 991–1044. Schweizerbarth, Stuttgart.

JAYARAM, K.C., 1981: The freshwater fishes of India, Pakistan, Bangladesh, Burma and Sri Lanka. – A handbook. The Zool. Survey of India, Calcutta. 475 pp., XIII pls.

KOTTELAT, M., 1989: Zoogeography of the fishes from Indochinese inland waters with an annotated check-list. – Bull. Zool. Mus. Univers. Amsterdam. **12** (1): 1–55.

SMITH, H. M., 1945: The fresh-water fishes of Siam or Thailand. – Bull. U.S. nat. Mus. 188: XI + 622 pp., 9 pls.

SYKES, W. H., 1840: The Fishes of Dukhun. – Trans. Zool. Soc. London **2:** 349–378.
TALWAR, P. K., and A. G. JHINGRAN, 1992: Inland fishes of India and adjacent countries. 1, A. A. Balkema, Rotterdam. 542 pp.

Addendum to Vol. 5/I

Romanogobio pentatrichus Naseka et Bogutskaya, 1998 [1)]

E = five-rayed gudgeon; Kuban' long-barbelled gudgeon; F = –, G = –,
R = Kubanskiy dlinnousyi peskar'

ALEXANDER M. NASEKA, NINA G. BOGUTSKAYA, and VLADIMIR G. POZNJAK

Synonyms

Gobio uranoscopus caucasicus (non Kamensky): Berg, 1913: 120 (Kuban', Laba); Berg, 1914: 468 (partim: Kuban')
Gobio ciscaucasicus (non Berg): Berg, 1932: 413 (partim: Kuban'); Berg, 1949: 655 (partim: Kuban'); Banarescu, 1961: 332 (Kuban'); Abdurakhmanov, 1962: 142 (partim: Kuban').
Gobio (Rheogobio) ciscaucasicus (non Berg): Banarescu and Nalbant, 1973: 141 (Kuban'); Banarescu, 1992: 320 (Kuban').
Romanogobio pentatrichus Naseka and Bogutskaya, 1998: 174 (Kuban', Laba)
Romanogobio pentatrichus: Naseka, 1998: 85 (Kuban', Laba)
Romanogobio pentatrichus: Naseka and Poznjak, 2000: 406 (Kuban', Laba)

Holotype

The holotype deposited in the Zoological Institute of the Russian Academy of Sciences (ZISP), St. Petersburg, under the number 15312 is 102.0 mm in *Sl**. It was caught in the Laba River near Labinskaya Stanitsa by D. VOLNUKHIN on May 4, 1911. The paratypes under the ZISP number 15309 include 14 specimens ranging in *Sl** from 66.5 to 82.5 mm, which were collected in the Kuban' River near Armavir by D. VOLNUKHIN on April 29, 1911, and under the ZISP number 51151 are 5

[1)] **Editor's remark**
Since the species *pentatrichus* lacks the main character of *Romanogobio* – the presence of epithelial crests on the dorsal scales – and has, unlike all species of *Gobio* sensu lato and all other members of Gobioninae, except *Microphysogobio obtusirostris,* five branched anal rays, it cannot be ascribed to *Romanogobio*, but to a fourth subgenus of *Gobio*, or even to a fourth genus oft the *Gobio* group.

specimens ranging from 72.0 to 118.0 mm in *Sl** from the Laba River near Labinskaya Stanitsa collected by D. VOLNUKHIN on May 4, 1911.

Etymology

The name *pentatrichus* refers to the main diagnostic feature, the presence of five branched rays in the anal fin: πεντε – five, and τριχιον – hair, ray in Greek.

Description

Diagnosis: *Du* 3, *Db* 7, *Au* 3, *Ab* 5, *Pu* 1, *Pb* 13–15, *Vu* 2, *Vb* 7, *l.l.* 42–46, *Squ.sup.* (5)6–7, *Squ.inf.* 4–5, *Sp.br.* 2–5, *D.ph.* usually 3.5–5.3, *Vert.* 40–43, *abd. vert.* 20, 21, *caud. vert.* 20–22, *preanal caud. vert.* (2)3–4; barbel long, reaching beyond posterior edge of the eye; caudal peduncle shallow with 10–12 scales around its narrowest part; epithelial crests absent from dorsal scales; supraorbital and infraorbital sensory canals are not connected; the base of the fifth ceratobranchial does not have more than four large holes; urohyal long, shallow, with rounded notch in the head of the bone. The fishes attain a standard length of 118 mm.

The genus *Gobio* sensu Banarescu, 1992, was divided by NASEKA (1996) into two genera: *Gobio* Cuvier, 1816, encompassing the species *G. gobio, G. uranoscopus, G. hettitorum, G. soldatovi,* and *G. coriparoides,* and *Romanogobio* Banarescu, 1961, which includes *R. kesslerii,* the type species, as well as *R. albipinnatus, R. ciscaucasicus, R. persus,* and six Asian species. A new species, *R. pentatrichus,* was assigned to the genus *Romanogobio* (Naseka and Bogutskaya, 1998; Naseka, 1998) because several of its characters differ from those typical for the genus *Gobio.* These include the shallow, elongate caudal peduncle; long barbel; anus relatively close to the base of the ventral fins; short, wide supraethmoid; high and oval rather than narrow, rod-shaped second infraorbital; large and round rather than slit-like notch in the head of the urohyal; relatively large number of vertebrae as well as more caudal than abdominal, relatively few predorsal, and relatively many preanal vertebrae. The vertebral complement in the genus *Romanogobio* can be expressed by the modal formula 42:(11)20(5)+(3)21(18), whereas that for the genus *Gobio* is 39:(11)21 (5)+(1)18(17), according to NASEKA (1996).

Romanogobio pentatrichus differs from most of the other gudgeon species in always possessing five branched rays in the anal fin. As a result, there are always seven branched rays in the dorsal and five in the anal fin.

Among the European species in the genus *Romanogobio, R. kesslerii* and *R. albipinnatus* differ most from *Romanogobio pentatrichus.* The supraorbital and infraorbital sensory canals of the two former species are commonly connected.

Their eyes are considerably larger with diameters as great as one to two interorbital widths. Their barbels are shorter than those of other species in the genus and do not usually extend past the posterior eye margin. Their neurocraniums are wide and relatively high in the ethmoid region. In addition, *R. kesslerii* has more rays in its dorsal and anal fins. There are usually eight branched rays in its dorsal and six in its anal fin. *R. albipinnatus* is characterized by an almost complete absence of spots on the unpaired fins.

Romanogobio persus differs from *R. pentatrichus* in possessing fewer scales in its lateral line: the average for *R. persus persus* (Gunther) is 40, while for *R. persus macropterus* (Kamensky), it is 42. It also has fewer vertebrae, which average 39 and 40 in the two subspecies, respectively. Furthermore, there is a connection between the supraorbital and infraorbital canals, which is lacking in *R. pentatrichus.*

R. pentatrichus is closest to *R. ciscaucasicus,* with which it is sympatrically distributed in the Kuban' Basin. These species are similar in such features as a small eye, the diameter of which is markedly less than the interorbital distance, as well as a large number of scales, usually more than 43, in the lateral line.

Below the key for identification of the three species of gudgeons present in the Kuban' drainage area is given (from NASEKA and POZNJAK 2000).

1(2) Barbel short, usually does not reach the vertical line through the posterior margin of the eye; caudal peduncle relatively deep, its length exceeds the minimum body depth by a factor of 1.8 to 2.2 . . . *Gobio gobio lepidolaemus*

2(1) Barbel long, usually reaches posteriad beyond the posterior margin of the eye often attaining the anterior to posterior margin of the preoperculum; caudal peduncle relatively shallow, its length exceeds the minimum body depth by a factor of 3.3 to 4.3.

3(4) Six anal branched rays; well-developed epithelial crests on the dorsal scales; connection between the supra- and infraorbital cephalic sensory canals present *Romanogobio ciscaucasicus*

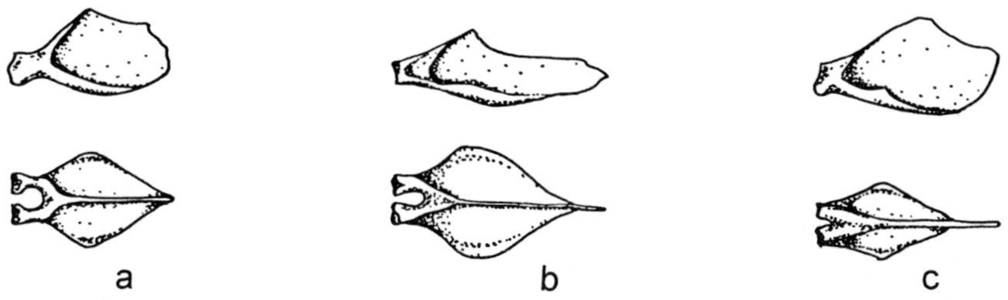

Fig. 38: The urohyal in lateral (above) and dorsal (below) view of *Romanogobio ciscaucasicus* **(a),** *R. pentatrichus* **(b),** and *Gobio gobio lepidolaemus* **(c).**

Fig. 39: The fifth ceratobranchial of *Romanogobio ciscaucasicus* **(a)**, *R. pentatrichus* **(b)** and *Gobio gobio lepidolaemus* **(c)**.

4(3) Five anal branched rays; no epithelial crests on the dorsal scales; no connection between the supra- and infraorbital cephalic sensory canals. *Romanogobio pentatrichus*

Besides the features mentioned in the key, the three Kuban' species differ in the shape of the urohyal (Fig. 38) and the fifth ceratobranchial (Fig. 39).

Morphology

The habitus is shown in Fig. 40. Morphometric data are provided in Table 34. The body is fusiform and slightly compressed laterally. The dorsal profile of the body is triangular, while the lower profile is slightly arched. The maximum body depth is along the vertical line through the dorsal fin origin. The caudal peduncle is slightly compressed and very shallow. The minimum body depth is 3.3 to 4.2 ($\bar{x} = 3.8$) times the length of the caudal peduncle. The anus is located about midway between the ventral and anal fin origins. There are 7 to 10 scales between the anus and the anal fin.

The head is wide, conical, and slightly flattened. The maximum head depth is considerably less than its maximum width, and its length is usually greater than its maximum body depth. The eye is small; its diameter is 1.3 to 1.8 ($\bar{x} = 1.5$) times the interorbital width. The mouth is inferior. Its upper lip is continuous, and its lower lip is widely interrupted in the middle. The barbel is long, reaching beyond the posterior margin of the eye and usually reaching the posterior margin of the preoperculum.

The dorsal fin is located at the midpoint of the body. The predorsal and postdorsal distances are approximately equal. The outer margins of the dorsal and anal fins are slightly concave. The paired fins are relatively long. The pectoral fin usually reaches a vertical line drawn through the dorsal fin origin. The caudal fin is deeply forked; its lobes are pointed. The upper lobe is slightly longer than the lower one.

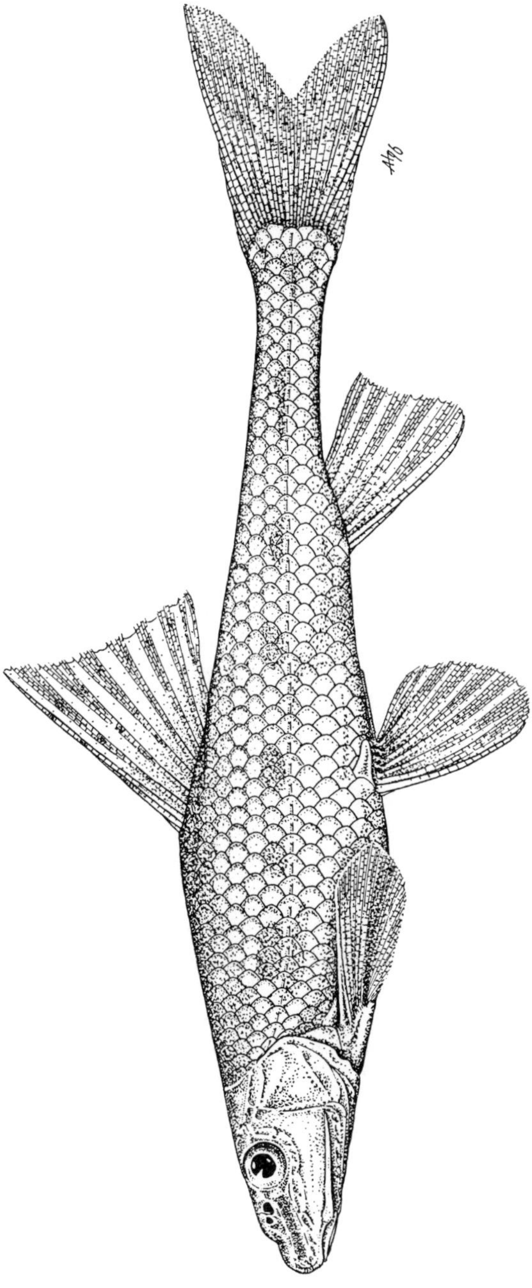

Fig. 40: The holotype of *Romanogobio pentatrichus*, ZIN 15312, *Sl** 102.0 mm, from the Laba River. Drawn by A. M. NASEKA.

Table 34: Morphometric features of *Romanogobio pentatrichus* from the Kuban' River basin.

	ZIN 15309; 15312; 51151; 51152 (n = 36)		
	Ranges	x̄	s
*Sl** (mm)	49.80–118.0	75.34	–
% of *Sl**			
lc	22.22–30.52	24.22	1.29
H	16.38–24.00	20.23	1.49
h	5.30– 6.62	6.02	0.28
max laco	12.93–17.71	15.08	1.21
min laco	3.03– 4.75	3.82	0.43
pD	42.44–47.09	44.49	1.09
poD	41.41–52.54	44.76	1.73
pV	34.55–52.04	47.33	2.74
pA	65.82–76.21	69.80	1.80
P-V	22.40–30.74	25.28	1.54
V-A	19.39–25.40	23.21	1.29
lpc	20.13–25.33	22.60	1.18
lD	12.67–14.88	13.74	0.70
hD	21.58–27.51	24.28	1.24
lA	7.22– 8.82	7.98	0.45
hA	17.13–20.63	19.00	0.94
lP	18.29–22.95	20.79	1.24
lV	15.00–19.83	17.36	1.04
V-an	9.47–13.80	11.32	1.04
% of *V-A*			
V-an	41.03–57.89	48.84	4.07
% of *lc*			
hc	48.35–62.38	56.24	2.51
hco	40.38–49.19	44.22	2.25
prO	31.97–44.44	39.08	2.84
Oh	15.29–25.00	21.61	1.94
poO	36.36–47.89	44.38	2.23
lac	50.48–68.21	59.39	3.86
io	24.79–34.38	29.56	2.17
lb	30.99–59.32	44.89	5.78

The lateral line is complete, having the scale formula (42)43–45(46). The most frequent number is 44, which was found in 61 % of the specimens. There are five to seven scales above the lateral line. The most common number is six. Four or five scales are located below the lateral line. There are 10 to 12 scales around the narrowest part of the caudal peduncle, where 2.5 scales are present above and 1.5 or 2.5 scales below the lateral line. The dorsal scales on the anterior part of the body lack epithelial crests, which are well developed above the lateral line on *R. ciscaucasicus*. The throat is scaleless.

The gill rakers are short. They number two to five, most commonly three, in 39 % of the specimens, or four, in 28 %, on the outer side of the entire first left branchial arch. There are 7 to 10, or rarely 11, on the inner side. The pharyngeal teeth are arranged in two rows, usually with a formula of 3.5–5.3, as found in 83 % of the specimens examined. Rarely, the formula is 2.5–5.3, 3.5–5.2, or 2.5–5.2. The teeth are hooked and possess smooth and narrow masticatory surfaces.

The cephalic sensory canals are complete. Their locations (Fig. 41) are almost identical with those described for *Romanogobio ciscaucasicus*. All specimens of *R. pentatrichus* examined lack a connection between the supraorbital and infraorbital canals. This feature distinguishes the new species from *Gobio uranoscopus, R. kesslerii,* and *R. albipinnatus,* as well as *R. ciscaucasicus* from the Kuban' River, which, as a rule, have connections between the supraorbital and infraorbital canals. There are eight pores in the supraortibal canal in 68 % of the specimens and nine or rarely ten in the others. Three canal openings are present in the nasal bone. The infraorbital canal has 15 to 20, usually 16 to 19, and in 53 % of the specimens, 17 pores. In the first infraorbital, there are five to seven openings, with six in 76 % of the speci-

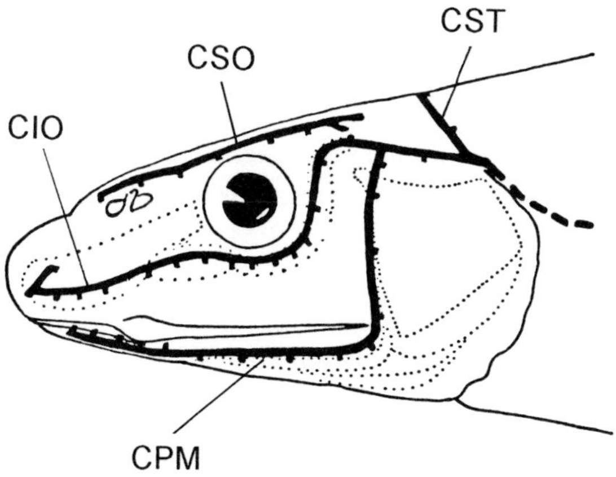

Fig. 41: The cephalic sensory canal pattern of *Romanogobio pentatrichus:* *CIO* = infraorbital canal, *CPM* = preopercular-mandibular canal, *CSO* = supraorbital canal, *CST* = supratemporal canal.

mens. There are 13 to 17 pores in the preopercular-mandibular canal, with 14 in 37 % and 15 in 35 % of the specimens. The number of canal openings in the dentary is usually four, which were found in 96 % of the specimens. The supratemporal canal had 5 pores in 77 % of the fishes examined. This canal in the other fishes had six or seven pores. It runs through the extrascapular and parietal and continues beneath the dorsal surface of the supraoccipital at the base of the supraoccipital crest, where a central pore is always present.

Osteological characters

Vertebrae: X-ray examinations of 36 specimens were completed. The total number of vertebrae varied from 40 to 43 with a mode of 41, the number in 50% of the specimens. There were 20 abdominal vertebrae in 33 % of the specimens and 21 in 67 %. The number of predorsal vertebrae was 10 to 12 with a mode of 11, which were present in 86 % of the fishes. There were four to seven intermediate vertebrae, with five present in 61% of the specimens. The number in the caudal region ranged from 20 to 22, but the most frequent number was 21, which were present in 56 % of the fishes examined. The number of preanal caudal vertebrae varied from two to four with the mode of three found in 69 % of the specimens. The following ratios were found between the abdominal and caudal vertebrae in the specimens examined: 1) 21:20, with fewer caudal than abdominal vertebrae, in 10 specimens or 28 % of the total; 2) 20:20 or 21:21, that is, equal numbers of caudal and abdominal vertebrae, in 14 specimens or 39 % of the total; 3) 20:21, 20:22, or 21:22, with fewer abdominal than caudal vertebrae, in 12 specimens or 33 % of the total. The formula for the mean number of vertebrae is 41.5:(10.9)20.7(5.1)+ (3.2)20.8(17.6). Thus, an approximately equal number of vertebrae in the abdominal and caudal regions is characteristic of the vertebral column of this species. The second and third vertebrae are completely fused.

A general view of the neurocranium is shown in Fig. 42. This is rather similar to that of *R. albipinnatus,* which was described in the preceeding book of this series. The neurocranium is relatively shallow and wide. Its depth is 16 % to 18 % of the

---

Fig. 42: The neurocranium of *Romanogobio pentatrichus, Sl** 75 mm, from the Kuban' River in lateral **(a)**, ventral **(b)**, and dorsal **(c)** view: *boc* = basioccipital, *eoc* = exoccipital, *epo* = epiotic, *eth.l* = lateral ethmoid, *f* = frontal, *f.dil.op* = dilatator operculi fossa, *f. st* = subtemporal fossa, *ic* = intercalar, *meth* = mesethmoid, *orbs* = orbitosphenoid, *orbs. sept* = interorbital septum of the orbitosphenoid, *p* = parietal; *peth* = preethmoid, *pr.ph* = pharyngeal process, *pr. rostr* = rostral process of the supraethmoid, *pr. Spho* = lateral sphenotic process, *pro* = prootic, *ps* = parasphenoid, *pto* = pterotic, *pts* = pterosphenoid, *seth* = supraethmoid, *soc* = supraoccipital, *spho* = sphenotic, *v* = vomer.

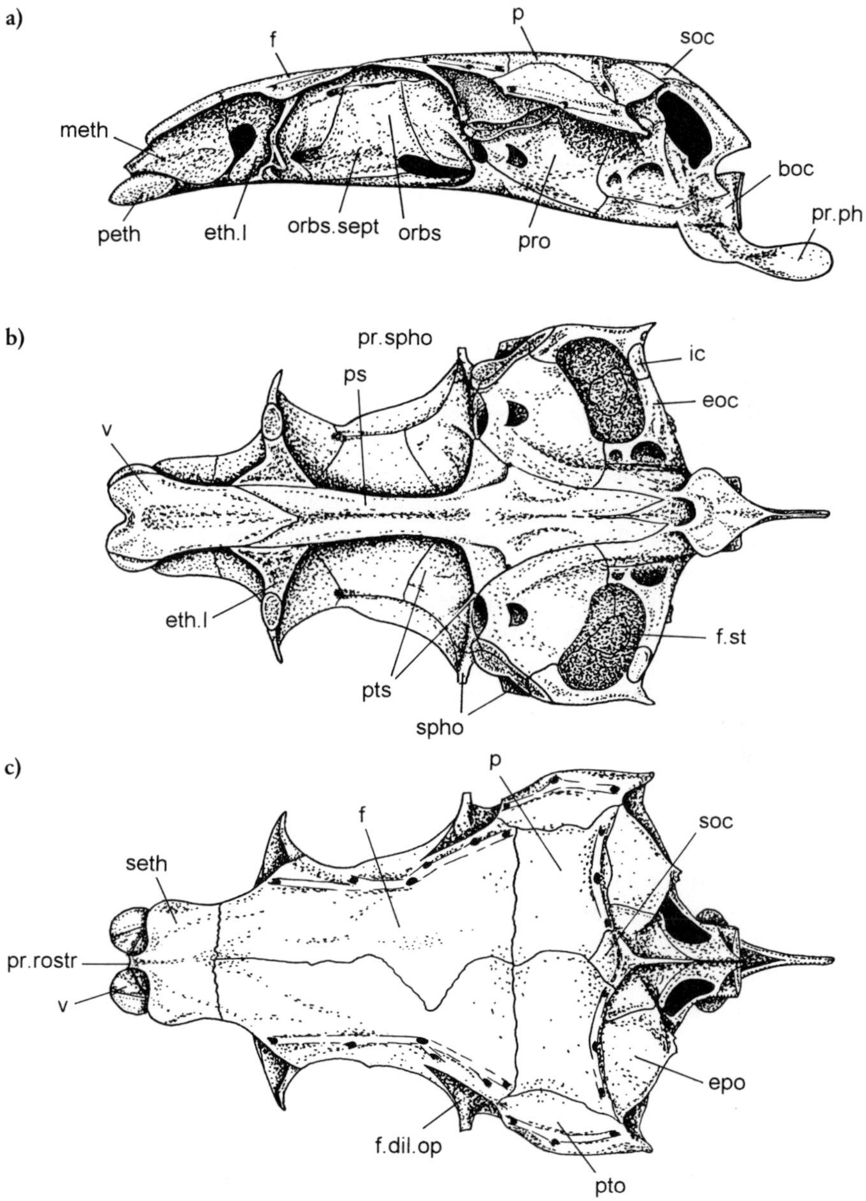
a)
p
f
soc
meth
boc
pr.ph
peth
eth.l
orbs.sept
orbs
pro
b)
pr.spho
ic
ps
eoc
v
eth.l
f.st
pts
spho
c)
p
f
soc
seth
pr.rostr
v
epo
f.dil.op
pto

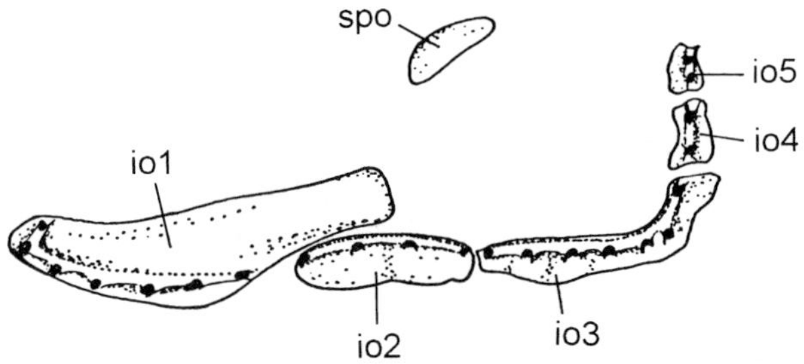

Fig. 43: Circumorbitals of the specimen in Fig. 42: *io* = infraorbitals, *spo* = supraorbital.

cranial base length (*L. bas. n.*) in the ethmoid region. In the sphenotic region, it is 21 % to 23 % of *L. bas. n.*, while in the occipital part, the depth is 25 % to 27 % of *L. bas. n.* The ranges for the width of the neurocranium between the lateral margins of the lateral ethmoids, sphenotics, and pterotics are 45 % to 47 %, 49 % to 51 %, and 54 % to 56 % of *L. bas. n.*, respectively. These measurements show that the neurocranium of *R. pentatrichus* is shallower and narrower than that of *R. ciscaucasicus.*

The supraethmoid is relatively short and wide like that of the other *Romanogobio* species. It differs from the elongated supraethmoid in the species of *Gobio.* The orbitosphenoid is shallow, but its interorbital septum is relatively well-defined. As in *R. ciscaucasicus* but unlike *G. gobio,* the pterosphenoid has a well-developed posterolateral process on the outer surface, which connects it to the margin of the ascending lateral wing of the parasphenoid and forms a narrow bony arch dividing the foramen of the trigeminal-facial canal into two separate apertures. The anterodorsal part of the supraoccipital, which is crossed by the central section of the supratemporal canal, extends between the parietals and is triangular in shape.

There are five infraorbitals (Fig. 43). The first is elongate, and a canal crosses only its anterior part. Its posterior part is located above the second infraorbital and is relatively long and high. The second infraorbital is approximately equal in depth to the posterior part of the first. In contrast, the first infraorbital of *G. gobio* and the other *Gobio* species has a deep posterior part, the depth of which considerably exceeds that of the shallow second infraorbital.

Some of the visceral bones are illustrated in Figure 44. The urohyal is long and shallow but relatively wide; its length is 27 % to 32 % of *L bas. n.* The head of the urohyal (Fig. 44, d) is divided by a deep, large, rounded notch. The base of the fifth

---

Fig. 44: Bones of the viscerocranium of the specimen in Fig. 42: opercular bones **(a),** jaws **(b),** suspensorium **(c),** urohyal **(d),** and fifth ceratobranchial **(e).**
aart = anguloarticular, *dn* = dentary, *ectpt* = ectopterygoid, *entpt* = entopterygoid, *iop* = interoperculum, *hm* = hyomandibular, *mtpt* = metapterygoid, *mx* = maxilla, *op* = operculum, *pal* = palatine; *pmx* = premaxilla, *pop* = preoperculum, *pr.cor* = coronoid process, *q* = quadrate, *rart* = retroarticular, *sop* = suboperculum, *sym* = symplectic.

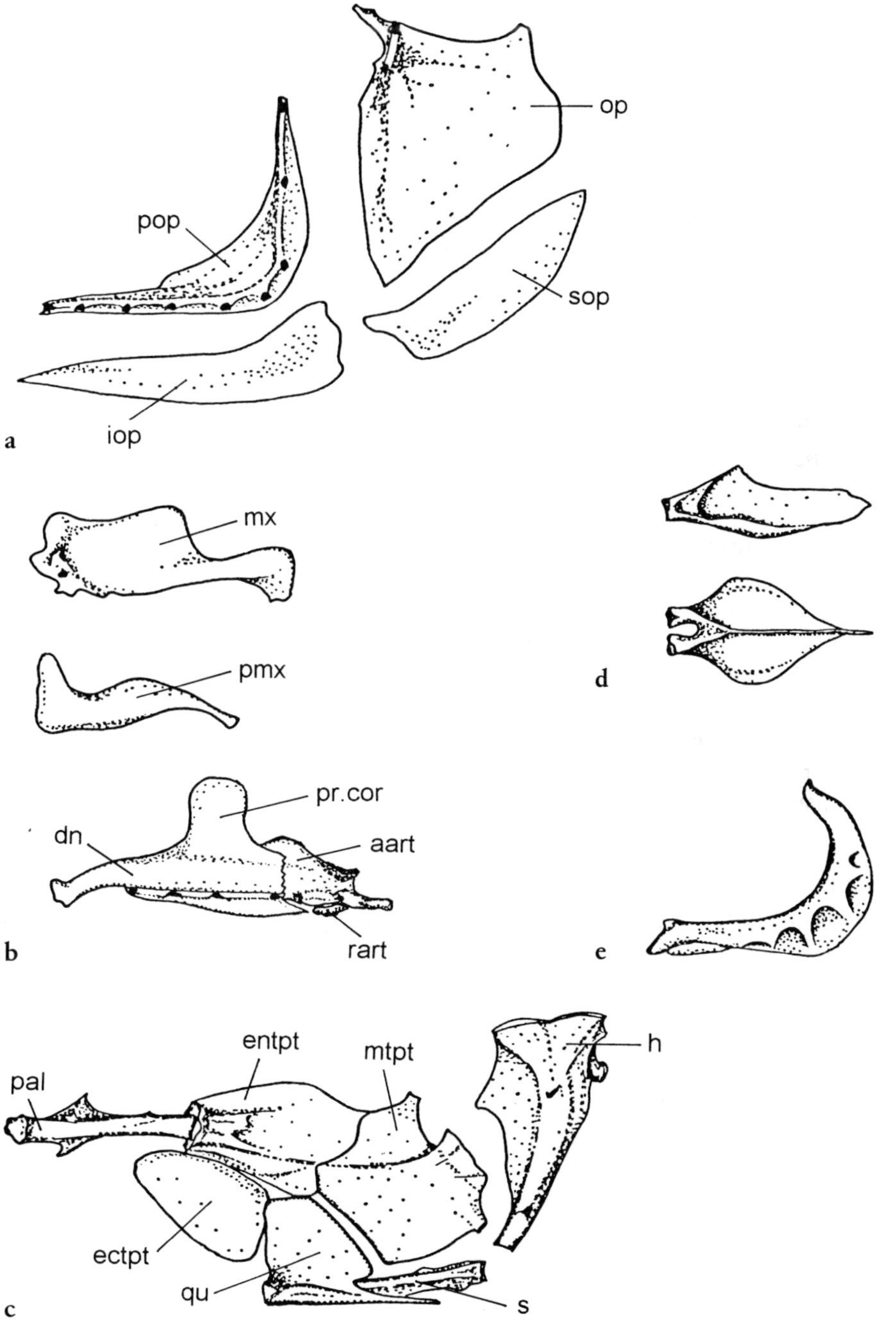
op
pop
sop
iop
a
mx
pmx
d
pr.cor
dn
aart
rart
b
e
entpt
mtpt
h
pal
ectpt
qu
s
c

ceratobranchial (Fig. 44, e) is relatively wide and has two to four very large deep holes. This feature is unlike that of *G. gobio*, which has shallow holes (Fig. 39, c), and *R. ciscaucasicus*, the narrow ceratobranchial of which has six to eight small holes (Fig. 39, c).

Colouration: The general body colour in alcohol is light yellowish. The upper part of the head and dorsum are darker. There is a row of dark, round spots along the dorsal mid-line, and 7 to 11 such spots are present along the lateral line. Pigment spots are grouped at the bases of the dorsal scales forming a pigmented meshwork on the upper part of the body. Up to four rows of small, intensely coloured spots are present on the dorsal and caudal fins, and rarely, there are similar rows on the pectoral fins.

Karyotype: Not investigated.

Protein specificity: Not studied.

Sexual dimorphism: Unknown.

Geographic variations: This endemic species with a limited range shows no geographic variation.

Age and size variability: Not studied.

Distribution

R. pentatrichus is known only from the Kuban' River and its tributaries, the Laba and the Belaya Rivers. Originally it is endemic to the Kuban' basin. However, two years ago the species was found in the Bol'shoy Egorlyk River, a tributary of the West Manych (the Don River system), now connected to the Kuban' River by an artificial canal.

Ecology

Habitat: This exclusively freshwater species prefers sections of rivers and streams with rapid currents and gravel or stony beds. It is commonly caught in shallow waters above sandy-gravel bottoms at a depth of 0.5 to 1.2 m and a current speed of 0.8 to 1.0 $m \cdot sec^{-1}$.

Migrations: Unknown.

Hardiness: Unknown.

Feeding habits: This fish feeds on benthic invertebrates.

Longevity: Unknown. As in the case of the other European gudgeons, it seems probable that the maximum age of *R. pentatrichus* exceeds five to six years.

Growth: Not studied.

Reproduction biology: Unknown.

Important parasites and diseases: No information is available.

Economic importance: This species has no commercial value.

Literature

ABDURAKHMANOV, Yu.A., 1962: (Freshwater fishes of Azerbajdzhan). – Izd. Akad. Nauk Azerb. SSR, Baku. 408 pp. (in Russian).

BĂNĂRESCU, P., 1961: Weitere systematische Studien uber die Gattung *Gobio* (Pisces, Cyprinidae), insbesondere im Donaubecken. – Vestn. Cesko-Slov. Zool. Spol. **25**(4): 318–346.

BĂNĂRESCU, P., 1992: A critical updated checklist of Gobioninae (Pisces, Cyprinidae). –Trav. Mus. Hist. Nat. "Grigore Antipa" **32:** 303-330.

BĂNĂRESCU, P., and T. NALBANT, 1973: Pisces, Teleostei, Cyprinidae (Gobioninae). – Tierreich **93:** 1-304.

BERG, L. S., 1913. (The fishes from the Kuban' basin.). – Ezhegodnik Zool. Mus. Imper. Akad. Nauk, St. Petersburg **17:** 116–122 (in Russian).

BERG, L. S., 1914: (Fishes. Vol. 3, Ostariophysi, Part. 2). – Izd. Imper. Akad. Nauk, St. Petersburg. pp. 337–704. (in Russian).

BERG, L. S., 1932: (Freshwater fishes of the USSR and adjacent countries. Part 1. 3rd ed.). Izd. Vses. Instituta Ozior. Retchn. Ryb. Khoz., Leningrad. 543 pp. (in Russian).

BERG, L. S., 1949: (Freshwater fishes of the USSR and adjacent countries. Part.2. 4th ed.). Izd. Akad. Nauk SSSR, Moscow-Leningrad. pp. 470–925 (in Russian).

NASEKA, A. M., 1996: Comparative study on the vertebral column in the Gobioninae (Pisces, Cyprinidae) with special reference to its systematics. – Publ. Spec. Inst. Esp. Oceanogr. **21:** 149–167.

NASEKA, A. M., 1998: Gobioninae. In: (Annotated check-list of Cyclostomata and fishes of the continental waters of Russia). Nauka, Moscow. pp. 81--87. (in Russian).

NASEKA, A. M., and N. G. BOGUTSKAYA, 1998: [A new gudgeon species *Romanogobio pentatrichus* (Gobioninae, Cyprinidae) from the basin of the Kuban' River.]. – Voprosy Ikhtiologii, Moscow **30:** 355–367. [in Russian; translated in J. Ichthyology **38**(3): 219–227.].

NASEKA, A. M., and V. G. POZNJAK, 2000: [Northcaucasian long-barbeled gudgeon, *Romanogobio ciscaucasicus* in the basin of the Kuban' River (Gobioninae, Cyprinidae)]. – Voprosy Ikhtiologii, Moscow **40:** 406–410. [in Russian].

Gasterosteidae Bonaparte, 1839

This section is dedicated to Dr. JOACHIM MÜNZING of Hamburg, born on August 21, 1932, and died on February 11, 2000. From 1959 through 1972, he measurably increased our knowledge of the European sticklebacks through numerous publications. Unfortunately, health problems prevented Dr. MÜNZING from writing the section on the sticklebacks for Freshwater Fishes of Europe, as originally planned. However, he provided the author with his valuable literature on the gasterosteids and other important records and supported the preparation of the manuscript through his critical comments. He was a good friend of the author for more than three decades.

HANS-JOACHIM PAEPKE

Gasterosteidae Bonaparte, 1839

Hans-Joachim Paepke

Members of this family are small fishes, 50 to 90 mm in *Tl*, high oval in cross section, and usually with short rostra and thin, fragile caudal peduncles. Only fishes in the genus *Spinachia* are very elongated, reaching 180 to 200 ml in *Tl*, and have extended, pipe-like rostra. The upper jaw is protrusible, and the upward extension of the premaxillary is well developed. There is no postcleithrum. Parietalia, nasalia, und circumorbital bones are present, but the eye ring is not fully closed. There are three branchiostegal rays. Epipleuralia are present. There are 28 to 42 vertebrae, of which the anterior ones are not elongated.

Anterior to the dorsal fin, there are 3 to 17 free anterodorsal spines. Each ventral fin is also armed with a strong spine and may bear one or two soft rays as well, or soft rays may be absent. The spines can be moved into rigid sockets and in this way maintained as defence weapons or spread to threaten members of the same species without the use of muscle power. The paired pelvic bones are stable and often have lateral processes, which are usually fused along the midline and externally visible.

The body lacks scales and is covered instead with more or less complete rows of body shields. The parts of the body not covered are completely bare. The pattern of shield arrangement is genetically determined and shows species-specific peculiarities and considerable intraspecific variability.

The species inhabit marine, brackish, and freshwater habitats in the Northern Hemisphere between 75°N at Baffin Island and about 35°N in southern Anatolia and Syria. Members of the southern populations remain small, have reduced shield cover, and are limited to fresh water. Many of the fishes inhabiting marine or brackish water habitats undertake an anadromous spawning migration into fresh water. The inhabitants of fresh water generally do not migrate. In freshwater bodies of Europe, the family is represented by only four species in two genera: *Gasterosteus* and *Pungitius*. However, the number of recognized species is still a matter of controversy (Kottelat 1997). The earliest paleontological finds of members of the genera *Pungitius* und *Gasterosteus* date from the Miocene (Bell 1984).

Highly specialized social and reproductive behaviour includes rituals related to establishing territories and defending them, building nests, courtship, spawning, and protecting the eggs and later, the larvae until they reach the free-swimming stage. The father cares for the young in a nest made from plant material, which is collected and pasted together with a secretion from the adrenal gland. Schools are formed, especially during the migration to the spawning sites but also during periods when reproduction does not take place (WOOTTON 1976, 1984).

The relationship between the Gasterosteidae and other fish groups has not been explained to everyone's satisfaction. However, according to BANISTER (1967), IDA (1976), PIETSCH (1978), FRITZSCHE (1984), ESCHMEYER (1990), NELSON (1994), and PAEPKE (1996), the family Gasterosteidae is more closely related to the Aulorhynchidae und Hypoptychidae than to other teleosts. Gasterosteidae can be defined as monophyletic and distinguished from the very similar Aulorhynchidae based on three apomorphic characteristics: a pair of submetacentric chromosomes; the presence of stable, extendable spines on the underside; and the collection of plant material for nest building. The relationship of the Hypoptychidae to the Gasterosteidae is still only hypothetical in nature. The Gasterosteidae encompasses seven species in five genera: *Apeltes*, *Culaea*, *Gasterosteus*, *Pungitius*, and *Spinachia*. *Spinachia spinachia* is the species with the most primitive, or plesiomorphic, features. *Gasterosteus aculeatus* seems to be the most advanced. *Apeltes* and *Culaea* each have 46 chromosomes, while *Pungitius* and *Gasterosteus* have only 42 (CHEN and REISMAN 1970, KLINKHARDT *et al.* 1995).

Key to the genera

1 (2) The body is very elongate and slender with *Sl* equal to about ten times *H*; the rostrum equals about 2.5 times one eye diameter; there are 13 to 17 very low dorsal spines. The fishes are found only in coastal waters of the North Atlantic from the Arctic seacoast of Europe to the Bay of Biscay, including the North and western Baltic Seas. . . . . *Spinachia* Cuvier, 1817

2 (3) Smaller, with a small rostrum and fewer dorsal spines.

3 (4) *Tl* 40–50 mm, rarely to 75 mm, slender with 7 to 12 low dorsal spines in closely spaced rows alternating left and right of the dorsal midline. Inhabits fresh and sometimes also brackish water bodies throughout a large part of Europe, Northeast Asia, and North America. . . . . . . *Pungitius* Coste, 1848

4 (5) There are fewer than seven dorsal spines

5 (6) The body size and shape is similar to those of *Pungitius*, but there are only 4 to 6 dorsal spines, which are inserted directly along the dorsal midline with considerable space between them. The membranes on the spines are dark, unlike those of *Pungitius*. It inhabits only freshwater bodies in continental North America. . . . . *Culaea* Whitley, 1950

6 (7) The body is thick-set as far as the slender caudal peduncle; *H* is about 1/4 of *Sl*. There are 3 to 5, usually 4, dorsal spines of different lengths. The relatively large anterior spines are inserted close together with about 1/4 of a spine length between them. The smaller posterior spines are separated from the

anterior ones by a larger space and are inserted immediately anterior to the dorsal fin. They occur only along the East Coast of North America. *Apeltes* Dekey, 1842

7 There are only 3, or rarely 4, dorsal spines. The two anterior spines are long and thick, while the posterior spine is very small and inserted immediately anterior to the dorsal fin. The spaces between the dorsal spines equal the length of one spine. In the ocean and in relatively large inland lakes, they are slender, with *Sl* equal to about 4.7 times *H*; in relatively small water bodies, the shape is more thick-set, with *Sl* about 3.5 to 4.2 times *H*. The maximal *Tl* is slightly more than 90 mm, but in fresh water, it often reaches only 55 mm. It is widely distributed in water bodies with various salinities in Europe, Northeast Asia, and North America. .... *Gasterosteus* Linné, 1758

Literature

BANISTER, K. E., 1967: The anatomy and classification of the order Gasterosteiformes (Pisces). Ph.D. Thesis, Univ. Newcastle-upon-Tyne, England. 211 pp.

BELL, M., 1984: Evolutionary phenetics and genetics. The threespined stickleback, *Gasterosteus aculeatus* and related species. In: B. J. TURNER (Ed.), Evolutionary Genetics of Fishes. Plenum Press, New York and London. pp. 431–528.

BERG, L. S., 1932: Übersicht der Verbreitung der Süßwasserfische Europas. – Zoogeographica **1:** 107–208.

BERG, L. S., 1958: System der rezenten und fossilen Fischartigen und Fische. VEB Deutscher Verlag der Wissenschaften, Berlin. 310 pp.

BERTIN, L., 1924: Classification variabilité, distribution geographique des poissons de la famille des Epinoches. – C.R Assoc. franc. Avanc. Sci. Paris **47:** 519–527.

BERTIN, L., 1925: Recherches bionomiques, biométriques et systématiques sur les Epinoches (Gastérosteides). – Ann. Inst. Oceanogr. Monaco **2:** 1–204.

CHEN, T. R., and H. M. REISMAN, 1970: A comparative chromosome study of the North American species of sticklebacks (Teleostei: Gasterosteidae). – Cytogenetics **9:** 321–332.

COAD, B. W.,1981: A bibliography of the sticklebacks (Gasterosteidae: Osteichthyes). Syllogeus No. 35. Nat. Mus. Canada, Ottawa. 142 pp.

GOSLINE, W. A., 1971: Functional morphology and classification of teleostean fishes. Univ. Press of Hawaií, Honolulu. 208 pp.

GREENWOOD, P. H., D. E. ROSEN, S. H. WEITZMAN, and G. S. MYERS, 1966: Phyletic studies of teleostean fishes, with a provisional classification of living forms. – Bull. Am. Mus. Nat. Hist. **131:** 339–456.

HALL, M. F., 1956: A comparative study of the reproductive behaviour of the sticklebacks (Gasterosteidae). Ph.D.dissertation. St. Hugh's College, Oxford Univ., Oxford, England.

IDA, H., 1976: Removal of the family Hypoptychidae from the suborder Ammodytoidei, order Perciformes, to the suborder Gasterosteodei, order Syngnathiformes. – Jap. Journ. Ichthyol. **23:** 33–42.

KLINKHARDT, M., M. TESCHE, and H. GREVEN, 1995: Databasis of Fish Chromosomes. Westarp Wissenschaften, Magdeburg. 237 pp.

KOTTELAT, M. 1997: European freshwater fishes. An heuristic checklist of the freshwater fishes of Europe (exclusive of former USSR), with an introduction for non-systematists and comments on nomenclature and conservation. – Biologia, Section Zool. **52**/Suppl. **5:** 1–271.

LEINER, M., 1934: Die drei europäischen Stichlinge (*Gasterosteus aculeatus* L., *Gasterosteus pungitius* L. und *Gasterosteus spinachia* L.) und ihre Kreuzungsprodukte. Vergleichende Betrachtungen ihrer Laichzeit, ihrer Körperformen und ihrer Brutpflegetätigkeit. – Z. Morphol. Ökol. Tiere **28:** 107–154.

MONOD, T., 1973: Gasterosteidae. In: Checklist of the fishes of the north-eastern Atlantic and of the Mediterranean. UNESCO, Paris. **1:** 280–286.

MURAL, R. J., 1973: The Pliocene sticklebacks of Nevada with a partial osteology of the Gasterosteidae. – Copeia 1973: 721–735.

NELSON, J. S., 1971: Comparison of the pectoral and pelvic skeletons and of some other bones and their phylogenetic implications in the Aulorhynchidae and Gasterosteidae (Pisces). – J. Fish. Res. Board Canada **28:** 427–442.

NELSON, J. S., 1980: A review of the family and generic relationship of the Gasterosteidae. Second. Int. Congr. Syst. Evol. Biol., Univ. British Columbia, Vancouver, July 17–24, 1980. 102 pp.

NELSON, J. S., 1984: Fishes of the World. 2nd ed., John Wiley and Sons, New York, Chichester, Brisbane, Toronto, Singapore. 523 pp.

PAEPKE, H.-J., 1996: Die Stichlinge, Gasterosteidae. Die Neue Brehm-Bücherei, Westarp Wissenschaften, Magdeburg. 175 pp.

SWINNERTON, H. H., 1907: The stickleback, its personal family history. 54th Ann. Rep. Nottingham Nat. Soc. **54:** 34–40.

WOOTTON, R. J., 1976: The biology of the sticklebacks. Academic Press, London. 387 pp.

WOOTTON, R. J., 1984: A Functional Biology of Sticklebacks. Croom Helm, London & Sydney, 265 pp.

Gasterosteus LINNAEUS, 1758

HANS-JOACHIM PAEPKE

Gasteracanthus PALLAS, 1814: 228
Leiurus SWAINSON, 1839: 175
Merriamella JORDAN, 1907: 131
Gladiunculus JORDAN & EVERMANN, 1927: 504

These are relatively small fishes, which have a high oval cross section. Their size and lateral covering of bony shields (*SL*) are subject to considerable variability, which is genetically determined but also ecologically influenced. Members of the marine and anadromous populations are usually larger and more slender than the small, thick-set individuals of the freshwater populations. The mouth is directed diagonally upward. It is about as large as an eye and smaller than the rostrum.

An obvious characteristic is the presence of three, or rarely four, dorsal spines (D_1), which are also highly variable in length and structure. They are located anterior to the soft-spined D_2. The first two dorsal spines are usually large and strong, while the third is much smaller and is therefore sometimes overlooked. Each ventral fin (*V*) is also armed with a strong spine. These ventral spines have a firm buttress in the massive pelvic girdle, which consists of a ventral shield with paired lateral wings. A shorter spine is located immediately anterior to the anal fin (*A*). The caudal fin (*C*) has a straight to slightly convex posterior margin

On each side, there is a lateral row of bony shields or plates (*SL*). They are expanded into a keel on the caudal peduncle. The shields in the pectoral region are greatly expanded vertically. Those anterior to the caudal keel are the smallest. Individuals from northern marine habitats usually have complete rows of plates and are therefore classified as the completely plated morph, formerly called forma *trachura*.

Those from southern freshwater habitats frequently have greatly reduced shield cover and are therefore classified as the low-plated morph, formerly called forma *leiura*. The many transitional individuals between the two extremes, which have gaps between the lateral rows of plates on the anterior part of the body and the caudal keel, were classified as the partially plated morph, formerly called forma *semiarmata*. More recently, a fourth morph has been recognized, which is low plated with a keel (ZIUGANOV 1983, BANBURA and BAKKER 1995).

There is an additional, anterodorsal row of up to eight plates, some of which are formed from fused pairs. They bear the dorsal spines (*Du*).

Within the family Gasterosteidae, the members of the genus *Gasterosteus* are more progressive in many characters than those in the other genera of the family. They possess optimally developed defensive modifications, including fewer but more effective spines and especially stable bony shields, which are most effective for protecting the body region. With such effective defences, the male can display itself with an eye-catching nuptial dress and inhabit benthic areas without places of concealment.

Their habitat is polyhaline and includes lakes and bodies of both brackish and fresh water widely distributed in the Northern Hemisphere. Many but not all marine and brackish water populations undertake anadromous spawning migrations. There are also permanent freshwater populations without obvious migrations for spawning. They are generally regarded to be the descendants of marine ancestors. The species do not penetrate as far into the Arctic as those of the genus *Pungitius*, thereby displaying a Siberian disjunction.

The three-spined stickleback in the genus *Gasterosteus* includes a large number of distinct morphologically divergent or parallel developing evolutionary lines, which are geographically or ecologically isolated and often genetically isolated, as well. In many cases, it is difficult to determine whether the identical appearance of populations is the result of a common origin or of adaptation to similar environmental conditions. The historically developed distribution structures of the phenotypes often seem to be masked by actual selection processes (GROSS, 1977). Among many developmental lines, there are no additional isolation mechanisms other than geographical ones. In other cases, the gene flow is more or less strongly limited. Although more than 40 morphologically identifiable groups have been described as nominal species, as shown in the list of European synonyms for *Gasterosteus aculeatus*, it has become accepted since the publication of BERTIN (1925) to treat all of these as conspecific, with the exception of *Gasterosteus wheatlandi* Putnam, 1866, and the so-called white stickleback from Alaska. As a result, the three-spined stickleback has been called a superspecies (BELL, 1976), and terms are generally used such as the *Gasterosteus aculeatus* species group or *Gasterosteus aculeatus* species complex (BELL 1984, 1995).

In contrast, in his remarkable heuristic checklist of the freshwater fishes of Europe, KOTTELAT (1997) placed the three-spined stickleback at the base of the phylogenetic species concept (PSC) according to CRACRAFT (1989), arranging a total of four species as follows: (1) *Gasterosteus aculeatus* Linnaeus, 1758; (2) *Gasterosteus crenobiontus* Bacescu & Mayer, 1956; (3) *Gasterosteus gymnurus* Cuvier, 1829; and (4) *Gasterosteus islandicus* Sauvage, 1874.

KOTTELAT considered *Gasterosteus gymnurus* to be the low-plated morph of *Gasterosteus aculeatus*, which is found mainly in Western European inland water bodies but also constitutes a small percentage of mixed populations along the seacoasts of Western Europe. Because all three-spined sticklebacks in fresh water invaded their habitats from the sea through river systems (MÜNZING 1959, WOTTON 1076, BELL 1984), *Gasterosteus gymnurus* is undoubtedly a polyphyletic taxon. It would therefore be difficult for it to fulfill the monophyletic criterion for a species demanded by the PSC. Furthermore, in the hybrid zones along the North Sea coast and in inland water of Central Europe, there is panmixia with the completely plated and the partially plated morphs (MÜNZING 1959, PAEPKE 1970, 1982). Similarly, it can be concluded that *Gasterosteus islandicus*, a low-plated morph that occurs in Iceland, has polyphyletic origins of similar phenotypes, since the sites of occurrence in different parts of Iceland listed by PENCZAK (1964) and checked by MÜNZING (unpubl. data) had to be colonized from the sea through different river systems.

In contrast to these, *Gasterosteus crenobiontus*, which was referred to as *Gasterosteus aculeatus* forma *crenobiontus* by BACESCU and MAYER (1956), has been described only from Lake Techirghiol in Romania and therefore is a taxon that undoubtedly fulfils the diagnosability and monophyletic grouping criteria of the PSC. However, in addition to this partially plated morph, there are similar partially plated morphs in Lake Iznik in Turkey and many hundreds of other more or less isolated freshwater populations of three-spined sticklebacks that could also be described as independent species according to the PSC. Whether doing this would contribute to a greater understanding of the phylogeny of the individual developmental lines within the genus *Gasterosteus* and their relationships to one another is subject to doubt. To decide this question, a consensus with other experts on the Gasterosteidae should be sought. These experts presently believe that a division of the *Gasterosteus aculeatus* species complex into separate taxonomic species would scarcely be practical and would not be very helpful (BELL 1995). Therefore, this account will assume that there is only one species of *Gasterosteus* distributed in Europe.

The **literature** follows the section on *Gasterosteus aculeatus*.

Gasterosteus aculeatus Linnaeus, 1758

E = three-spined stickleback, F = épinoche, G = Dreistachliger Stichling, R = trechiglaya kolyushka

HANS-JOACHIM PAEPKE

Synonyms
[only for the European populations, some of which were listed by MÜNZING (1959) and KOTTELAT (1997).]

Gasterosteus teraculeatus Lacepède, 1801
Gasterosteus trachurus Cuvier, 1829: 170
Gasterosteus gymnurus Cuvier, 1829: 170
Gasterosteus leiurus Cuvier in Cuvier and Valenciennes, 1829: 487
Gasterosteus semiarmatus Cuvier in Cuvier and Valenciennes, 1829: 493
Gasterosteus semiloricatus Cuvier in Cuvier and Valenciennes, 1829: 494
Gasterosteus argyropomus Cuvier in Cuvier and Valenciennes, 1829: 498
Gasterosteus brachycentrus Cuvier in Cuvier and Valenciennes, 1829: 499
Gasterosteus tetracanthus Cuvier in Cuvier and Valenciennes, 1829: 499
Gasterosteus spinosus Jenyns, 1835: 350
Gasterosteus quadrispinosa Crespon, 1844: 283
Gasterosteus nemausensis Crespon, 1844: 283
Gasterosteus ponticus Nordmann, 1840: 380
Gasterosteus biarmatus Nordmann, 1840: 381
Gasterosteus neustrianus Blanchard, 1866: 220
Gasterosteus bailloni Blanchard, 1866: 231
Gasterosteus argentatissimus Blanchard, 1866: 232
Gasterosteus elegans Blanchard, 1866: 234
Gasterosteus islandicus Sauvage, 1874: 20
Gastrosteus (sic) *hologymnus* Regan, 1909: 436
Gasterosteus aculeatus f. *crenobionta* Bacescu and Mayer, 1956: 19
Gasterosteus aculeatus v. *messinicus* Stephanidis, 1971: 202

Holotype

No holotype was designated. In the fish collection of the Linnean Society of London, there is a skin from the left side of a fish under No. 29 and two skins from a right side under the Nos. 30 and 31, which had been in the personal possession of C. v. LINNAEUS. These should be considered the syntypes (WHEELER 1985).

Ethymology

Gasterosteus means bony stomach. Gaster refers to the stomach in Latin, while osseus means bones or bone-like in Greek.

Description

D_1 III–IV, usually III; D_2 10–11, mean 11.2; *A* I/6–10, mean 8.3; *C* 10–12, mean 11.2; *P* 9–11, mean 10.0; *V* I/1; *Sp. br.* 14–23, mean 18.6; *Sl** 40.2–73.0, mean 54.7; *SL* 5–33, mean 23.6.

Morphology. The body is laterally compressed and higher than those of species in the genus *Pungitius*. *H* 24.8–36.1 % of *Sl**, mean 26.3 %. Head length: 8.8–16.9 % of *Sl**, mean 12.4 %. The eyes are relatively large; the mouth is small and directed diagonally upward. It does not extend posteriad as far as the anterior margin of the eye. Due to the diagonally upward orientation of the mouth, the lower jaw is clearly shifted anteriad. D_1 usually consists of two long and a third, very short spine. The middle spine is longest, reaching a height 7.3 % to 14.6 % and averaging 11.0 % of *Sl**. D_2 consists exclusively of segmented soft rays, the anterior of which reach a height of 5.4 % to 14.4 % and averages 11.3 % of *Sl**. Posterior to

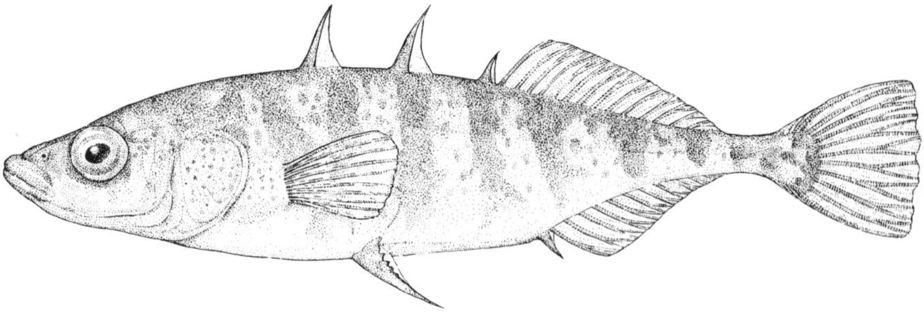

Fig. 45: *Gasterosteus aculeatus* Linnaeus, 1758 (drawn by H.-J. BEHRENDT).

that ray, the others decrease continuously in length. The rays of *A* are similar in arrangement. Their greatest length reaches 6.1 % to 14.4 % and averages 10.4 % of *Sl**. *A* is inserted in the middle beneath D_2. It is always armed with a short spine that is often curved. The posterior margin of C is staight or slightly concave. Both of the outermost segmented *C* rays are unbranched, while all others are forked. The number of unforked *C* rays is therefore one or two, usually two. The forked *C* rays number 8 to 11 and average 9.9. The *P* are fan-like and have a straight posterior margin. Their length is 11.1 % to 23.7 % of *Sl**, with a mean of 17.4 %. *V* is

Table 35: Morphometric and meristic features of *Gasterosteus aculeatus*, determined for two resident freshwater populations in Germany.

	Stöckerfliess Brook near Potsdam				Lake Ruppin in Neuruppin			
	n	average	s	ranges	n	average	s	ranges
*Sl** in mm	32	48.18	2.92	40.20-59.00	32	47.24	2.03	41.00-58.00
% of *Sl**								
lc	32	29.93	2.17	26.50-35.10	32	29.06	2.17	25.30-32.20
H	32	28.39	1.53	25.20-31.30	32	26.81	1.72	23.30-29.80
lpc	32	12.03	1.15	10.40-14.80	32	13.62	1.83	10.30-16.90
pD	32	38.04	2.22	35.10-42.40	32	37.78	1.79	33.50-40.90
pA	32	73.65	3.09	67.80-80.00	32	71.83	2.06	68.10-76.40
lP	32	16.47	1.69	13.55-20.00	32	15.49	1.88	11.06-19.02
hD_1 [x)]	32	10.40	0.97	7.71-12.04	32	9.83	1.46	7.32-12.16
hD_2	32	10.56	0.92	8.78-12.40	32	10.93	1.67	5.37-12.45
hA	32	9.84	1.35	6.12-12.40	32	8.71	1.56	6.83-12.25
lV	32	15.74	1.61	12.88-18.37	32	14.28	3.41	7.80-18.60
pelvic plate [xx)]	32	24.74	2.24	20.80-30.00	32	22.28	1.59	18.58-25.59
% of *lc*								
prO	32	29.43	4.95	25.90-36.20	32	29.18	2.26	23.80-33.30
Or	32	26.45	2.78	20.70-32.00	32	27.19	2.70	21.50-33.30
Du	32	3	0.00	3	32	3	0.00	3
Db	32	11.09	0.73	10-12	32	11.31	0.61	10-12
Au	32	1.00	0.00	1	32	1.00	0.00	1
Ab	32	8.03	0.22	7-9	32	8.25	0.25	7-9
P	32	10.03	0.18	10-11	32	9.97	0.18	9-10
Vu+Vb	32	I/1	0.00	I/1	32	I/1	0.00	I/1
Cb	32	10.00	0.00	10	32	9.88	0.43	8-10
Cu	32	2.00	0.00	2	32	2.00	0.00	2
SL	32	19.06	10.20	6-32	32	7.65	1.56	5-11
Sp.br.	32	19.13	2.50	15-22	32	18.09	1.44	15-21

[x)] the second dorsal spine was measured, [xx)] length of the pelvic bone plate.

Table 36: Morphometric and meristic features of three anadromous populations of *Gasterosteus aculeatus*.

	North Sea, Eiderstedt Peninsula Germany				Baltic Sea, Oder Estuary German-Polish border				Black Sea, Lake Varna, Bulgaria			
	n	$\bar{x}$	s	ranges	n	$\bar{x}$	s	ranges	n	$\bar{x}$	s	ranges
Sl	6	69.23	3.44	63.00–73.00	24	51.73	2.27	45.00–57.00	25	56.28	3.82	50.20.–65.00
% of *Sl*												
lc	7	26.54	2.20	25.70–29.10	26	27.40	1.27	24.89–30.70	25	31.12	2.37	25.70–36.10
H	7	27.26	1.69	25.70–30.00	26	25.25	1.27	22.70–27.00	23	23.79	1.25	21.10–25.70
lpc	7	12.69	1.07	11.00–14.30	26	12.82	1.56	10.70–15.60	25	10.88	1.36	8.80–13.80
pD	7	35.68	1.81	32.90–38.60	26	36.40	1.45	32.40–37.70	24	37.03	1.59	34.40–40.20
pA	7	70.83	1.25	69.40–72.10	26	71.94	2.20	65.70–76.10	25	70.28	3.21	64.60–77.10
lP	7	17.60	1.07	16.44–18.20	26	17.44	1.66	14.62–20.65	25	20.24	1.30	17.19–23.72
hDu [x]	7	11.69	1.36	10.00–13.02	26	12.05	1.48	8.13–14.60	25	11.56	1.70	9.44–13.75
hDb	7	11.83	1.00	10.27–13.60	26	10.78	1.33	8.68–13.67	24	12.23	0.92	10.72–14.04
hAb	7	11.11	0.76	10.14–12.06	26	10.75	1.19	9.30–12.30	25	11.60	1.60	8.79–14.39
hV	7	14.97	1.20	13.43–16.83	26	16.02	0.85	13.82–20.00	25	14.68	3.72	8.77–17.45
lPb [xx]	7	24.40	1.31	22.90–26.30	26	25.02	1.78	21.50–28.01	25	21.56	1.89	16.30–25.70
% of *lc*												
prO	7	28.97	1.77	26.00–31.00	26	27.98	2.95	20.30–32.10	25	28.92	2.27	24.60–34.10
Or	7	27.69	2.24	24.70–30.30	26	28.35	2.39	24.10–33.33	25	26.24	2.02	22.80–29.40

[x] The second D spine was measured, [xx] length of pelvic bone plate

Table 36: (cont.)

	North Sea, Eiderstedt Peninsula Germany				Baltic Sea, Oder Estuary German-Polish border				Black Sea, Lake Varna, Bulgaria			
	n	$\bar{x}$	s	ranges	n	$\bar{x}$	s	ranges	n	$\bar{x}$	s	ranges
Du	7	3.00	0.00	3	26	3.00	0.00	3	25	3.00	0.00	3
Db	7	10.86	0.85	10–12	26	10.65	0.71	10–12	25	11.60	0.63	11–12
Au	7	1.00	0.00	1	26	1.00	0.00	1	25	1.00	0.00	1
Ab	7	7.86	0.66	7–9	26	7.69	0.62	6–9	25	8.36	0.68	7–10
P	7	9.86	0.38	9–10	26	9.92	0.28	9.10	25	10.00	0.00	10
Vu+b [xxx]	7	I/1	0.00	I/1	26	I/1	0.00	I/1	25	I/1	0.00	I/1
Cb	7	9.86	0.38	10–11	26	9.96	0.20	9–10	25	9.80	0.20	9–10
Cu	7	1.86	0.38	1–2	26	2.00	0.00	2	25	2.00	0.00	2
SL	7	29.11	5.52	21–33	26	31.73	0.86	30–33	25	30.44	1.51	27–33
Sp.br.	7	17.43	1.00	16–19	26	19.27	1.40	17–23	25	19.00	1.92	14–23

[xxx] ventral spine and ray

supported by a strong spine and a short, delicate ray. The highly variable spine reaches a length from 8.8 % to 20.0 % of *Sl**, with a mean of 15.1 %. The *pD* varies from 32.4 % to 42.4 % of *Sl** with a mean of 37.0 %. The *pA* ranges from 64.6 % to 80 % of *Sl** with a mean of 71.5 %. The length of the short, fragile caudal peduncle (*lpc*) ranges from 8.8 % to 16.9 % of *Sl** and averages 12.2 %.

The body colouration varies according to the environment. The inhabitants of the sea and large lakes are light silvery with the irregular pattern of dark spots strongly suppressed. In contrast, this darker pattern is generally prominent on inhabitants of lotic water and is called "stream-bed markings". Often, the dorsum is darkened by a greenish, iridescent coloration. All fins are colourless and transparent. During the reproductive period, the males develop a prominent "wedding suit", which will be discussed in detail below.

Karyotype: According to KLINKHARDT and BUNK (1990) and KLINKHARDT *et al.* (1995), the diploid number of chromosomes is obviously 42. Hetermorphic sex chromosomes could not be recognized, although SWARUP (1956, 1959a) reported them. The karyotype consists of the following: 2n = 8 metacentric + 12 submetacentric + 14 subtelocentric + 8 acrocentric chromosomes. This complement corresponds to the total number of 42 reported in other publications, although other totals have been reported for the karyotype of some non-European *Gasterosteus aculeatus* populations (CHEN and REISMAN 1970, MURAMATO and IGARASHI 1969, MURAMATO *et al.* 1969). SWARUP (1956, 1959c) was able to produce triploid specimens by exposing zygotes 0.5 to 3.0 hours after fertilization to temperatures of 0 °C or 33 °C. The triploids had shorter bodies and longer tails than normal individuals, and although their cells were larger, their organs were not.

Protein specificity: The most important electrophoretic studies of the enzymes conducted in Europe are those of the Polish authors, RAFINSKI *et al.* (1989). They invesigated the variability of 13 presumptive enzyme loci of *G. aculeatus*. The specimens they examined were obtained from 6 marine and 13 freshwater habitats in Poland. The marine sticklebacks show a greater enzyme polymorphism than those from fresh water with p = 0.38 for the proportion of polymorphic loci compared to p = 0.12 for those in fresh water. Similarly, the average number of alleles per locus were significantly higher for individuals from the sea, n = 1.54, than for those from fresh water, n = 1.14. Nevertheless, there is no significant difference between the average values for heterozygosity of the two groups investigated. Several alleles were found to be less abundant only in the marine sticklebacks. The genetic difference between the inhabitants of seawater and freshwater was slight, fluctuating between D = 0.000053 and D = 0.031410. The value of *D* varies among catches from the sea less than it does among catches from freshwater habitats. Apparently, the inhabitants of the sea are more panmixic than the more isolated freshwater populations.

However, TANIGUCHI *et al.* (1990) used similar methods to study the three-spined sticklebacks inhabiting marine and freshwater habitats in Japan and determined that the opposite was true. While the genetic differences between the two ecotypes were very great with D = 0.6746, those among the individual populations in each group were much less with D = 0.0006 to 0.0015. The authors concluded that the freshwater populations in northern Japan are monophylic and are descended from common ancestors that lived before the postglacial colonization of the freshwater habitats. HAGLUND *et al.* (1992) studied the allozyme variation of 16 North American, European, and Asiatic populations of *G. aculeatus* and reached the same phylogenetic conclusions. It is possible that they also worked to some extent with the same Japanese populations as TANIGUCHI *et al.* (1990), which would help to explain their results. HAGLUND *et al.* (1992) distinguished an Atlantic group, encompassing populations from Europe and eastern North America, and a Pacific group, which consists of a large aggregation of populations from western North America and several from Japan together with a small number of deviant ones, also from Japan, which might possibly form a species distinct from *G. aculeatus.*

Electrophoretic studies of the phylogenetic relationships of *Gasterosteus aculeatus, G. wheatlandi, Pungitius pungitius,* and *Apeltes quadracus* were conducted by HUDSON and GUDERLEY (1984). Their locus by locus analysis of interspecific allozyme variation as well as band counting analysis of soluble protein in the muscles confirmed their assumptions about the relationships among these four gasterosteid species, which had been reached based on anatomical, karyological, and ethological findings. They assume that the two *Gasterosteus* species are the most closely related, which should be expected. *Apeltes quadracus* retained the most primitive features, while *Pungitius pungitius* assumes an intermediate position within the group. Serological distances between *Gasterosteus aculeatus* and *Pungitius pungitius* have already been determined by PENCZAK (1961).

Differences in other biochemical features, such as the activities of aspartate aminotransferase and glutamate aminotransferase, and other peculiarities of the proteins were found by JÜRSS *et al.* (1982, 1983, 1985) among the varieties of *Gasterosteus aculeatus* near the mouth of the Warnow in the Baltic Sea, that were fully or partially described above, and those in freshwater habitats farther upstream, that have scarcely been mentioned. These differences can be correlated with the fact that the three varieties discussed are adapted in different ways to the temperature and salinity conditions in their preferred habitats (GUTZ 1970).

Osteological characteristics: Information on the osteology of *Gasterosteus aculeatus* was supplied by BERTIN (1925), and specific reference to the skull was made by GREGORY (1959), who characterized the skull of *Gasterosteus* as being in a primitive stage prior to the elongation of the preorbital part that we find in *Spinachia*. At the present time, however, the opposite would seem to be true. The mouth is small, and the articulation of the quadrate is clearly located anterior to the preorbital margin. The premaxillary process is, compared to other species, intermediate in length. The relatively short maxillary is firmly joined to the distal part of the premaxillary and attached to the narrow dentale by a ligament so that the mouth can be extended forward. Only the premaxillary and dentale bear teeth. The anterior teeth are arranged in three or four rows, while those at the posterior margin of the jaw are arranged in only one. The bony ring around the eye is incomplete. The lacrimal bone is large. The third suborbital bone covers the cheek and makes contact with the preoperculum.

The number of vertebrae varies between 29 and 33. Of these, 14 are usually somatic vertebrae, and 18, caudal vertebrae. The flattened ventral or haemal arches of the caudal vertebrae have been modified to form a two-part hypural plate.

In this respect, *Gasterosteus* is actually more primitive than the species in the genera *Apeltes, Culaea,* and *Pungitius*, which all have fully fused, one-piece hypural plates (NELSON 1971). A characteristic of the gasterosteids is that parts of their pectoral and pelvic girdles lie directly beneath the skin and are therefore externally visible. This is also true of the massive ectocoracoid, or interclaviculare, which is considerably narrower in other gasterosteids. It is firmly joined by musculature to the basibranchiale and also joined to the cleithrum and coracoid. It extends caudad as far as the pelvic girdle but does not join with it.

Both of the ossa pubes, which are formed from the pelvic girdle, are visible. They mesh together along the midline to form a massive abdominal shield, from which this fish taxon derives its name. To support the large abdominal spines, this bony shield has ascending branches, which are longer and wider than those of other sticklebacks. The ventral and dorsal spines can be set in special locking sockets, which considerably increases their effectiveness as defensive weapons. The dorsal spines are sufficiently anchored by a row of broadened, scutiform pterygophores.

The greatest attention, however, has been given to the lateral bony scutes, which fully or partially protect the sides of the body of *Gasterosteus aculeatus.* In the plesiomorphic condition, the shields are arranged in single bilateral rows, and their number corresponds to the number of myomeres and consequently also to the number of vertebrae. Their number and arrangement are genetically determined and are subject to considerable variability (HEUTS 1947b, MÜNZING 1959, ZIUGANOV 1983, BELL 1984).

The individual varieties of scute arrangement were formerly designated and determined in different ways, which occasionally resulted is misunderstandings. Recently, BELL (1984), WOOTTON (1976), and BACKER and SEVENSTER (1988) proposed eliminating the commonly used designations, *leiurus, semiarmatus,* and *trachurus*, and substituting the following verbal descriptions, which are unfortunately somewhat exhaustive: low plate morph, corresponding to *leiurus,* with the body sides covered by a row of five to eight plates confined to the trunk region and without a caudal keel; partially plated morph, corresponding to *semiarmatus,* with a gap of

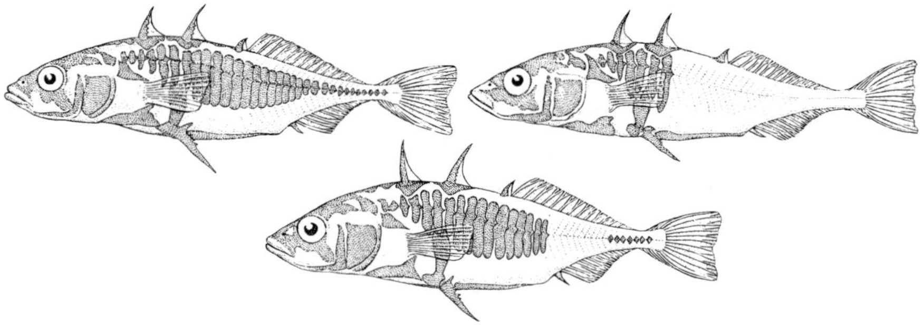

Fig. 46: The three lateral plate morphs of *Gasterosteus aculeatus*: the completely plated morph (left), low plate morph (right), and partially plated morph (below). The fourth morph, the low plate with keels, is not depicted (drawn by H.-J. BEHRENDT).

more than one plate between the anterior row of bony plates and the caudal keel; completely plated morph, corresponding to *trachurus*, with a complete row of 29 to 33 plates, including the posteriormost, which form a raised caudal keel. ZIUGANOV (1983), as well as BANBURA and BAKKER (1995), described a fourth variant: low plate with a keel. This designation only serves to characterize the variants phenotypically. They have no primary ecological significance other than the fact that the completely plated morph is more frequently encountered among the marine migratory forms than the low plate morph, which is found primarily among non-migratory freshwater populations.

Table 37: The best available two locus genetic models for the inheritance of the lateral plate morph of *Gasterosteus aculeatus* (BANBURA 1995).

A) Model by ZIUGANOV (1983): Two unlinked loci, each locus with two alleles (A,a and C,c) that interact epistatically		
AAC-, AAcc, Aacc	=	completely plated morph
AaC-	=	partially plated morph
aaC- aacc	=	low plate morph
B) Model by BANBURA (1994): Two unlinked loci, one with three alleles, in order of decreasing dominance: A, a, a_k, and the other with two alleles (C, c), with epistatic interaction		
AAC-, AAcc, Aacc, Aa$_k$cc	=	completely plated morph
AaC-, Aa$_k$C-	=	partially plated morph
aaC-, aacc, aa$_k$C-, aa$_k$cc	=	low plate morph
A$_k$a$_k$C-, a$_k$a$_k$cc	=	low plate morph with keel

Sexual dimorphism: The most striking sexual dimorphism is observed in the colouration of the males during the period of reproduction. While the females maintain the same simple colour pattern throughout the year, the males become bright orange on the entire underside, or just on the underside of the head and pectoral region of some individuals. Even more spectacular are the radiant, turquoise-coloured eyes of the males. The dorsal surface also becomes brighter, taking on a transparent light blue hue, which is considerably lighter than the dark pattern that prevails during the rest of the year. During the reproductive cycle, the individual elements of the nuptial colouration pattern vary in a way that changes the intraspecific signals they give (McLennan and McPhail 1989). Titschack (1922), in his study of the secondary sexual characteristics of *Gasterosteus aculeatus,* described the nuptial colouration pattern and mentioned additional peculiarities. For instance, the dorsal musculature and the muscles of the pectoral fins are better developed in the males. Similarly, males show a better development of parts of the brain and the spinal cord. In addition, the kidneys of the two sexes show morphological differences related to function. Those of the male are also greatly enlarged. This is related to the production of a sticky secretion that hardens quickly in water and is used for building the nest. This is accomplished along with the primary function of the organ. In contrast, the liver of the female is enlarged. Furthermore, females are often larger and, during the spawning period, more swollen than males in the same age class (Heuts 1947a, Münzing 1959, Reimchen and Nelson 1987). Reimchen and Nelson (1987) also found that the number of vertebrae in the males is greater, but this had not been reported by Münzing (1959). Penczak (1962) and Aneer (1973) noted a significant difference between the sexes in the relative head length. According to Penczak (1962), the mouth opening of the males is deeper, and their snouts are longer. The report of Bertin (1925) that females bear fewer lateral shields than the males could not be generally confirmed by Heuts (1947a). Münzing (1959) found no difference at all. Titschack (1922) summarized this subject by reporting that the two sexes use the energy available to them in very different ways: the females almost exclusively for the production of eggs, and the males for a highly developed form of parental care, which permits them to produce considerably less sperm than free-spawning fishes. The energy saved on sperm production is then fully consumed by the greater muscle activity required for territorial defence and care of the young, by the kidney secretion, and generally "through the more active temperament."

Variability: *Gasterosteus aculeatus* is among the most variable fishes known. This has resulted in the unusually large number of synonyms, more than 40, that have been given to this species.

Most obvious is the variability in the pattern of the lateral scutes, which was shown through cross-breeding experiments to be genetically determined (Heuts 1947, Münzing 1959, Hagen and Gilbertson 1973a, Avise 1976, Ziuganov 1978,

1983, BANBURA and BAKKER 1995). ZIUGANOV's genetic model has the great advantage of providing an explanation for the spontaneous appearance of the low plate morph, which can be produced by mating two members of the completely plated morph. It also explains how populations of individuals with few scutes in inland water bodies arise from members of the fully shielded migratory marine form. The model of BANBURA and BAKKER (1995) explains how the relatively rare variant of the low plate form with keels arises. Aside from the four genetically determined varieties, there are variations in the number of scutes within the individual morphs. The least variability is shown by the low plate morph (MÜNZING 1959, PAEPKE 1996, BELL 1984).

Furthermore, the number of lateral bony scutes is not the same on different sides of the same fish, as BERTIN (1925) reported. Individuals with asymmetrical scute patterns occur among all three varieties. They are rarest among members of the low plate morph and most abundant in the partially plated morph (PAEPKE 1970). Such asymmetrical fishes are considered to be hybrids between members of the other two varieties, but they are able to form monomorphic populations, such as those in Lake Iznik, Turkey (MÜNZING 1962), and in Lake Terkirghiol, Romania (BACESCU and MAYER 1956, BANARESCU 1966). ANEER (1973) found that among the three-spined sticklebacks in the northern Baltic Sea proper, there is a great predominance of the partially plated morph, which accounts for as much as 80 % of the individuals. A large proportion of these display an asymmetrical arrangement of the lateral scutes. The greatest difference between the number of lateral scutes on the right and left sides of an individual was reported to be 18 by PAEPKE (1970). In addition to genetic control, this is influenced by geographical and ecological factors, as well as by natural selection mechanisms, such as predation (GROSS 1978, BELL 1984).

The variability in the gill rakers is also genetically controlled (HAGEN and GILBERTSON 1972, HAGEN 1973). Usually, three-spined sticklebacks from flowing water have fewer gill rakers than those from large lakes, which can be attributed to a long-term adaptation to the prevailing benthic or pelagic food items.

The size and shape of the body also varies greatly. Along the Pacific Coast of North America, the body size seems to decrease from north to south (MYERS 1930). Such a tendency might also be found in other regions. In addition, members of the migratory marine form are larger than those inhabiting small freshwater habitats (MÜNZING 1959), which are often more truncated and have higher dorsal profiles than members of marine populations (PAEPKE 1983).

Among the features investigated by HEUTS (1949) was the variability in the number of fin rays. A more obvious variation, which is of great significance for evolutionary biology, is that of the dorsal spines. More than 97.6 % of all three-spined sticklebacks have three dorsal spines. Within single populations, the percentage of individuals with only two spines can reach a maximum of 11.1 %, while a maximum of 17.2 % can have four spines (HEINCKE 1889, PENCZAK 1965, GROSS

1978a, Bell 1984). The number of dorsal spines is determined genetically but can also be influenced by environmental factors (Lindsey 1962, Campbell 1979, Bell 1984). The dorsal spines are anchored in shield-like pterygophores, which are arranged along the dorsal midline anterior to the D_2. Usually, six plates are present, but occasionally they may number as many as eight due to the splitting of the third and fourth plates. On these eight plates, the three, or rarely two or four spines take various positions. Seven different combinations are possible. This suggests that *Gasterosteus aculeatus* is descended from ancestors that possessed more than three spines (Heincke 1889, Paepke 1981, 1996).

The size and shape of the pelvic girdle also varies. Consequently, the relative proportions of the ventral shields may be sexually dimorphic (Penczak 1965). The distal margins on both lateral ascending branches of the pelvis are especially variable. As in other sticklebacks, there is a tendency toward the reduction of the entire pelvic girdle, including the ventral spines, of *Gasterosteus aculeatus*. This is an interesting case of parallel evolution (Nelson 1971, Moodi and Reimchen 1973, Bell 1974, Reimchen 1980).

Münzing (1959) determined that the inhabitants of isolated freshwater bodies possess considerably more vertebrae than members of the migratory marine form.

The body colouration varies considerably. This is due just to the extent of the orange portion of the nuptial colouration of the male (Titschack 1922, Bell 1984). The dark spotted markings of the normal colour pattern are also variable. Among the population at Treuenbrietzen, Germany, individuals have appeared for several years with a banded pattern, in most cases among otherwise irregular markings. This suggests that heredity might be involved (Paepke 1982a, 1996). Albinotic and xanthoristic individuals have also been reported several times (Arnold 1984, Stevens 1987, Bakker *et al.* 1988, Bößneck 1999). Populations with melanistic males are known, mainly from the Northwest of North America (McPhail 1969, Moodie 1972, Hagen and Moodi 1979, Bell 1982), but none are yet known from Europe. Recently, there have been reports from Canada of "white sticklebacks", which display a bright iridescence dorsally and lack the characteristic colour patterns. Because these colour variants also displayed a fully distinct behaviour in caring for their young, they seem to belong to an independent species (Jamison *et al.* 1992).

Specimens belonging to about 100 populations from all over Europe were examined by Gross (1977) during his study of the geographic variation of ten meristic characters, including the lateral plates, vertebrae, fin rays, and pterygophores. He reported interesting clinal trends in the changes in these characters, some of which are gradual and others stepwise. He believes that abiotic environmental factors or historical events are less likely to have caused this than "the cline in the magnitude of the selection pressure from predators".

Age and size variability: The most obvious change in growing juveniles is the successive formation of the lateral scutes. They are first formed on juveniles that have reached a length of at least 25 mm *Sl**, according to the particular genotype. First to form are the large scutes in the pectoral region, followed in the partially and fully shielded varieties by the raised caudal keel and then the small scutes that develop in the spaces between the other two groups of scutes (MÜNZING 1959, BELL 1981). Other changes, such as the increase in the number of vertebrae, occur more discretely during the course of the individual development. Pre-adults have more vertebrae than younger individuals and fewer than the great majority of specimens longer than 80 mm *Sl** (REIMCHEN and NELSON 1987). The width of the toothed portions of the jaws shows a similar change. The teeth of juveniles about 27 mm in *Sl** are arranged in at most two rows, while those of adults can form three or four rows, as determined by personal observations not previously published.

Subspecies

Gasterosteus aculeatus varies greatly within its large geographic and ecologically diverse range. Identical variants are often polytopically distributed. Their intraspecific subdivision has still not been satisfactorily completed. KOTTELAT (1997) generally disputed the value of subspecific categories in phylogenetic systematics and preferred treating all discrete groups that could be diagnosed through their own natural history as species.

Hybrids

LEINER (1934a, b) artificially crossed *Gasterosteus aculeatus* with *Pungitius pungitius* and *Spinachia spinachia*. The hybrids of *G. aculeatus* x *S. spinachia* stagnated in their embryonic development prior to hatching. Similarly, 50 % to 75 % of the hybrids of *G. aculeatus* x *P. pungitius* perished within the eggs, or the yolk-stage larvae died eight to ten hours after hatching. The few survivors "had malformed eyes with tiny, slit-like pupils and bulging corneas, giving them extremely poor vision or making them blind. Their entire bodies, especially their abdominal regions, were covered with large blisters filled with water. The pericardial space was expanded, and the dorsal, anal, and caudal fins were poorly developed or malformed." The few survivors showed a remarkable degree of variation. However, none reached sexual maturity. PAEPKE (1996) also artificially crossed *G. aculeatus* with *P. pungitius* and was able to confirm the findings of LEINER (1934a, b). Among polymorphic populations of *Gasterosteus aculeatus*, the partially plated morph is considered to be a hybrid between the other two morphs.

Distribution

The three-spined stickleback inhabits a vast range in the Palearctic and Nearctic Regions that is greatly differentiated ecologically and includes not only the sea and brackish water but also freshwater habitats.

In Europe, the range extends from the coastal regions of the Barent Sea and waters around Iceland southward past the Faroe, Shetland, and Orkney Islands, across Great Britain and Ireland, except for the Scottish Highlands and Wales, and throughout Scandinavia to the North and Baltic Seas as well as the lowlands in large parts of Northeast, Central, and Western Europe (ANDRIJASHEV 1954, BERG 1932, 1949, MÜNZING 1959, 1961, 1962a, 1963, 1964, 1972, PENCZAK 1962a, b, 1964, 1965, POTAPOVA 1967). The southern border of its range is located approximately along the northern roof of the Wolynian-Podolic plate, the Beskydy Range near Krakow and Wroclaw, the Sudeten, the Erzgebirge near Dresden, and the German Mittelgebirge. The region near the headwaters of the Vistula, Oder, and Elbe Rivers are beyond the range. The species was originally absent from Czechia, Slovakia, Austria, Hungary, and Switzerland (HECKEL und KNER 1858, SIEBOLD 1863). An introduced population is present near Prague, and another has been known near Bratislava since 1966 (BALON 1967). In the 1880s, members of the species were released near Vienna and Munich (GROTE *et al.* 1909, GASCHOTT 1941). The many locations of populations in the Danube catchment area of Austria today were discussed in detail by AHNELT (1986) and AHNELT and AMANN (1994), who considered them to be the descendents of introduced fishes. The species was first reported from Hungary in 1956 (BERINKEY 1960). In the Rhine, it has moved upstream as far as the vicinity of Basel (HEINCKE 1889, AHNELT and AMANN 1994, Ahnelt *et al.* 1994). It was released near Geneva in 1872 (FATIO 1882), and today it is widely distributed in Switzerland (KIRCHHOFER *et al.* 1990). It is also found throughout much of France, although it is absent from the high southern and northwestern plateaus, the Ardenne Region, and the Vogesen (BERTIN 1925). It has been reported only sporadically from Portugal, Spain (LOBON-CERVIA *et al.* 1988), Italy (BIANCO 1979, 1981), and Yugoslavia, where it inhabits mainly freshwater bodies near the coast or brackish water. The situation is similar in Greece (ECONOMIDIS 1991). The isolated, relict-like populations along the Mediterranean Coast of southwestern Anatolia (KOSSWIG 1954, 1956, MÜNZING 1962b) and in Syria (KRUPP and COAD 1985) deserve special attention. The similar population on the Mitidia Plain south of Algiers (PELLEGRIN 1921, BERTIN 1925) seems to have died out (HEUTS 1956).

In the Mediterranean itself, *Gasterosteus aculeatus* seems no longer to occur. However, it is still present in the Black Sea and the Sea of Azov, as well as the rivers flowing into them, including the lower Danube as far as the Iron Gate, the mouths of the Don and Kuban, and the lower Dnieper and Prypyat. It does not inhabit the Caspian Sea, the Volga, or the Aral Sea (BERG 1932, 1949, BĂNĂRESCU 1966, MÜNZING 1963a, 1963b, 1964, SVETOVIDOV 1964, WOOTTON 1976).

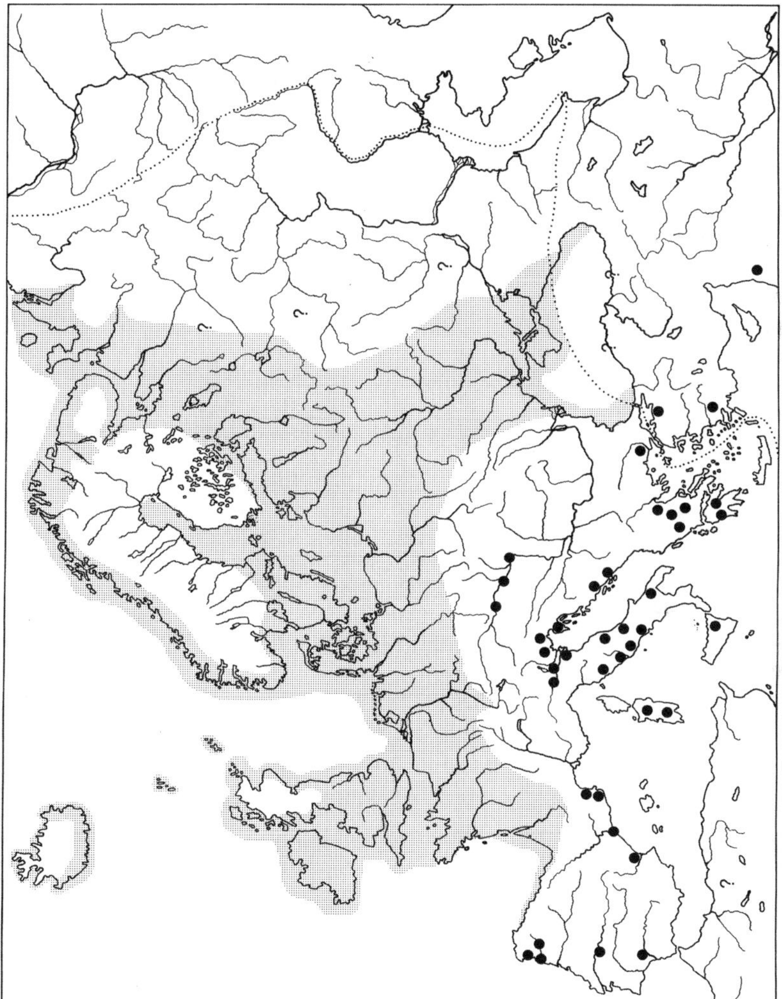

Fig. 47: The geographical distribution of *Gasterosteus aculeatus*. Dots in Western and Southern Europe indicate the locations of partially or fully isolated populations beyond the main area of the range. These include introduced populations in the upper Danube drainage area.

Between the European and Northeast Asian populations is a large disjunction caused by the Ice Age. The species is not encountered eastward from Europe as far as the eastern enclave running from the Bering Sea and Aleutian Islands southward to the East Coast of Korea at 36°6′N and to Nagasaki on Kyushu at 32°44′N (WOOTTON 1976).

In North America, its range extends along the Pacific Coast from the Alaska Peninsula to about the northern border of Mexico at 32°N and along the Atlantic Coast from the Hudson Straits south to the vicinity of Cape Hatteras at about 35°N. Here, as well, the species penetrates inland through the rivers. Isolated populations exist along the southern coast of Greenland (BERTIN 1925, MÜNZING 1959, WOOTTON 1976).

The distinct morphs have the following ranges in Europe:

The completely plated morph inhabits mainly coastal water bodies along the White Sea and North Atlantic, including those in Iceland, northern Ireland, and Scotland, as well as large parts of the Baltic and Black Seas. It is also very abundant in the inland waters of Eastern Europe to the east of the Oder watershed (PENCZAK 1965, POTAPOWA 1967, 1972, ZIUGANOV 1978, BANBURA 1994). Along the coasts of the English Channel, Irish Sea, southern North Sea, Skagerak, Kattekatt, and Baltic Sea in Poland, polymorphic populations occur that include all three morphs. Within these populations, the percentage of the low plate morph decreases from southwest to northeast (MÜNZING 1962a, 1972, BANBURA 1994).

Polymorphic populations including all three morphs are also predominant in the large Central European river valleys of the Elbe and Oder and possibly also the Weser and Rhine. Their characteristic is a bimodal distribution of the variants corresponding to the low plate and completely plated morphs, which constitute high percentages of the individuals present. In contrast, the partially plated morph is less frequently represented. These mixed populations characterize a region of intergradation between the predominant completely plated morph of Eastern European inland water bodies and the low plate morph, which is predominant in the inland water bodies of Western Europe. The region of intergradation is located mainly in Brandenburg, Germany, from which it extends eastward into Silesia in southwestern Poland (PAEPKE 1970, 1982a, 1996, BANBURA 1994). In contrast to the polymorphic populations inhabiting the main courses of the Elbe and Oder, those in the minor tributaries can consist up to 100 % of members of the low plate morph. Pure populations of the low plate morph are found mainly throughout large areas of France (BERTIN 1925), Spain (LOBON-CERVIA *et al.* 1988), Italy (BIANCO 1981), Syria (KRUPP and COAD 1985), and formerly Algeria (PELLEGRIN 1921). It also inhabits freshwater bodies in Finland and Iceland (SAUVAGE 1874, PENCZAK 1964). The partially plated morph constitutes varying proportions of the polymorphic populations. In contrast, nearly monomorphic populations of them are found only in a few lakes near the seacoast, such as Lake Iznik in Turkey and Lake Tekirghiol in Romania (BACESCU and MAYER 1956, MÜNZING 1962b, BĂNĂRESCU 1966).

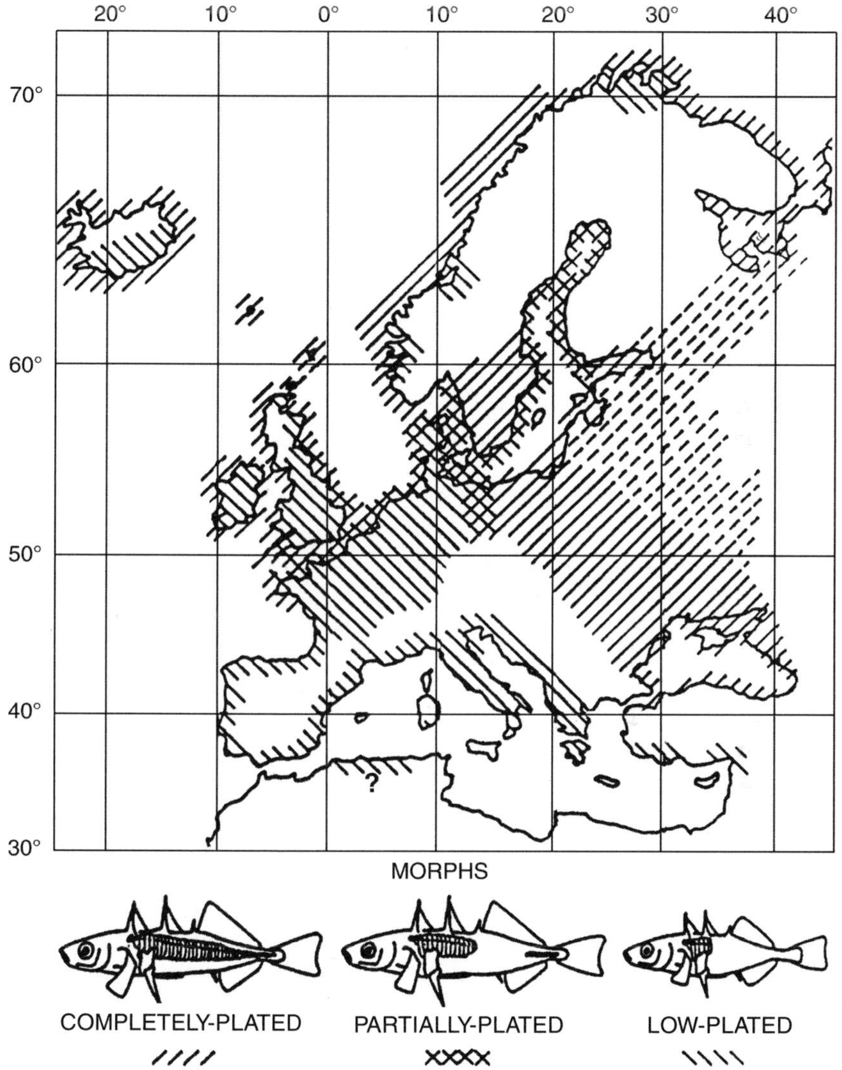

Fig. 48: The distribution of the different *Gasterosteus aculeatus* plate morphs in Europe. Where the line patterns intersect, intergrades occur. Broken lines in the patterns indicate that the composition of the populations is not satisfactorily known (GROSS 1977).

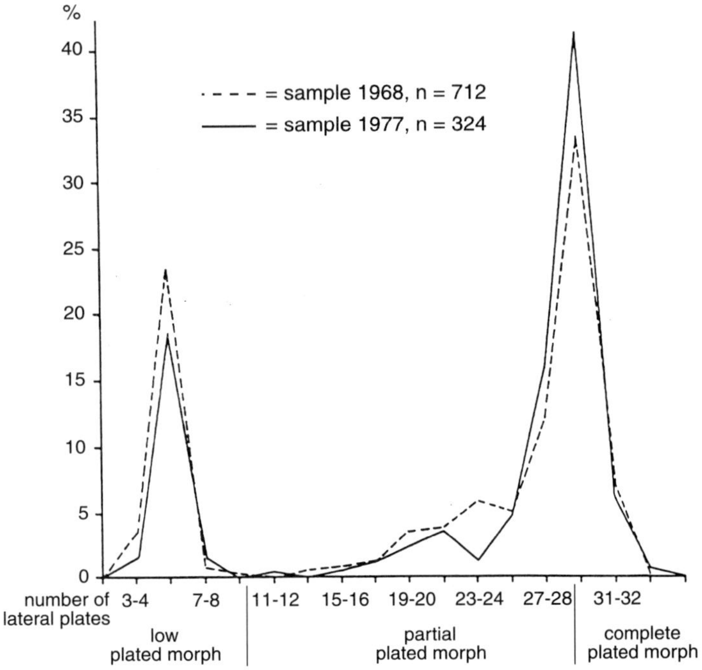

Fig. 49: Relative abundances of the different plate morphs near Pillgram, Germany. These values are typical for polymorphic populations of *Gasterosteus aculeatus* in freshwater habitats of the Elbe and Oder drainage areas (Paepke 1982a).

Different hypotheses have been proposed to explain the European distribution of the three varieties that are recognized from their scutes, namely the low plate, partially plated, and completely plated morphs (Heuts 1947a, 1947b, Münzing 1959, 1961, 1962a, 1962b, 1963a, b, 1964, 1972, Gross 1977, 1978). Heuts (1947a, b) attributed the distribution pattern to a physiological adaptation to the regional temperature and salinity conditions. Münzing (1959, 1961, 1962a, 1962b, 1963a, b, 1964, 1972) stressed aspects of the natural history of the fauna. He viewed the modern distribution pattern of the morphs as the result of their postglacial spread and subsequent intergradation processes. According to Gross (1977, 1978), the distribution of the individual variants resulted from a recent selection process, mainly by predators. All three hypotheses have certain weak points, showing that the multifaceted cause and effect combinations affecting the sticklebacks are not simple and should not be interpreted in a one-sided manner.

Ecology

Without doubt, the three-spined stickleback was originally an inhabitant of the sea (MÜNZING 1959) that developed three ecologically distinct life history modes in the course of its evolution. Marine populations remain permanently in seawater. In contrast, freshwater populations remain in fresh water throughout their entire life cycle. Anadromous populations that are members of the migratory marine form sensu MÜNZING (1959) hatch in fresh water but spend most of their lives prior to reaching sexual maturity in the sea. Like the salmon and sea-trout, they enter rivers and streams along the coast, where they spawn in the downstream sections (BELL 1984). Therefore, the three-spined stickleback shows the broadest ecological valence of all gasterosteids. It is plainly polyhaline, and each ecotype, or perhaps, better stated, ecologial adaptation form, is adapted to the special conditions in its habitat. The body fluids of the sea-dwelling fishes are hypotonic to the environment. They must drink in order to avoid the permanent danger of desiccation. The excess salt that is taken in is eliminated by special chloride cells. This requires a high consumption of energy for active ion transport from cell to cell. In contrast, the inhabitants of fresh water are exposed to the opposite osmotic pressure. Their blood is hypertonic to the environment. They must continuously excrete the water that enters while retaining the vital salts. This is accomplished passively without a requirement for energy by maintaining a certain level of amino acids. This kind of osmoregulation is inadequate in seawater. The salt water exerts a toxic effect, especially at low water temperatures, and the fishes show symptoms of poisoning (GUTZ 1970, WOOTTON 1976, GUDERLEY 1994). That is probably the reason that the already small percentage of the low plate morph within the polymorphic populations along the North Sea Coast of The Netherlands and Germany decreases from southwest to northeast (MÜNZING 1962a). It also explains why a marine population consisting only of the low plate morph is known only from the warm coast of Southern California (BELL 1979). However, SCHAARSCHMIDT *et al.* (1999) failed to find fundamental differences in the osmoregulatory systems of freshwater populations of three-spined sticklebacks and brackish water populations in the Baltic Sea equivalent to those described by GUTZ (1970) and GUDERLEY (1994).

Generally, the temperature gradient represents a limiting factor for all varieties. To the north, *Gasterosteus aculeatus* does not range into the Arctic regions as far as *Pungitius pungitius*. Without considering this general range information, LACHANCE *et al.* (1987) independently determined under laboratory conditions that the temperature preference of G. *aculeatus* from Quebec is 9 °C to 12 °C, while *P. pungitius* seemed to have two different preferred ranges of temperature, between 9 °C to 10 °C and 15 °C to 16 °C. The southern limits of the range of *Gasterosteus aculeatus* are located between 32°N and 35°N latitude, according to the mean temperature for the year. At the southern borders of the range in Europe and North America, reductions in the spine length and number of lateral plates are especially frequent signs of the deteriorating living conditions (REGAN 1909, MILLER and HUBBS 1969, ROSS 1973, GROSS 1978, BIANCO 1979, 1981).

Habitat: The greatest local abundance of the three-spined stickleback is reached along shallow seacoasts. Examples of such locations are the Haff regions of the Baltic Sea, where a modest fish oil industry formerly depended on the great masses of sticklebacks for its existence (HEILBORN 1931, "Stt." 1933, 1934), and in the lower stretches of the great northward-flowing rivers, including the Oderhaff, Elbe and Weser Estuaries, and Rhine Delta. However, all naturally standing and flowing waters as well as artificial ditch and canal systems are also colonized, as long as they can be reached by the sticklebacks. However, the genetic impoverishment increases toward the peripheral parts of the river systems, where isolation effects increase (PAEPKE 1982, 1996). Well protected by its strong spines, *Gasterosteus aculeatus* is able to colonize more open habitats with fewer protective structures than *Pungitius pungitius*. Thus, it is frequently found to be the first settler of newly constructed artificial water bodies.

Migrations: The most prominent migrations are undertaken by members of the anadromous populations. In Europe, they were subjected to special studies by MÜNZING (1959), VAN MULLEM and VAN DER VLUGT (1964), and DANIEL (1965, 1985). The freshwater populations migrate in spring only from the deep waters of their winter quarters to the shallow waters near shores and into adjacent ditches (WOOTTON 1976). There is almost nothing known about the movements of the marine populations.

In the Norderbootfahrt on the German Eiderstedt Peninsula, isolated immigrants have already begun to appear in November, but it is not until March that the invasion reaches its maximum. The fishes enter in large groups, each arriving at times the water temperature drops (DANIEL 1965, 1985). Similarly, in the Elbe Estuary at Hamburg, under normal weather conditions, large numbers of adults assemble after the beginning of March. Soon after, a mass migration into brackish and fresh water takes place. During normal years, the great bulk of the individuals move upstream during a relatively short period of time. After cold or long winters, the migration begins later, and its progress is correspondingly delayed (MÜNZING 1959). Thus, the spawning grounds near Bremen were first occupied in mid-April after a hard winter (LEINER 1930). After the long and cold winter of 1962–63, the immigration was delayed for two months (DANIEL 1965). In Japan, the immigration into fresh water occurs from the middle of February to the end of March (IKEDA 1933). In contrast, in the coastal rivers of British Columbia, the first immigrants arrive in the middle of May (HAGEN 1976).

Among the essential factors that stimulate the migratory behaviour of the anadromous populations are the spring increases in water temperature and illumination (BAGGERMANN 1957, 1969). They activate the endocrine glands, which secrete the hormones that produce the changes in the salinity preference. Thus, the environmental conditions in spring first influence the hypopysis, which secretes paralactin and thyrotropin. These hormones promote metabolic processes and also stimulate the thyroid, gonads, and adrenal glands to secrete their own hormones. A higher level of paralactin permits osmoregulation by sea-dwelling fishes in fresh water. Furthermore, the thyroxin that is secreted in greater amounts by the thyroid gland reduces the capability of the sticklebacks to continue regulating the ion concentrations of their body fluids in salt water. In addition, it increases the locomotory activity, thereby producing the necessary requirements for strenuous anadromous migrations (Gutz 1970, Wootton 1976). Gonadotropin and the adrenocorticotropic hormone (ACTH) stimulate the release of hormones from the gonads. Van Iersel (1953) was able to show that ACTH contributes to produce the excitement that the sticklebacks show shortly before the spawning season. In this way, these hormones also apparently contribute to the initiation of the migratory behaviour. That would explain why the schools of the anadromous populations moving upstream consist almost exclusively of sexually mature individuals. The return migration of the adults to the sea in Germany occurs mainly in June and July. The females begin to travel seaward earlier than the males, which remain to guard the young. Lam and Hoar (1967) as well as Wootton (1976) suggested that decreases in the temperature and illumination may be causes of the reversal in salinity preferences, but the exhaustion of the fishes at the end of the spawning period may also play an important role since the return migration begins long before the summer solstice. After completion of spawning, the fishes permit themselves to be carried downstream by the current, scarcely showing signs of life. Most of them die before reaching the sea (Münzing 1959, Daniel 1965, 1985). In contrast to the adults, the juvenile fishes leave the spawning sites and head toward the sea in schools or smaller groups. According to Daniel (1965, 1985), schools of juveniles headed for the sea have already reached the upstream end of the Norderbootfahrt on the German North Sea Coast in May. The number of these fishes increases enormously in June, and in July, the schools reach the mouth of the main river. Münzing (1959) reported similar periods of time for the migrations. Large schools migrate faster than smaller ones. For example, 200 individuals require only 8 seconds to travel 2 m, while 25 individuals require 19 seconds. When the juveniles leave the spawning places in the tidal ditches along the lower Elbe, they are 18 mm in *Sl**. By the time they have reached 20 mm several weeks later, they have already left the fresh water (Münzing 1959).

In contrast to the anadromous populations, the non-migratory freshwater populations spend the entire year in water of the same salinity. Thus, there are certain differences in the osmoregulatory abilities the members of these two ecologically adapted forms (Gutz 1970). Recent investigations have shown, however, that these differences are not as great as previously assumed (Schaarschmidt *et al.* 1999). In the lower courses of rivers, stationary freshwater populations can spawn syntopically with members of anadromous populations, but after spawning, they do not leave the spawning site (Daniel 1985, personal observations).

It is remarkable that "pelagic sticklebacks" have been found hundreds of miles from the coast in the Barent Sea (Andriyashev 1954), the Atlantic (Jones and John 1978), and the open North Pacific (Quinn and Light 1989).

Hardiness: *Gasterosteus aculeatus* does not seem to be able to tolerate ice cover and a lack of oxygen as well as *Pungitius pungitius* (Daniel 1985), which is also suggested by the differences in their geographical distribution.

Feeding habits: The three-spined stickleback is an opportunisic feeder and not a specialist. It nourishes itself mainly as a carnivore. Its diet consists preferably of aquatic invertebrates, which are visually found and investigated by snapping and chewing prior to selection (Thomas 1977). The diet is determined primarily by the temporal and spatial availability of the prey. The behaviour during feeding varies from individual to individual according to the degree of fullness of the stomach and other factors (Salvanes and Hart 1998). The seasonal variations in the nutrient components have been determined through numerous long-term analyses (Hynes 1950, Zander *et al.* 1984, Behnke *et al.* 1986, Ukegbu and Huntingford 1988). Thus, in the Birket River in England, cladocerans are predominant in early summer, while copepods are the dominant food item during all other seasons (Hynes 1950). In the Barther Bodden, Germany, the three-spined sticklebacks feed mainly on chironomid larvae, ostracods, and other benthic prey in the shallow littoral zone during the spawning season, while in the pelagic zone during the rest of the year, they prefer copepods (Behnke *et al.* 1986). Generally, in flowing water, sticklebacks prefer benthic food items, including chironomid larvae, tubificids, and young gammarids. In contrast, they prefer zooplankton in large inland lakes and in the haffs of the Baltic Sea. Hagen (1967) found differences in the feeding behaviour of members of the low plate and completely plated morphs in British Columbia. While the low plate morph consumed mainly benthic prey, the completely plated morph nourished itself mainly from plankton. These varieties also differed from one another in the number, form, and length of the gill rakers. Finally, Zander *et al.* (1984) reported that the quantity of the foods consumed is influenced by the temperature, while the quality is determined partially by the salinity and perhaps also by the nitrate concentration in the environment. More food is eaten at higher than at lower temperatures. A three-spined stickleback weighing 1 g must eat 2.4 % of its own weight per day at a temperature of 10 °C, 4.8 % at 14 °C, and 13.3 % at 18 °C (Rajasilta 1980). In addition to the copepods that are predominant everywhere, Zander *et al.* (1984) frequently found polychaetes and amphipods in the stomachs of sticklebacks at sites on the seaward side of the mouth of the Elbe, while oligochaetes were temporary supplements to the diet that had been more important for the sticklebacks caught upstream. Behnke *et al.*

(1986) determined that the mean length of *Eurytemora affinis* from the stomachs of sticklebacks exceeded the mean length of this copepod in plankton samples. The entire spectrum of the diet is extraordinarily species-rich, as determined from the numerous analyses of the diet (SAUNDERS 1914, BLEGVAD 1917, MARKLEY 1940, HARTLEY 1948, HYNES 1950, VAINU 1961, HAGEN 1967, LEMMETYINEN and MANKKI 1975, RAJASITTA 1980, ZANDER *et al.* 1984, PAEPKE 1996, GISLASON *et al.* 1998).

Generally, copepods constitute the most important food items among the crustaceans, while chironomid larvae are most important among the insects. In addition to the foods available in the open water and on the sediment, food items that have fallen into the water, including tiny collembolans, are also consumed (PAEPKE 1996). During the spawning season, eggs and occasionally larvae of their own species become regular components of the diet (HYATT and RINGLER 1989). They account for increased proportions of the diet as the population density increases and during periods of food shortage and are consumed by members of both sexes. They are consumed mainly during attacks on the nests of other individuals. In the village pond at Pillgram, Germany, the stomachs of 31 % of the sticklebacks examined contained stickleback eggs. The average number was 12, and the maximum was 55. One male had eaten 45 larvae (PAEPKE 1996).

The great interest in the feeding habits of the three-spined stickleback has resulted in part from the concern that this fish is a potential competitor for the food of economically valuable fish species (WOOTTON 1976, BEHNKE *et al.* 1986). An individual weighing 1 g requires 9 to 19 g of food before it is capable of reproduction. A female can utilize 14 to 24 g of food during one spawning season (WOOTTON 1976). When massive numbers of these fishes appear during the so-called "stickleback years", they have an especially strong effect on the available food supply. The relationship between the food consumption and egg production was investigated by WOOTTON (1986). In addition, the three-spined stickleback can occasionally feed upon the offspring of other fish species, such as juvenile eels returning to western Europe (DANIEL 1965). It is also considered to be a predator on amphibian larvae (MCLEE *et al.* 1992/93). However, laboratory tests have shown that toad larvae are only occasionally killed and eaten as long as a sufficient supply of other foods, such as *Tubifex*, is available. Predation on tadpoles of the grass frog, *Rana arvalis*, also remains within tolerable limits (GLANDT 1983, 1984).

Longevity: Corresponding to the different living conditions to which *Gasterosteus aculeatus* is exposed in the many water bodies within its wide range, the life expectancies of the individual populations differ. Based on his extensive statistical studies, BERTIN (1925) believed that the three-spined stickleback survives for two summers in France and The Netherlands, dying after spawning during the second summer. This would give them a life expectancy of somewhat more than one year. GISLASON *et al.* (1998) reported similar conclusions for the three-spined sticklebacks in Lake Mývatn on Iceland, where they died after their first breeding season. In contrast, Jones and Hynes (1950) assumed that the maximum age for members of all European populations is about 3.5 years. Mann (1971) and van Mullem and van der Vlugt (1964) reported the presence of strictly annual reproductive communities with life cycles completed almost without exception in about one year from South England and The Netherlands, respectively. Similarly, the population in Lake Bara, Alaska, also has an annual reproductive cycle. Only a few survive the breeding season to reach a maximum age of 2.25 years. A similar longevity pattern has been observed in the Martamek River System of Quebec (Greenbank and Nelson 1959, Coad and Power 1974). In contrast, members of the polymorphic anadromous populations along the German North Sea Coast reproduce during their second year of life, but with few exceptions, they cannot survive to an age older than 2.5 years (Münzing 1959). The population in the Darss-Zingster Boddenkette behaves similarly, as personally observed by the author. Members of the freshwater populations in the German State of Brandenburg, which include all three varieties, reproduce as one-year-olds in the second summer of life and usually die immediately afterwards. Only a few participate in reproduction a second time or become even older. The maximum age in this region was reached by two females from Angermünde, which were captured on October 7, 1969, at lengths of 58 mm and 61 mm *Sl**. These had already survived for the fourth summer of their lives and must have been from 39 to 40 months old. It was noticed that members of the completely plated morph appear more frequently among the relatively old individuals than members of the low plate and partially plated morphs (Paepke 1971, 1982b, 1996). Very early, it was noticed that three-spined sticklebacks kept in captivity usually can reach a greater age than most of those in the wild. Bock (1928) estimated that it is possible for them to reach an age of five years, while the estimate by Wunder (1928, 1930) was only four. Crucial for the life expectancy within the individual populations may be whether the sticklebacks become sexually mature during the first or second year of life.

Growth: Like the potential age of the fishes, their growth in body length and weight is assumed to be dependent upon ecological factors, such as the food supply, climatic pattern, photoperiods, water temperature, and salinity, just as it probably depends upon a genetically controlled pattern of life within each of the individual populations, which may be anadromous or resident freshwater just as they may live for only one or more than one year. Making general statements concerning the growth is made even more difficult by the fact that many older publications fail to report whether the length data provided refer to standard or total length. Furthermore, there are fast-growing and slow-growing individuals within each reproductive community. It is possible for fast growers to reach the length of slow growers from the previous year within the first three months of their lives. Finally, the growth rate decreases more and more with increasing age, so it is difficult to assign older individuals to any particular age class (Jones and Hynes 1950,

MÜNZING 1959, WOOTTON 1976). Under the best circumstances, it is possible to establish a relationship between the dimensions and weight recorded and findings from examinations of the otoliths. That was first attempted for the three-spined stickleback by BOCK (1928). For members of the low plate morph from Germany, he calculated the following relationships: first year length 35 to 40 mm, second year length 45 mm, third year length 50 to 55 mm. The maximum length was about 60 mm. However, the most comprehensive growth and age analysis employing otolith examinations was completed by JONES and HYNES (1950) in England. The results for this relationship among members of the Birket population were provided as follows: first year length 42.7 mm, second year length 45.6 mm, third year length 51.4 mm. Similar growth was achieved by the sticklebacks in Priddy Pool, southwestern England (PENNYCUICK 1971). MÜNZING (1959) as well as VAN MULLEM and VAN DER VLUGT (1964) confirmed the findings of HEUTS (1947a) that after an *Sl** of 35 mm is reached, the males and females begin to grow at different rates, with the females becoming 5 mm longer than the males during the course of their lives. According to the calculations of MÜNZING (1959), the following relationships between length and age were observed within non-migratory, freshwater populations: first year males – 40 mm, females – 45 mm; second year – males 50 mm, females – 55 mm; third year - females only – 55 to 66 mm. In contrast, the members of the anadromous populations first migrate into the lower Elbe at an age of two years, when the males are 53 mm, and the females, 58 mm. Examinations of the otoliths generally confirmed the results of the length measurements, although MÜNZING (1959) emphasized the fact that the methods employed in the study of the anadromous populations did not provide unambiguous results in all cases. For Polish populations, PENCZAK (1965) was able to provide the following results, which are similar to those of MÜNZING (1959): first year – 42 mm in a range from 39 to 45 mm, second year – 46 mm in a range from 44 to 60 mm, third year, mainly females – not less than 58 mm. PAEPKE (1971) used fishes from the region of Brandenburg, Germany, to show that the growth in fresh water can vary according to the locality. Investigations similar to those in Europe were undertaken on North American populations by such authors as GREENBANK and NELSON (1959), COAD and POWER (1973, 1974), and HAGEN (1967). HAGEN (1967) found that members of the completely plated morph grow faster than those of the low plate morph. Investigations of polymorphic populations along the Belgian, Dutch, and German North Sea Coasts, however, could not confirm these findings (HEUTS 1947a, MÜNZING 1959, VAN Mullem and VAN DER VLUGT 1964).

The growth in weight of the three-spined stickleback was investigated, among others, by PENNYCUICK (1971), who calculated the following average values for weight increase: during the first year the average weight of fishes was 0.514 g; during the second, 0.888 g; during the third, 1.224 g. In southern England, MANN (1971) found that growth was more rapid, already reaching 0.86 g at an age of one year. Equations to determine the relationship between length and weight and also the increase in weight were proposed by such authors as PENNYCUICK (1971). As determined by VAN MULLEM (1967), growth rates of 45 mm in *Sl** and more than 1 g in weight can be reached in only four months. Three-spined sticklebacks kept in aquaria by WUNDER (1928, 1930) grew to a length of 7 cm in only six months.

Population dynamics: It has been reported several times that females are numerically predominant in anadromous populations (MÜNZING 1959, DANIEL 1965). DANIEL (1965) determined that from January to June, the ratio of males to females was 1:11.25 but attributed this finding to a selective effect of the mesh size of his gill nets. Females are also predominant in freshwater populations. PAEPKE (1971) examined 4253 individuals, most of which had been caught in April and May, and determined that 41.2 % of them were males and 58.8 %, females.

The age structure of the populations varied from time to time and place to place. According to MÜNZING (1959), the spawning schools in the lower Elbe consist primarily of two-year-olds but also include a few three-year-olds. DANIEL (1965) modified this report by explaining that the fishes entering freshwater included members of the three previous year classes, of which those almost two years old were predominant. The freshwater populations in Brandenburg were similarly composed predominantly of the two to three previous age classes, although for a certain period of time each year, members of four year classes would be present together. Thus, the standard length of fishes in samples taken at Wiesenau, Germany, on May 22, 1969, varied from 10 to 74 mm *Sl**. Among them were the 1969 hatchlings that had just left the eggs, many members of the 1968 and 1967 year classes, and a few survivors of the 1966 year class (PAEPKE 1971, 1996). The size of this reproductive community was dependent to a large degree on the size and trophic potential of the individual habitat. PAEPKE (1971, 1996) calculated that the population density in small streams in the state of Brandenburg, Germany, varied between 1 and 12 individuals per 10 m^2. KROKHIN (1970) estimated that the population in the 136 ha Lake Dal'neye, Kazakistan, consisted of 2.5 million individuals, which formed sub-populations of 100,000 to 200,000 individuals, each. BOLDUC and FITZGERALD (1989) estimated the maximal density in tidal pools in Quebec to be 16 individuals per m^2. DANIEL (1965) calculated the total number of sticklebacks that immigrated each year into the Norderbootfahrt, Germany, an inlet with a length of 15 km and an area of 75,000 m^2, to be 10^6. The number of nests they constructed was estimated at 10,000, and the number of hatchlings at about 13.5 million. It is even more difficult to estimate the mass of the sticklebacks along the seacoasts. According to HEILBORN (1931) and "STT" (1933, 1934), about 2000 tons, corresponding to about 2,000,000,000 individuals, were caught each year at the beginning of the 1930s in the Pillauer Deep between the Frischem Haff at Kaliningrad Saliv and the Baltic Sea. ANWAND (1964) noted considerable fluctuations in the freshwater populations. RANTA and LINDSTROM (1990) showed how increasing predator pressure forced the individuals to form larger schools of individuals of about the same size and how increasing population density caused their tendency to congregate in discrete size groups to increase. PEUHKURI (1997, 1998a, 1998b) studied the problems of size-assortative shoaling and food-patch finding by stickleback shoals.

Reproductive Biology

Maturity: Depending upon when the three-spined sticklebacks reach a standard length of 35 mm (MÜNZING 1959) or 40 mm (CRAIG-BENNETT 1931, BAGGERMAN 1957), they spawn at the end of either their first or their second year of life. PAEPKE (1971) found that ripeness within any spawning population is reached on an individual basis, and its onset can occur even before the minimum size reported above is reached. He observed two females from freshwater populations in Brandenburg that already contained eggs in advanced stages of ripeness while still only 30 mm and 31 mm in *Sl**. They were found on April 26 and April 17, 1968, respectively. According to SNYDER and DINGLE (1990), sticklebacks in populations farther upstream in rivers become sexually mature before those in estuaries.

Gonads: The weight of the gonads accounts for 0.57 % of the body weight of the males and 25.6 % of the female weight (TITSCHACK 1922), but in extreme cases, it can reach 56.4 % of a female's body weight (DANIEL 1965). WOOTTON (1976) reported the weight of the testes to be 2.0 % of the body weight. The number of eggs produced and number of spawning acts during a given season depend upon the size of the female. Young females with a live weight of 0.4 g release 20 to 40 eggs during each spawning act, while older individuals weighing 3.0 g can release from 300 to 400 eggs each time (WOOTTON 1973, 1976). SNYDER (1990) reported the average initial clutch size in California to be 56.6 eggs for inland populations, 57.0 eggs for populations living upstream in rivers, and 70.2 eggs for populations in estuaries. DANIEL (1965) counted about 1475 eggs in one female weighing 8.18 g. He reported that $F = 1475 + 68\ Sl^*$ and $F = 1475 + 8.18\ W$, where F = the absolute number of eggs, Sl^* = standard length, and W = weight of the female in g. Other formulas for the relationship between the length of the female and the number of eggs were proposed by HAGEN (1967) and WOOTTON (1973, 1976). Personal, previously unpublished measurements revealed that the ovaries of a female 59.5 mm in Sl* were 24.0 mm long and 8.2 mm in diameter.

Spawn: The diameter of an egg cell is about 1.5 mm, the range being 1.1 to 1.8 mm. Its fresh weight is between 0.0024 and 0.0025 g. For the production of egg cells containing 30 calories, about 100 calories of food must be consumed (WOOTTON 1976). According to other calculations, a clutch of eggs contains about 200 % of the daily energy requirements of a ripe female (WOOTTON 1976, WOOTTON *et al.*

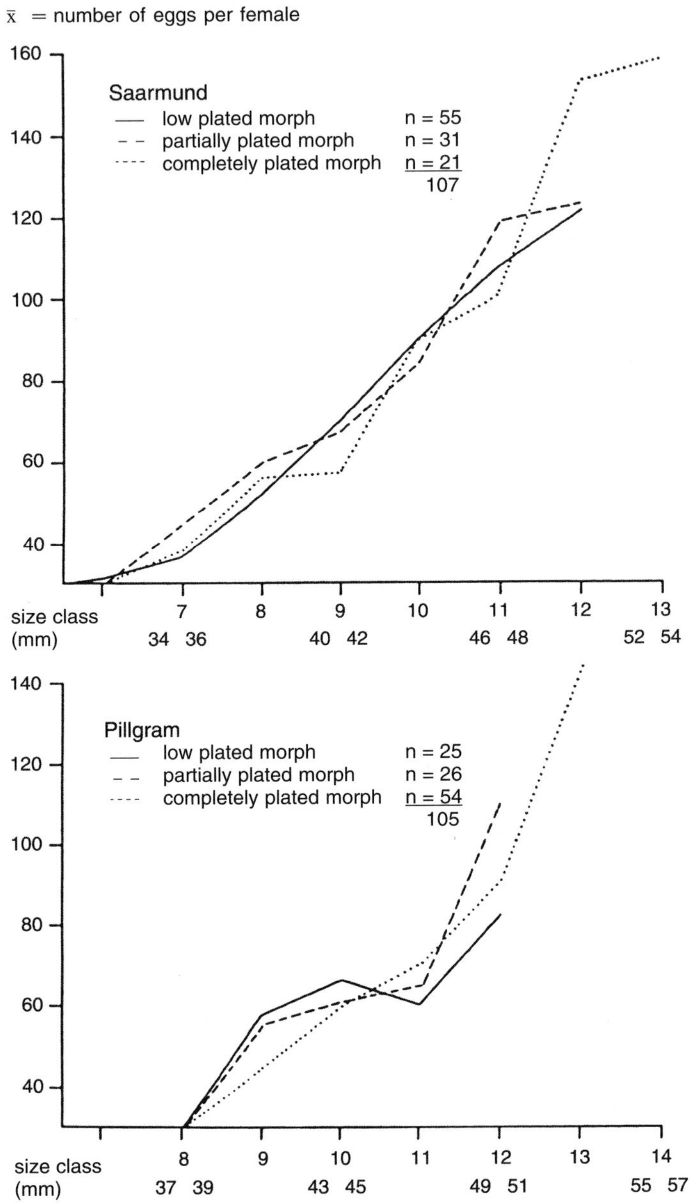

Fig. 50: Relationships between the number of eggs and standard length of females from two polymorphic populations of *Gasterosteus aculeatus* in Saarmund and Pillgram near Berlin, Germany (PAEPKE 1982a).

1995, ALI and WOOTTON 1999). According to SNYDER (1990), the eggs from estuarine populations in California are smaller than those of typical freshwater populations. The spawn is usually opaque but colorless, although it can also be yellowish to bright orange (EHRENBAUM 1904, VRAT 1949, WOOTTON 1973 b).

Spawning period: As in the case of the spawning migration, the beginning and duration of the spawning period depends on the weather conditions during the winter and spring months, especially the change in the photoperiod from short to long days and the temperature amplitude (BAGGERMAN 1969). When the days are short, the sticklebacks do not become ripe for spawning (BORNESTAF *et al.* 1997). In Central Europe, the main spawning period occurs during the months of April and May (MÜNZING 1959, DANIEL 1965, PAEPKE 1996). In Brandenburg, most spawning sites are first occupied during the second half of April (PAEPKE 1996). In June, the breeding activity is renewed. On June 19, 1996, the research group of WINKLER, SCHRÖDER, TSCHIESCHE, and PAEPKE personally collected about 700 *Gasterosteus aculeatus* using a push net at Bodden in Barhöft, northwest of Stralsund. As many as about 80 % of these fishes were two-year-old females, some of which were still releasing eggs. Another 7 % were spent males, while 13 % were one-year-olds, several of which were sexually mature females. The great proportion of the males in the two-year-old spawning generation had already died by this time. The young fishes captured among the reeds had already reached an *Sl** between 13 and 17 mm. However, DANIEL (1965) could still find stickleback eggs in the stomachs of eels during July and August. On August 6, 1981, the author personally observed a male several years old building a nest in an inland water body and two females ready to spawn, which were about the same age (PAEPKE 1983). MÜNZING (1959) reported similar observations among freshwater populations, leading him to believe that their spawning season generally lasts longer than that of anadromous populations. While the spawned out members of the anadromous populations had already left the spawning grounds by the beginning of June, the spawning of the freshwater populations generally continued at least until the end of July. In the Bay of Finland, the three spined sticklebacks generally spawn in June and July, which is about two months later than normal in Central Europe (SCHNEIDER and LEVANDER 1900). GISLASON et al. (1998) reported a similar situation in Lake Mývatn on Iceland.

Spawning sites: Three-spined sticklebacks prefer to spawn in fresh and brackish water, although they can also reproduce in the sea (EHRENBAUM 1904, BELL 1979, 1984). The nests are often encountered in shallow water near the shore, small inlets along the shore, streams flowing into the sea, and even artificial ditches. In the Little Campbell River in Canada, HAGEN (1967) observed that males of the completely plated morph prefer sandy bottoms, while those of the low plate morph would rather spawn in muddy regions. Where only stony bottoms are available, as in gravel quarries, the males placed their nests in the narrow spaces between the stones. Lentic water regions are preferred, but in flowing water, the nests are built in the sheltered water downstream from beds of plants, as personally observed. The importance of water plants and other objects as "landmarks" for establishing territorial borders between competing neighbors and to help the fishes find their way home has been reported several times (WUNDER 1928, 1930, TER PELKWIJK and TINBERGEN 1937, VAN DEN ASSEM 1967). Numerous experiments and field observations have shown that spawning sites with more objects present are more densely settled than barer habitats (WUNDER 1930, THER PELKWIJK and TINBERGEN 1937, TINBERGEN 1951, 1957, VAN IERSEL 1958, JENNI 1972, VAN DEN Assem 1967). However, even sandy bottoms without many objects can be settled by *Gasterosteus.* Because of its effective defenses against predators, it is the only gasterosteid that is able to spawn at sites such as the shallow, fine sandy landing zones along the Baltic Sea Coast of Mecklenburg, as personally observed by the author. At such spawning places, isolated landmarks are especially attractive for the males. One male even constructed its nest in a shoe on the bottom of Parsteiner Lake (PAEPKE 1983). Two others in the River Wear in Durham City built nests in a tin can and an empty milk bottle (WOOTTON 1972).

Mating habits: There is probably no other fish with a reproductive behavior studied as intensively as that of the three-spined stickleback. It is possible only to refer the reader to some important authors on the subject: WUNDER (1928, 1930), LEINER (1929, 1930, 1931), TER PELKWIJK and TINBERGEN (1937), TINBERGEN (1951, 1957), VAN IERSEL (1953, 1958); BAGGERMANN (1966), VAN DEN ASSEM (1967), WOOTTON (1972). An especially thorough summary of all details was provided by WOOTTON (1976).

After arrival at the spawning site, the males choose and begin to defend a territory (VAN DEN ASSEM 1967), and the success of reproduction very strongly depends upon the size of this territory (GOLDSCHMIDT and BAKKER 1990). After the territory is established, a shallow, round spawning depression is dug in the sand. This can have a diameter of as much as 10 cm. Parts of plants are carried into the depression for construction of the nest. These are often carried considerable distances or occasionally stolen from the nests of neighbors. The material is pasted together with a secretion from the adrenal gland that hardens in water. This secreted protein contains a special glycoprotein called spiggin, the production of which is induced by the androgen 11-ketotestosterone (JACOBSSON *et al.* 1999). By entering the nest and pushing, a cavity in the nest is created, first with only one entrance and exit. Frequently, a prominent stem extends from this location to mark the entrance. As the structure is improved, its strength is tested by rapid movements of the pectoral fins, and, according to the security demands of the builder, it is more or less completely decorated with sand as camouflage. The establishment of a territory and construction of the first nest is completed within a few days (WOOTTON 1976), during which a great deal of the time is spent in conflict with the neighbors. These battles for territory involve ritualized actions to a considerable degree (SEVENSTER 1961). The size of the territory decreases with an increasing density of males at the spawning site. The minimum size under natural conditions is apparently from 0.20 to 0.25 m^2, as reported by BERTIN (1925) and confirmed experimentally by ASSEM (1967). In the Norderbootfahrt, DANIEL (1965) reported that the territory encompasses about 4 m^2, while PAEPKE (1996) reported territory sizes averaging from 2.0 to 2.5 m^2 in ditches located in Brandenburg. However, the size decreased to 1.4 m^2 at ideal locations with plenty of sunlight and dense plant growth.

A special form of cannibalism occurs when neighbors steal not only nest material but also eggs and larvae (WOOTTON 1976, 1984, PAEPKE 1982b, FORSTER 1995).

Spawning habits: Only after the construction of the nest is nearly completed does the territorial male show the full splendor of his nuptial colouration. The underside of the head and body are more or less bright orange red, while the dorsum and the eyes take on a glowing turquoise hue. The usual pattern of black flecks largely disappears. The intensity of the nuptial colour pattern varies in intensity within a population just as does the preference of females for especially strongly red coloured males or those with particularly varied colour patterns. Both of these have been shown to be genetically correlated (BAKKER 1993, BAUBE et al. 1995). The males cease to react only aggressively to females nearly ready for spawning. Instead, they either swim directly toward her or perform the famous zig-zag dance, as defined by LEINER (1929). This dance has been interpreted as an expression of the conflict between aggression and sexuality. The presence of predators and competitors can inhibit the intensity of the dance (CANDOLIN 1997). Females identify themselves to the males by their simple colouration consisting of a dark pattern on a bright silvery background, the position of their heads against their swollen bellies, and their slow swimming in the direction of the male. The phylogenetic diversification of the male and female colour signals have been subjected to different selection pressures throughout their evolutionary histories (MCLENNAN 1996). In the ideal situation, the male swims back toward the nest, and the female follows. When they arrive at the nest, the male turns on his side to place his dorsal spines in the direction of the female and indicates the location of the entrance to the nest with his snout. The female swims past her partner and enters the nest and spawns, while the male continually prods her on the dorsum and caudal peduncle. The female then leaves the nest by boring a hole in the rear wall and is usually driven away by the male immediately. Only after the female is gone does the male enter the nest to fertilize the eggs. The entire procedure can be completed in 0.5 to 1.5 minutes, although complications frequently arise during the sequence of events that can cause delays or even prevent completion of the spawning. Such complications come about due to an insufficient stage of ripeness of the female, too great a difference between the sizes of the partners, too much aggression by the male, or disturbances caused by another stickleback. The parental care itself involves the protection of the offspring from predators, especially those of the same species, as well as the production of water circulation over the eggs with the fins, repair of the nest after spawning, cutting additional ventilation holes in the nest to facilitate water circulation as the size of the hatchlings increases, and catching the larvae that leave prematurely and returning them to the nest.

As many as seven spawning acts can take place in one nest within a period of one to two days (Moodi 1972). However, as many as 40 clutches of eggs can found in a single nest (Bakker and Sevenster 1995). During one spawning season, a male can build as many as four nests at various sites. Although a maximum of 500 to 800 eggs per nest is possible, and Winnicki *et al.* (1998) even found 1810 eggs in one nest, Wootton (1976) assumed an average of 400 eggs per nest and calculated that as many as 1600 offspring can be raised by one male within 60 days. However, the number of zygotes is usually decimated by fungi and predators. During a two to three month spawning season, assuming a schedule of 16 hours daylight and 8 hours darkness and a water temperature of 20 °C, one female can spawn as many as 15 times (Baggerman 1957). The intervals between spawning depend upon the food supply and may vary from 60 hours to three or four days or may be as much as 15 days.

Early ontogeny: The embryonic development depends on the temperature. It requires 40 days at 8 °C, 20 days at 12 °C, and only 6 days at 25 °C (Heuts 1956). Recalculated in degree-days, the embryonic development prior to hatching requires 108 to 120 at higher temperatures and 216 to 262 at lower temperatures (Winnicki *et al.* 1998). At 18 to 19 °C, the cytoplasm begins to move toward the animal pole 30 minutes after fertilization. Within the next 25 minutes, the blastodermal disc or germinal disc has developed. According to Winnicki *et al.* (1998), these are located laterally in the equatorial zone. Oil drops, which are responsible for shifting the germinal disc of other species to the upper part of the zygote, do not have this effect in the sticklebacks. Invagination of the blastoderm occurs 22 hours after fertilization. The movement of the musculature in the tail begins 106 hours after fertilization, and after 160 hours, the development of the heart has been completed. Eight days after fertilization, the yolk sac larva hatches with a length of 3 to 5 mm. After an additional four days, the yolk in the sac has been consumed, the mouth can be opened and closed, and the swim bladder is already present but must be filled with air at the surface of the water (Vrat 1949, Swarup 1958a, b). At a body length of 6 to 8 mm, the first fin rays appear, and when 9 to 10 mm is reached, the dorsal and ventral spines become visible (Ehrenbaum 1904). The dorsal spines are still not joined to the vestiges of the embryonic fin seam, and they remain only along the ventral margin. By the time an *Sl** of 15.0 mm is reached, the vestiges have also disappeared. In the meantime, the young fishes have long since left the nest and largely resemble their parents. At an *Sl** of 15.5 mm, the first scutes on the body can be recognized, and by the time 16.0 mm in *Sl** is reached, traces of a raised caudal keel become evident, as first reported here by the author. However, the typical number of scutes on the completely plated morph is only attained at an *Sl** of 25 mm (MÜNZING 1959) or 30 mm (HEUTS 1947a, HAGEN and MCPHAIL 1970, IGARASHI 1970a, b, WOTTON 1976).

Important Parasites and Diseases

The fauna and flora parasitizing *Gasterosteus aculeatus* is very diverse. In the former Soviet Union, 48 parasites have been identified (BYKOHVSKY 1962). WOOTTON (1976) even counted 94: 2 Fungi, 3 Flagellata, 10 Sporozoa, 11 Ciliata, 11 monogenetic and 21 digenetic Trematoda, 11 Cestoda, 6 Nematoda, 7 Acanthocephala, 1 Annelida, 10 Crustacea, and the larvae of the mollusc, *Unio* sp. About 30 species are ectoparasites, and about 50 are endoparasites. The rest encompass unspecialized forms. At the end of the spawning period, many three-spined sticklebacks are infested with *Saprolegnia* sp. because their energy reserves and therefore also their natural resistance has been depleted due to the exertions of the spawning migration and reproduction. More serious *Saprolegnia* infestations usually lead to the death of the weakened fishes. *Glugea anomala* produces bright spherical growths 3 to 4 mm in diameter in the connective tissue. Greater infestations can greatly limit the ability of the sticklebacks to move around. In 1981, the entire *Gasterosteus* population in the isolated village pond in Osmünde was infested by this sporozoan (PAEPKE 1996). In a Scottish lake, *Ichthyophthirius multifiliis* had weakened numerous adults and almost fully destroyed their offspring (HOPKINS 1959). In lotic waters, attacks by *Ichthyophthirius* are less harmful because most of their swarmers drift away. An extraordinarily heavy infestation of sticklebacks by the fish louse, *Argulus* sp., was found by WINKLER, SCHRÖDER, TSCHIESCHE, and PAEPKE on June 19, 1996, at Bodden bei Barhöft, northwest of Stralsund, but not previously reported. As many as four to five parasites per fish were present. The best known parasite of the stickleback is undoubtedly the cestode, *Schistocephalus solidus*. There are an exceptionally large number of publications on its developmental cycle, encompassing an egg called the coracidium, which floats freely in the water; a procercoid living in a copepod; a pleoceroid inhabiting the body cavity of a stickleback; and its adult stage, which inhabits the intestinal tract of a fish-eating bird, as well as on its effects on local stickleback populations. The pleocercoid, which has white or ivory coloured rings, can reach a length of 4 cm. As many as 106 small pleoceroids have been found in a single fish. These parasites exert pressure in the body cavity, forcing organs out of their normal positions. The maturation of eggs, spermatogenesis, formation of the kidney secretions, and nutrition can be mildly to seriously impaired. The hosts are hindered in their activity and suffer from respiratory difficiencies, so they are more likely to fall victim to gulls, terns, loons, and mergansers than other sticklebacks. The infestation of the sticklebacks is dependent upon the presence of these birds as well as of the copepod species *Diacyclops bicusidatus*, *Eucyclops serrulatus*, und *Megacyclops viridis* (ARME and OWEN 1967, HOPKINS and SMYTH 1951, BARBER and HUNTINGFORD 1995, BARBER and G. D. RUXTON 1998).

Economic Importance

Economic exploitation of the three-spined stickleback is known to occur only temporarily and within geographically limited regions. In 1890, the Seefischerei-Gesellschaft Germania built a factory for producing fish meal and fish oil at Alt Pilau, which is now called Baltisk on the Kaliningrad Saliv. The factory exploited the enormous stickleback schools that migrated annually between August and December from the Frische Haff through the Pillauer Deep into the Baltic Sea. Each day, catches weighing from 1000 to 2000 hundredweight were reported, and the annual catch amounted to between 40,000 and 50,000 hundredweight ("STT."1933,1934, HEILBORN 1931). Even today, considerable catches are possible in the eastern Baltic Sea. For example, 394.7 tons were caught in Estonia during 1974, predominantly consisting of *Gasterosteus aculeatus* together with considerably fewer *Pungitius pungitius* (MIKELSAAR 1984). During times of emergency, the inhabitants of regions along the Tiber used three-spined sticklebacks as food (GESSNER 1583), just as the inhabitants of Danzig did while their city was besieged by Napoleon's troops during the year 1807 (HEILBORN 1931). During the war year 1943, the species was used in The Netherlands for the manufacture of fish meal used as animal food (BAGGERMAN 1957). According to WOOTTON (1976), during the 18th century, an English worker caught 96 sheffels, which equals 350 buckets, of sticklebacks every day for fertilizer and to produce fish oil. DANIEL (1965) reported that while he was still going to school, the pupils could earn pockey money by catching sticklebacks. From oil produced from the sticklebacks, a salve for treating burns was obtained (NIKOLSKI 1957).

On the negative side, *Gasterosteus aculeatus* has been denounced as a predator on the eggs and larvae of other fish species (REGAN 1911), leading KÜHLMANN (1963) and ANWAND (1964) to contemplate suitable methods for combating it. In England, the sticklebacks have been found to prey upon great crested newts (BAKER 1999). Rotenone has been used to eliminate the sticklebacks from a newt pond (MCLEE *et al.* 1992/93). DANIEL (1965) showed that they also feed on glass eels. The three-spined stickleback also acts as a competitor for the food of other fish species, especially in the Black Sea and Sea of Azov (NIKOLSKI 1957). Sticklebacks and young salmon use nearly identical food resources, but the damage caused by this competiton is difficult to estimate since the sticklebacks themselves serve as prey for older salmonids (MAITLAND 1965, ROGERS 1968, KROKHIN 1970, WOOTTON 1976).

At least their spawn is also consumed by eels (DANIEL 1965, SAAT in prep.). In Europe, it is mainly the pike, perch, and trout that regularly use sticklebacks as a food source (WOOTTON 1976). SAAT (in prep.) mentioned other fish species that consume sticklebacks, including cod, Baltic herring, small sand eels, turbot, flounder, bulltrout, and sea scorpions. When threatened by predators, sticklebacks produce stickleback alarm pheromones, which had formerly been identified only from ostariophysan fishes (BROWN and GODIN 1997). The sticklebacks also play an important role as nourishment for birds, at least for limited periods during the year. For example, they are consumed by herons and black-headed gulls (CREUTZ 1965, DANIEL 1965), red-breasted mergansers and goosanders, Arctic terns and common terns, coots, and dipper and little grebes (BENGSTON 1971, PENCZAK 1968, LEMMETYINEN 1973, WASSMANN 1987, DIERSCHKE and KLUMANN 1988, PAEPKE 1996). Even spoonbills were reported to consume sticklebacks as food (KEMPER 1987). They have therefore achieved a certain significance for maintaining the diversity of the avifauna. This is understandable in view of the fact that both stickleback species inhabiting fresh water are often the only fish species that can be captured by herons in small ditches. In 1989, a special fish ladder was constructed for anadromous stickleback species in order to improve the food reserves of the spoonbills that were nesting in the polder areas (KEMPER 1995). It should not be forgotten that an ideal profit can be made by using the three-spined sticklebacks as laboratory animals and experimental animals in the field for ecological, ethological, physiological, population genetic, and evolotionary biological studies (PAEPKE 1996, BELL 1984).

Literature

AHNELT, H., 1986: Zum Vorkommen des Dreistachligen Stichlings *Gasterosteus aculeatus* (Pisces, Gasterosteidae) im österreichischen Donauraum. – Ann. Naturhist. Mus. Wien 88/89 B: 309–314.

AHNELT, H., and E. AMANN, 1994: *Gasterosteus aculeatus* (Pisces, Gasterosteidae) in Österreich – eine Lanze brechen für den Dreistachligen Stichling? – Österreichs Fischerei **47**(1): 19–23.

AHNELT, H., H.-J. PAEPKE, and E. AMANN, 1994: "Vierstachlige" Stichlinge aus dem Rheintal in Vorarlberg (*Gasterosteus aculeatus*: Pisces, Gasterosteidae). – Österreichs Fischerei **47**(4/5): 125–132.

ALI, M., and R. J. WOOTTON, 1998: Do random fluctuations in the intervals between feeding affect the growth rate in juvenile three spined ticklebacks? – J. Fish. Biol. **53**(5): 1006–1014.

ALLEN, J. R. M., and R. J. WOOTTON, 1983: Rate of food consumption in a population of threespine stickleback, *Gasterosteus aculeatus*, estimated from the faecal production. – Environ. Biol. Fish. **8**(2): 157–162.

ANDRIYASHEV, A. P., 1954: (Fishes of the northern seas in the U.S.S.R.). – Izv. Akad. Nauk. SSSR Moskow-Leningrad **1**: 566. (in Russian).

Aneer, G., 1973: Biometric characteristics of the three-spined stickleback (*Gasterosteus aculeatus* L.) from the northern Baltic proper. – Zool. Scripta **2:** 157–162.

Anwand, K., 1964: Zum Massenauftreten des Stichlings. – Deutsch. Fisch. Z. **11**(4): 111–112.

Arme C., and R.W. Owen, 1967: Infections of the three-spined stickleback (*Gasterosteus aculeatus* L.) with the pleocercoid larvae of *Schistocephalus solidus* (Müller, 1776) with special reference to pathological effects. – Parasitology **57:** 301–314.

Arnold, A., 1984: Besonderheiten eines Stichlingsvorkommens: Goldstichlinge und vierstachlige Stichlinge. – Aquarien Terrarien **31**(11): 381.

Avise, J. C., 1976: Genetics of plate morphology in an unusual population of threespine sticklebacks (*Gasterosteus aculeatus*). – Gent. Res. Camb. **27:** 33–46.

Bacescu, M., and R. Mayer, 1956: Certari asupra Gludrinilor (*Gasterosteus aculeatus*) din apele Rominesti. – Bul. Inst. Ceret. Piscicole **15**(2): 19–36.

Baggermann, B., 1957: An experimental study on the timing of breeding and migration in the three-spined stickleback (*Gasterosteus aculeatus*). –Arch. Neérl. Zool.**12**(2): 105–317.

Baggermann, B., 1959: The role of external factors and hormones in migration of sticklebacks and juvenile salmon. In: Aubrey Grobman (Ed.), Comparative Endocrinology. J. Wiley and Sons, New York **16:** 24–37.

Baggermann, B., 1966: On the endocrinology control of reproductive behaviour in the male three-spined stickleback *Gasterosteus aculeatus* L. – Sympos. Soc. Exp. Biol. **20:** 427–456.

Baggermann, B., 1969: Influence of photoperiod and temperature on the timing of the breeding season in the stickleback, *Gasterosteus aculeatus*. – Gen. Comp. Endocrinol. **13:** 491.

Baker, J. M. R., 1999: Abundance and survival rates of great crested newts (*Triturus cristatus*) at a pond in Central England: Monitoring individuals. – Herpetol. J. **9:** 1–8.

Bakker, T. C. M., 1993: Positive genetic correlation between female preference and preferred male ornament in sticklebacks. – Letters to Nature **363:** 255–257.

Bakker, T. C. M., and E. Feuth-De-Bruijn, 1988: Juvenile territoriality in stickleback, *Gasterosteus aculeatus* L. – Anim. Behav. **36**(5): 1556–1558.

Bakker, T. C. M., E. Feuth-De-Bruijn, and P. Sevenster, 1988: Albinism in threespine stickleback, *Gasterosteus aculeatus*. – Copeia 1988(1): 236-238.

Bakker, T. C. M., and P. Sevenster, 1988: Plate morph of *Gasterosteus aculeatus* Linnaeus (Pisces: Gasterosteidae): comments on terminology. – Copeia 1988(3): 659-663.

Bakker, T. C. M., and B. Mundwiler, 1994: Female mate choice and male red coloration in a natural three-spined stickleback (*Gasterosteus aculeatus*) population. – Behav. Ecol. **5**(1): 74–80.

Bakker, T. C. M., and P. Sevenster, 1995: Preface (of papers read at the Second Intern. Symposium on Stickleback Behaviour). – Behaviour **132:** 907–913.

Bakker, T. C. M., and W. J. Rowland, 1995: Male mating preference in sticklebacks: effects of repeated testing and own attractiveness. – Evolution **132:** 935–949.

Balon, E. K., 1967: Three-spined stickleback – *Gasterosteus aculeatus* Linnaeus, 1758 in the Danube near Bratislava. – Sb. Slov. národ. Muz. (Prir. Ved.) **13:** 127–134.

Bănărescu, P., 1966: Intraspecific variation, subvariation and speciation in Romanian freshwater fishes. – Z. zool. Syst. u. Evolut. Forsch. **4**(3/4): 378–396.

Banbura J., 1994: Lateral plate morph differentiation of freshwater and marine populations of the three-spined stickleback, *Gasterosteus aculeatus* in Poland. – J. Fish Biol. **44:** 773–784.

Banbura, J., and T. C. M. Bakker, 1995: Lateral plate morph genetics revisited: Evidence for a fourth morph in three-spined sticklebacks. – Behaviour **132:** 1153–1171.

BARBER, I., and F. A. HUNTINGFORD, 1995: The effect of *Schistocephalus solidus* (Cestoda: Pseudophyllidae) on the foraging and shoaling behaviour of three-spined sticklebacks, *Gasterosteus aculeatus*. – Behaviour **132:** 1223–1240.

BARBER, I., and G. D. RUXTON, 1998: Temporal prey distribution affects the competitive ability of parasitized sticklebacks. – Animal Behaviour **56:** 1477–1483.

BAUBE, C. L., W. J. ROWLAND, and J. B. FOWLER, 1995: The mechanisms of colour-based mate choice in female threespine sticklebacks: hue, contrast and configurational cues. –Behaviour **132:** 979–996.

BEHNKE, H., H. M. WINKLER, and L. DEBUS, 1986: Untersuchungen zum Bestand und zur Nahrungsökologie der Stichlinge (*Gasterosteus aculeatus* und *Pungitius pungitius*) im Freiwasser des Barther Boddens. – Wiss. Z. W.-Pieck-Univ. Rostock **35** Naturwiss. 5: 54–60.

BELL, M. A.1976: Evolution of phenotypic diversity in *Gasterosteus aculeatus* superspecies of the Pacific Coast of North America. – Syst. Zool. **25:** 211–227.

BELL, M. A., 1979: Low plate morph of the threespine stickleback breeding in salt water. – Copeia 1979: 529–533.

BELL, M. A., 1981: Lateral plate polymorphism and ontogeny of the complete plated morph of threespine sticklebacks (*Gasterosteus aculeatus*). – Evolution **35:** 67–74.

BELL, A., 1982: Melanism in a high elevation population of *Gasterosteus aculeatus*. – Copeia 1982: 829–835.

BELL, M. A., 1984: Evolutionary phenetics and genetics. The threespine stickleback, *Gasterosteus aculeatus* and related species. In: TURNER, B .J. (Ed.), Evolutionary Genetics of Fishes. New York and London. pp. 431–528.

BELL, M. A., 1995: Intraspecific systematic of *Gasterosteus aculeatus* populations: implications for behavioral ecology. – Behaviour **132:** 1131–1152.

BELL, M. A., and S. A. FORSTER, 1994: The evolutionary biology of the threespine stickleback. Oxford Univ. Press., Oxford, 571 pp.

BENGSTON, S., 1971: Food and feeding of diving ducks breeding at Lake Myvatn, Island. – Ornis Fenn. **48:** 77–92.

BERG, L. S., 1932: Übersicht der Verbreitung der Süßwasserfische Europas. – Zoogeographica **1:** 107–208.

BERG, L. S., 1949: (Freshwater fishes of the U.S.S.R. and adjacent countries.). Vol. 3. Izd. Akad. Nauk SSSR Moskow-Leningrad. pp. 971--976. (in Russian)

BERINKEY, L., 1960: The stickleback (*Gasterosteus aculeatus* L.) a new fish species from Hungary. – Vertebr. Hungarica **2**(1): 1–10.

BERTIN, L., 1925: Recherches bionomiques, biométriques et systematiques sur les épinoches. – Ann. Inst. Océanogr. Monaco **2:** 1–204.

BIANCO, P. G., 1979: Sulla presenzia di *Gasterosteus aculeatus* nella penisola Salentina. –Natura – Soc. ital. sci. nat. Museo civ. stor. nat. e Acquario civ. Milano **70**(1/2): 76–80.

BIANCO, P. G., 1981: Areale Italico, kinvenimento in Calabria e origini della popolazioni mediterranee di *Gasterosteus aculeatus* L. – Boll. Mus. civ. st. nat. Verona **7:** 197–216.

BLANCHARD, E., 1866: Les poissons des eaux douces de la France. Paris. pp. 220–234.

BLEGVAD, H., 1917: On the food of fish in the Danish waters within the Skaw. – Rep. Danish Biol. Sta. **24:** 19–72.

BOCK, F., 1928: Die Hypophyse des Stichlings (*Gasterosteus aculeatus* L.) unter besonderer Berücksichtigung der jahreszyklischen Veränderungen. – Z. wiss. Zool. **131:** 645–710.

Bolduc, F., and G. FitzGerald, 1989: The role of selected environmental factors and sex ratio upon egg production in threespine sticklebacks, *Gasterosteus aculeatus*. – Can. J. Zool. **67:** 2013–2020.

Bornestaf, C., E. Antonopoulou, I. Mayer, and B. Borg, 1997: Effects of aromatise inhibitors on reproduction in male three-spined sticklebacks, *Gasterosteus aculeatus*, exposed to long and short photoperiods. – Fish Physiol. and Biochemistry **16**(5): 419–423.

Bößneck, U., 1999: Freiland-Nachweis eines teilalbinotischen Dreistachligen Stichlings (*Gasterosteus aculeatus* Linnaeus, 1758) bei Erfurt (Thüringen). – Veröff. Naturkundemus. Erfurt 1999: 109–110.

Brown, G. E., and J.-G. J. Godin, 1997: Anti-predator responses to conspecific and heterospecific skin extracts by threespine sticklebacks: Alarm pheromones revisited. – Behaviour **134:** 1123–1134.

Bykohvsky, B. E. (ed.), 1962: Key to the parasites of the USSR (in Russian). Isd. Akad.Nauk SSR, Moscow, Leningrad [in cooperation with B. Bykovskaja, A. V. Shekin, S. S. Shulman, and V. M. Epishtejn], 866pp.

Campbell, R. N., 1979: Sticklebacks [*Gasterosteus aculeatus* (L.) and *Pungitius pungitius* (L.)] in the Outher Hebrides, Scotland. – Hebredian Nat. **3:** 8–15.

Candolin, U., 1997: Predation risk affects courtship and attractiveness of competing threespine stickleback males. – Behav. Ecol. Socialbiol. **41:** 81–87.

Chen, T. R., and H. M. Reisman, 1970: A comparative chromosome study of the North American species of sticklebacks (Teleostei: Gasterosteidae). – Cytogenetics **9:** 321–332.

Coad, B. W., 1981: A bibliography of the sticklebacks (Gasterosteidae: Osteichthyes). Syllogeus series 35, Nat. Mus. Can. 142 pp.

Coad, B. W., and G. Power, 1973: Observation on the ecology and phenotypic variation of the threespine stickleback, *Gasterosteus aculeatus* L., 1758, and the blackspotted stickleback, *G. wheatlandi* Putnam, 1867 (Osteichthyes: Gasterosteidae) in Amory Cove, Quebec. – Can. Field. Nat. **87:** 113–122.

Coad, B. W., and G. Power, 1974: Meristic variation in the threespine stickleback, *Gasterosteus aculeatus*, in the Matamek River System, Quebec. – J. Fish Res. Board Can. **31:** 1155–1157.

Cracraft J., 1989: Speciation and its ontology: the empirical consequences of alternative species concepts for understanding patterns and processes of differentiation. In: O. Otte and J. A. Endler (Eds.), Speciation and its Consequences. Sinauer Associates, Sunderland, MA. pp. 28–59.

Craig-Bennet, A., 1931: The reproductive cycle of the three-spined stickleback, *Gasterosteus aculeatus*. – Linn. Philos. Trans. London ser. **B 219:** 197–279.

Crespon, J., 1844: Faune meridionale du description de tous les animaux vertébrés virants et fossiles, sauvage ou domestiques qui se rencontrent toute l'année ou quine sont que des passage dans la plus grande partie de la France. Montpellier **2:** 283.

Creutz, G., 1965: Stichlinge als Vogelnahrung. – Orn. Beob. Bern **62:** 24–26.

Cuvier, G., 1829: La règne animal distribué d'après son organisation, pour servir de base à l'histoire naturelle des animaux et d'introduction a l'anatomie comparée. Nov. éd. Paris **2:** 122–406.

Cuvier, G., and A. Valenciennes, 1829: Histoire naturelle des poissons. Paris-Strasbourg, Vol. **4:** 479–498.

Daniel, W., 1965: Beiträge zur Biologie des Dreistachligen Stichlings (*Gasterosteus aculeatus* L.) Beobachtungen in einem Marschsielzug. – Faun. Mitt. Norddeutschl. **2**(10–12): 289–307.

DANIEL, W., 1985: Fragen zum Wanderverhalten des Dreistachligen Stichlings (*Gasterosteus aculeatus* L.). – Faun.-Ökol. Mitt. **5:** 419–429.

DIERSCHKE, V., and H. KLUMANN, 1988: Zur Nahrung nestjunger Küstenseeschwalben (*Sterna paradisaea*) im NSG Oehe-Schleimündung. – Seevogel **9**(2): 26–27.

DUNCKER, G., 1926: Die Verbreitung der Süßwasserfische der Nordmark. -Arch. Hydrobiol. **17**(4): 736–750.

DUNCKER, G., and W. LADIGES, 1960: Die Fische der Nordmark. – Abh. Verh. Naturwiss. Ver. Hamburg N.F. 3 (Suppl.): 178–181.

ECONOMIDIS, P. S., 1991: Checklist of freshwater fishes of Greece. Hellenic Soc. Protect. Nat. Athen 30–31.

EHRENBAUM, E., 1904: Eier und Larven von Fischen der Deutschen Bucht. III. Fische mit festsitzenden Eiern. – Wiss. Meeresunters. Helgoland **6**(2): 127–200.

FATIO, V., 1882: Faune des vertebres de la Suisse. Histoire Naturelle des Poissons. Geneva-Bale. Vol. **4:** 1–806.

FELDMANN, R., 1980: Zur Verbreitung und Ökologie des Dreistachligen Stichlings und des Zwergstichlings in Westfalen. – Nat. u. Heimat **40:** 99–109.

FORSTER, S. A., 1995:Understanding the evolution of behaviour in threespine stickleback: the value of geographical variation. – Behaviour **132:** 1107–1129.

FORSTER, S. A., J. A. BAKKER, and M. A. BELL, 1992: Phenotypic integration of life history and morphology: an example from the three-spined stickleback, *Gasterosteus aculeatus* L. – J. Fish Biol. **41** (Suppl. B) Dec. 1992: 21–35.

GASCHOTT, O., 1929: Die Stichlinge (Gasterosteidae). In: R. DEMOLL and H. N. MAIER (ed.): Handbuch der Binnenfischerei Mitteleuropas. Stuttgart **3**: 131–141.

GESSNER, C., 1583: Thierbuch. Zürich.

GISLASON, G. M., A. GUDMUNDSSON, and A. EINARSSON, 1998: Population densities of the three-spined stickleback (*Gasterosteus aculeatus* L.) in a shallow lake. – Verh. Intern. Verein. Limnol. **26:** 2244–2250.

GLANDT, D., 1983: Experimentelle Untersuchungen zum Beute-Räuber-Verhältnis zwischen Stichlingen, *Gasterosteus aculeatus* L. und *Pungitius pungitius* (L.) (Teleostei) und Grasfroschlarven, *Rana temporaria* L. (Amphibia). – Zool. Anz. **211:** 277–284.

GLANDT, D., 1984: Laborexperiment zum Beute-Räuber-Verhältnis zwischen Dreistachligen Stichlingen, *Gasterosteus aculeatus* L. (Teleostei), und Erdkrötenlarven, *Bufo bufo* (L.) (Amphibia). – Zool. Anz. **213:** 12–16.

GOLDSCHMIDT, T., and T. C. M. BAKKER, 1990: Determination of reproductive success of male sticklebacks in the field and in the laboratory. – Netherl. J. Zool. **40**(4): 664–687.

GREENBANK, J., and P. R. NELSON, 1959: Life history of the threespine stickleback *Gasterosteus aculeatus* Linnaeus in Karluk Lake und Bare Lake, Kodiak Island, Alaska. – Fish Bull. Fish U. S. Wildl. Service **59:** 537–559.

GREGORY, W.K., 1959: Fish skulls. A study of the evolution of natural mechanisms. Laurel, Florida. 481 pp.

GROCHMALICKI, J., 1920: Notes sur la zoogeographia polonaise *Citellus guttatus* Pall. et *Gasterosteus aculeatus* L. Extrait Kosmos Bull. Soc. Polon. Nature. Leopol. (Cited in MÜNZING 1959).

GROSS, H. P., 1977: Adaptive trends of environmentally sensitive traits in the three-spined stickleback, *Gasterosteus aculeatus* L. – Z. zool. Syst. Evol. Forsch. **15:** 252–278.

GROSS, H. P., 1978: Natural selection by predators on the defensive apparatus of the three-spined stickleback, *Gasterosteus aculeatus* L. – Can. J. Zool. **56:** 398–413.

Grote, W., C. Vogt, and B. Hofer (Ed.), 1909: Die Süsswasser-Fische von Mittel-Europa. Commissionsverlag Engelmann, Frankfurt a. M. und Leipzig, 558 pp.

Guderley H., 1994: Physiological ecology and evolution of the threespine stickleback. In: M. A. Bell and S. A. Forster (Eds), The Evolutionary Biology of the Threespine Stickleback. Oxford Univ. Press. Oxford, pp. 85–113.

Gutz, M., 1970: Experimentelle Untersuchungen zur Salzadaption verschiedener Rassen des Dreistachligen Stichlings (*Gasterosteus aculeatus* L.). – Intern. Rev. ges. Hydrobiol. **55**(6): 845–894.

Hagen, D. W., 1967: Isolating mechanisms in three-spine sticklebacks (*Gasterosteus aculeatus*). – J. Fish. Res. Board Can. **24:** 1637–1692.

Hagen, D. W., and J. D. McPhall, 1970: The species problem within *Gasterosteus aculeatus* on the Pacific coast of North America. – J. Fish. Res. Board Can. **27:** 147–155.

Hagen, D. W., and L. G. Gilbertson, 1973: The genetics of plate morphs in freshwater threespine sticklebacks. – Heredity **31:** 75–84.

Hagen, D. W., and G. E. E. Moodi, 1979: Polymorphism for breeding colours in *Gasterosteus aculeatus* I. Their genetics and geographical distribution. – Evolution **33:** 641–648.

Hagen, D. W., and G. E. E. Moodi, 1982: Polymorphism for plate morphs in *Gasterosteus aculeatus* on the East Coast of Canada and an (sic!) hypothesis for their global distribution. – Canad. J. of Zoology **60:** 1032–1042.

Haglund, T. R., D. G. Buth, and R. Lawson, 1992: Allozyme variation and phylogenetic relationship of Asian, North American, and European populations of the threespine stickleback, *Gasterosteus aculeatus*. – Copeia 1992(2): 432–443.

Hartley, P. H. T., 1948: Food and feeding relationships in a community of freshwater fishes. – J. Anim. Ecol. **17:** 1–14.

Heckel, J., and R. Kner, 1858: Die Süßwasserfische der Österreichischen Monarchie mit Rücksicht auf die angränzenden Länder. Commissionsverlag Engelmann, Leipzig. 388 pp.

Heilborn, A., 1931: Der Stichling. Die Brehm-Bücher. Vol. 8. Brehm, Berlin. 32 pp.

Heincke, F., 1889: Untersuchungen über die Stichlinge. – Öfversigt af k. Vetensk.-Akad. Förhandlingar 1889(6): 395–410.

Heutz, M. L., 1947a: The phenotypical variability of *Gasterosteus aculeatus* (L.) populations in Belgium. – Its bearing on the general geographical variability of the species. – Verh. van de Koninklijke Vlaamse Acad. voor Wetensch., Lett. an schone Kunsten van Belgie. Kl. Wetensch. **9**(25): 1–63.

Heutz, M. L., 1947b: Experimental studies on adaptive evolution in *Gasterosteus aculeatus*. – Evolution **1**(1–2): 89–102.

Heutz, M. L., 1949: Racial divergence in fin ray variation patterns in *Gasterosteus aculeatus*. – J. Genetics **49:** 183–191.

Heutz, M. L., 1956: Temperature adaption in *Gasterosteus aculeatus* L. – Pubbl. Staz. Zool. Napoli **28:** 44–61.

Honma, J., and E. Tamura, 1984: Anatomical and behavioural differences among threespine sticklebacks: the marine form, the landlocked form and their hybrids. – Acta zool. Stockholm **65**(2): 79–87.

Hopkins, C.A., 1959: Seasonal variation in the incidence and development of the cestode *Proteocephalus filicollis* (Rud. 1810) in *Gasterosteus aculeatus* (L.1776). – Parasitology **49:** 529–542.

Hopkins, C. A., and J. D. Smyth, 1951: Notes on the morphology and life history of *Schistocephalus solidus* (Cestoda: Diphyllobothriidae). – Parasitology **41:** 283–291.

HUDSON, J., and H. GUDERLEY, 1984: An electrophoretic study of the phenotypic relationships among four species of sticklebacks (Pisces, Gasterosteidae). – Can. J. Zool. **62**(11): 2313–2316.

HYATT, K. D., and N. H. RINGLER, 1989: Egg canibalism and the reproductive strategies of threespine sticklebacks (*Gasterosteus aculeatus*) in a coastal British Columbia lake. – Can. J. Zool. **67**(8): 2036–2046.

HYNES, H. B. N., 1950: The food of freshwater sticklebacks (*Gasterosteus aculeatus* and *Pygosteus pungitius*), with a review of methods used in studies of the food of fishes. – J. Anim. Ecol. **19:** 36–58.

IGARASHI, K., 1970a: Formation of the scutes in the marine form of the three-spined stickleback, *Gasterosteus aculeatus* (Linnaeus) from Nasu area, Tochigi-ken. – Annot. Zool. Jap. **43:** 43–49.

IGARASHI, K., 1970b: On the variation of the scute in the three-spined stickleback, *Gasterosteus aculeatus* (Linnaeus) from Nasu area, Tochigi-ken. – Annot. Zool. Jap. **43:** 43–49.

IKEDA, K., 1933: Effects of castration on the sexual characters of anadromous three spine stickleback, *Gasterosteus aculeatus* (L.). – Jap. J. Zool. **5:** 135–157.

JACOBSEN, P., and T. KAPAKER, 1985: Freshwater tolerance in the trachurus form of three-spined stickleback *Gasterosteus aculeatus* L. in west Norwey. – Faun. norv. (A)**6:** 9–12.

JACOBSSON, S., B. BORG, C. HAUX, and S. J. HYLLNER, 1999: An 11-ketotestosterone induced kidney-secreted protein: the nest building glue from male three-spined stickleback, *Gasterosteus aculeatus*. – Fish Physiol. and Biochemystry **20:** 79–85.

JAMIESON, I. G., D. W. BLOUW, and P. W. COLGAN, 1992: Field observations on the reproductive biology of a newly discovered stickleback (*Gasterosteus*). – Can. J. Zool. **70:** 1057–1063.

JENNI, D. A., 1972: Effects of conspecifics and vegetation on nest site selection in *Gasterosteus aculeatus* L. – Behaviour **42:** 97–118.

JENYNS, L., 1835: A manual of British vertebrate animals. Cambridge and London. Part **32:** 350.

JONES, J. W., and H. B. N. HYNES, 1950: The age and growth of *Gasterosteus aculeatus*, *Pygosteus pungitius*, and *Spinachia vulgaris*, as shown by their otoliths. – J. Anim. Ecol. **19**(1): 59–73.

JONES, D. H., and A. W. G. JOHN, 1978: A threespined stickleback, *Gasterosteus aculeatus* L. from the North Atlantic. – J. Fish Biol. **13:** 231–236.

JÜRSS, K., T. BITTORF, T. VÖKLER, and R. WACKE, 1982: Experimental studies on biochemical and physiological differences between the three morphs of the three-spined stickleback, *Gasterosteus aculeatus* L. I. Gill NA/K-ATPase, muscle alanine aminotransferase and muscle aspartate aminotransferase activities. – Zool. Jb. Physiol. **86:** 267–272.

JÜRSS, K., T. BITTORF, T. VÖKLER, AND R. WACKE, 1983: Experimental studies on biochemical and physiological differences between the three morphs of the three-spined stickleback, *Gasterosteus aculeatus* L. II. Alanine aminotransferase, aspartate aminotransferase and glutamate dehydrogenase activities of the liver. – Zool. Jb. Physiol. **87:** 1–7.

JÜRSS, K., T. BITTORF, T. VÖKLER, R. WACKE, and R. BASTROP, 1985: Experimental studies on biochemical and physiological differences between the three morphs of the three-spined stickleback, *Gasterosteus aculeatus* L. III. Liver-somatic index and enzyme activities at different times of year. – Zool. Jb. Physiol. **89:** 441–451.

KEMPER, J. H., 1987: Large prey for the spoonbill in theory and practice. – Limosa **60**(4): 209–210.

KEMPER, J. H., 1995: Role of the three-spined stickleback, *Gasterosteus aculeatus* L. in the food ecology of the spoonbill, *Platalea leucorodia*. – Behaviour **132:** 1285–1299.

KIRCHHOFER, A., B. ZAUGG, and J.-C. PEDROLI, 1990: Rote Liste der Fische und Rundmäuler der Schweiz. – Documenta Faunistica Helvetiae, CSCF, Neuchatel **9:** 1–23.

KLINKHARDT, M.B., and B. BUNK, 1990: Karyologische Studien an verschiedenen Süßwasserfischen aus brackigen Küstengewässern der südwestlichen Ostsee. 3. Der Dreistachlige Stichling, (*Gasterosteus aculeatus* Linnaeus, 1758) und der Neunstachlige Stichling [*Pungitius pungitius* (Linnaeus, 1758)]. – Zool. Anz. **225**(5–6): 341–352.

KLINKHARDT, M. B., M. TESCHE, and H. GREVEN, 1995: Database of fish chromosomes. Westarp Wiss., Magdeburg. 132 pp.

KOSSWIG, C., 1954: Beitrag zur Zoogeographie der Seen im Marmaragebiet. – Hidrobiologi B. **1:** 4.

KOSSWIG, C., 1956: Beitrag zur Faunengeschichte des Mittelmeeres. – Publ. Staz. zool. Napoli **28:** 78–88.

KOTTELAT, M., 1997: European freshwater fishes. An heuristic checklist of the freshwater fishes of Europe (exclusive of former USSR), with an introduction for non-systematists and comments on nomenclature and conservation. – Biologia, Section Zool. **52**/Suppl. **5:** 1–271.

KROHKIN, Y. M., 1970: Estimation of the biomass and abundance of the three-spine stickleback (*Gasterosteus aculeatus* L.) in lake Dal'neye based on the food consumption. – Vopr. ikhtiol **10:** 637–642.

KRUPP, F., and B. W. COAD, 1985: Notes on a population of the threespine stickleback, *Gasterosteus aculeatus*, from Syria. – Senckenbergiana biol. **66**(1/3): 35--39.

KÜHLMANN, D. H. H., 1963: Beobachtungen zur Biologie von *Gasterosteus aculeatus* L. und Gedanken zu seiner Bekämpfung als Schädling. – Z. Binnenfischerei (N.F.) **11**(3/4): 301–309.

LACHANCE, S., P. MAGNAN, and G. J. FITZGERALD, 1987: Temperature preferences of three sympatric sticklebacks (Gasterosteidae). – Can. J. Zool. **65:** 1573–1576.

LAM, T. J., and W. S. HOAR, 1967: Seasonal effects of prolactin on freshwater osmoregulation of the marine form (*trachurus*) of the stickleback *Gasterosteus aculeatus*. – Can. J. Zool. **45:** 509–516.

LEINER, M., 1929: Ökologische Studien an *Gasterosteus aculeatus*. – Z. Morphol. Ökol. Tiere **14**(2): 360–399.

LEINER, M., 1930: Fortsetzung der ökologischen Studien an *Gasterosteus aculeatus*. – Z. Morphol. Ökol. Tiere **16**(3/4): 499–540.

LEINER, M., 1931: Ökologisches von *Gasterosteus aculeatus*. – Zool. Anz. **93:** 317–333.

LEINER, M., 1934a: Die drei europäischen Stichlinge (*Gasterosteus aculeatus* L., *Gasterosteus pungitius* L. und *Gasterosteus spinachia* L.) und ihre Kreuzungsprodukte. Vergleichende Betrachtung ihrer Laichzeit, ihrer Körperform und ihrer Brutpflegetätigkeit. – Z. Morphol. Ökol. Tiere **28:** 107–154.

LEINER, M., 1934b: Beiträge zur ontogenetischen Entwicklung der drei europäischen Stichlingsarten und ihrer Kreuzungsprodukte. – Z. Wiss. Zool. **145:** 366–388.

LEMMETYINEN, R., 1973: Feeding ecology of *Sterna paradisaea* Pontopp. and *S. hirundo* L. in the archipelago of southwestern Finland. – Ann. Zool. Fenn. **10:** 507–525.

LEMMETYINEN, R., and J. MANKKI, 1975: The three-spined stickleback (*Gasterosteus aculeatus*) in the food chains of the northern Baltic. – Merentutkimuslait. Julk./Havsforskn. Skr. No. **239:** 155–161.

LINDSEY, C. C., 1962: Experimental studies of meristic variation in a population of threespine sticklebacks, *Gasterosteus aculeatus*. – Can. J. Zool. **40:** 271–312.

LOBON-CERVIA, J., T. PENCZAK, and A. DE SOSTOA, 1988: Morphological variability and distribution of stickleback (*Gasterosteus aculeatus* L.) in Spain. – Cybium **12**(3): 219–227.
MAITLAND, P. S., 1965: The feeding relationship of salmon, trout, minnows, stone loach and three-spined sticklebacks in the river Endrick. – Scotland J. Anim. Ecol. **34:** 109–133.
MANN, R. H. K., 1971: The populations, growth and production of fish in four small streams in southern England. – J. Anim. Ecol. **40:** 155–190.
MARKLEY, M. H., 1940: Notes on the food habits and parasites of the stickleback, *Gasterosteus aculeatus* (Linnaeus), in the Sacramento river, California. – Copeia 1940: 223–225.
MCLEE, A. G., and R. W. SCAIFE, 1992/93: The colonization by great crested newts (*Triturus cristatus*) of the water body following treatment with a piscicide to remove a large population of sticklebacks (*Gasterosteus aculeatus*). – Brit. Herpet. Soc. Bull. No. **42:** 6–8.
MCLENNAN, D.A ., 1996: Integrating phylogenetic and experimental analyses: The evolution of male and female nuptial coloration in the stickleback fishes (Gasterosteidae). – Syst. Biol. **45:** 261–277.
MCLENNAN, D. A., and J. D. MCPHAIL, 1989: Experimental investigations of the evolutionary significance of sexually dimorphic nuptial colouration in *Gasterosteus aculeatus* (L.): the relationship between male colour and male behaviour. – Can. J. Zool. **67**(7): 1778–1782.
MCPHAIL, J. D., 1969: Predation and the evolution of a stickleback (*Gasterosteus*). – J. Fish. Res. Board Can. **26:** 3183–3208.
MEHNER, T., 1990: Zur Bestimmung der Beutefischarten aus Fragmenten der Wirbelsäule bei der Nahrungsanalyse (Osteichthyes, Teleostei). – Zool. Anz. **225**(3/4): 210–222.
MIKELSAAR, N., 1984: Fishes of the Estonian SSR. Valgus, Tallin. 432 pp.
MILLER, R. R., and C. L. HUBBS, 1969: Systematics of *Gasterosteus aculeatus*, with particular reference to intergradation and introgression along the Pacific Coast of North America: A commentary on a recent contribution. – Copeia 1969: 52–69.
MOODI, G. E. E., 1972: Morphology, life history and ecology of an unusual stickleback (*Gasterosteus aculeatus*) in the Queen Charlotte Islands, Canada. – Can. J. Zool. **50:** 721–732.
MOODI, G. E. E., and T. E. REIMCHEN, 1973: Endemism and conservation of sticklebacks in the Queen Charlotte Islands. – Can. Field. Nat. **87:** 173–175.
MÜNZING, J., 1959: Biologie, Variabilität und Genetik von *Gasterosteus aculeatus* (Pisces). Untersuchungen im Elbegebiet. – Intern. Rev. Hydrobiol. **44:** 317–382.
MÜNZING, J., 1961: *Gasterosteus aculeatus* L. (Pisces) im Ostseeraum (Das heutige Bild der Verbreitung und postglaziale Entwicklung). – Mitt. Hamb. Zool. Mus. Inst. **59:** 61–72.
MÜNZING, J., 1962a: Die Populationen der marinen Wanderform von *Gasterosteus aculeatus* L. (Pisces) an den holländischen und deutschen Nordseeküsten. Netherl., J. Sea Res. **1**(4): 508–525.
MÜNZING, J., 1962b: Ein neuer *semiarmatus*-Typ von *Gasterosteus aculeatus* L. (Pisces) aus dem Iznik See. – Mitt. Hamburg. Zool. Mus. Inst. **60:** 181–194.
MÜNZING, J., 1963a: The evolution of variation and distributional patterns in European populations of the three-spined stickleback, *Gasterosteus aculeatus*. – Evolution **17**(3): 320–332.
MÜNZING, J., 1963b: Die Verbreitung des dreistachlichen Stichlings in Europa. – Natur und Museum **93:** 284–290.
MÜNZING, J., 1964: Variabilität und Verbreitung von *Gasterosteus aculeatus* (Pisces), in Europa. – Verh. Ver. naturw. Heimatforsch. Hamburg **36:** 1–22.

MÜNZING, J., 1972: Polymorphe Populationen von *Gasterosteus aculeatus* L. (Pisces, Gasterosteidae) in sekundären Intergradationszonen der Deutschen Bucht und benachbarter Gebiete. – Faun.-Ökol. Mitt. **4:** 69–44.

MURAMATO, J., and K. IGARASHI, 1969: A preliminary note on the chromosomes and enzymatic patterns of three forms of sticklebacks. – Jap. Fac. Sci. Hokkaido-Univ. (ser. Zool.) **17:** 266–270.

MURAMATO, J., K. IGARASHI, M. ITOH, and S. MAKINO, 1969: A study of the chromosomes and enzymatic patterns of sticklebacks in Japan. – Proc. Jap. Acad. **45:** 803–807.

MYERS, G. S., 1930: The killifish of San Ignacio and the stickleback of San Ramon, Lower California. – Proc. Calif. Acad. Sci. **19:** 95–104.

NELSON, J. S., 1971: Comparison of the pectoral and pelvic skeletons and some other bones and their phylogenetic implications in the Aulorhynchidae and Gasterosteidae. – J. Fish. Res. Board Can. **28**(1): 427–442.

NIKOLSKY, G. V., 1957: Spezielle Fischkunde. VEB Deutsch. Verl. Wiss. Berlin. 632 pp.

NORDMANN, A. VON, 1840: Observations sur la fauna pontique. In: A. DE DÉMIDOFF (Ed.), Voyage dans la Russie méridionale et la Crimée. Vol. **3:** 380–381.

PAEPKE, H.-J., 1968: Über eine Mischpopulation des Dreistachligen Stichlings (*Gasterosteus aculeatus* L.) in der Umgegend von Potsdam. – Veröff. Bez. Mus. Potsdam **16:** 23–34.

PAEPKE, H.-J., 1970: Studien zur Ökologie, Variabilität und Populationsstruktur des Dreistachligen und Neunstachligen Stichlings. II. Die Variabilität der Lateralbeschildung von *Gasterosteus aculeatus* in einer brandenburgischen Intergradationszone und ihre zoogeographisch-historischen Hintergründe. – Veröff. Bez.-Mus. Potsdam **21:** 5–48.

PAEPKE, H.-J., 1971: Studien zur Ökologie, Variabilität und Populationsstruktur des Dreistachligen und Neunstachligen Stichlings. III. Größen-, Alters- und Geschlechtsverhältnisse in brandenburgischen Populationen von *Gasterosteus aculeatus* L. – Veröff. Bez.-Mus. Potsdam **23/24:** 5–22.

PAEPKE, H.-J., 1981: "Vierstachlige" Stichlinge und ihre evolutionsbiologische Aussage. – Aquarien Terrarien, Leipzig/Jena/Berlin **28**(11): 384–385.

PAEPKE, H.-J., 1982a: Phänogeographische Strukturen in den Süßwasserpopulationen von *Gasterosteus aculeatus* L. (Pisces, Gasterosteidae) in der DDR und ihre evolutionsbiologischen Aspekte. – Mitt. Zool. Mus. Berlin **38**(2): 269–328.

PAEPKE, H.-J., 1982b: Nestüberfälle bei Stichlingen – Erscheinungsformen und Deutungsversuche. – Aquarien Terrarien **29:** 202–203.

PAEPKE, H.-J., 1983: Über die innerartliche Variabilität von *Gasterosteus aculeatus* in der DDR. – Aquarien Terrarien **30:** 318–320.

PAEPKE, H.-J., 1996: Die Stichlinge. Die Neue Brehm-Bücherei, Vol. 10. Westarp Wissenschaften, Magdeburg. 2nd ed., 175 pp.

PELKWIJK, J.J. THER, and N. TINBERGEN, 1937: Eine reizbiologische Analyse einiger Verhaltensweisen von *Gasterosteus aculeatus* L. – Z. Tierpsychol. **1:** 193–204.

PELLEGRIN, J., 1921: Les poissons des eaux douces de l'Afrique du Nord Francaise; Maroc, Algerie, Tunisie, Sahara. – Mem. Soc. Sci. Nat., Maroc **1:** 1–216.

PENCZAK, T., 1961: Serological relationship between the nine-spine and the three-spined stickleback. Nature **192**(4803): 673–674.

PENCZAK, T., 1962a: Biometria ciernika *Gasterosteus aculeatus* L. – z rzeki Ner. Fragm. Faun. **10**(10): 138–161.

PENCZAK, T., 1962b: Morphometry of the stickleback, *Gasterosteus aculeatus* L., from the Baltic Sea. – Zool. Poloniae **12**(3): 239-246.

PENCZAK, T., 1964: Three-spined stickleback from Iceland *Gasterosteus aculeatus islandicus* Sauvage. – Ann. Zool. Warsz. **22:** 441–449.
PENCZAK, T., 1965: Morphological variation of the stickleback (*Gasterosteus aculeatus* L.) in Poland. – Zool. Poloniae **15**(1): 3–49.
PENCZAK, T., 1967: Comments on the taxonomy of the three-spined stickleback, *Gasterosteus aculeatus* Linnaeus. – The Ohio J. Sci. **66**(1): 81–87.
PENCZAK, T., 1968: *Gasterosteus aculeatus* L. as food of *Mergus merganser*. – Przegl. Zool. **12:** 357–358.
PENNYCUICK, L., 1971: Seasonal variations in the parasite infections in a population of three-spined sticklebacks, *Gasterosteus aculeatus* L. – Parasitology **63:** 373–388.
PEUHKURI, N., 1997: Size-assortative shoaling in fish: the effect of oddity on foraging behaviour. – Anim. Behav. **54:** 271–278.
PEUHKURI, N., 1998a: Body size and food-patch finding in a stickleback shoal. – J. of Fish Biol. **53:** 687v689.
PEUHKURI, N., 1998b: Shoal composition, body size and foraging in sticklebacks. – Behav. Ecol. Sociobiol. **43:** 333–337.
POTAPOVA, T. L., 1967: Distribution of the different forms of the three-spined stickleback in the Soviet-Union. – Vestn. Mosk. Univ. Ser. 6, Biol. Pochvoved **1:** 19–27.
POTAPOVA, T. L., 1972: Intraspecific variability of the threespine stickleback [*Gasterosteus aculeatus* (L.)]. – J. Ichthyol. **12**(1): 20–33.
QUINN, T. P., and J. T. LIGHT, 1989: Occurrence of threespine sticklebacks (*Gasterosteus aculeatus*) in the open North Pacific Ocean: migration or drift? – Can. J. Zool. **67:** 2850–2852.
RAFINSKI; J., J. BANBURA, and M. PRZYBYLSKI, 1989: Genetic differentiation of freshwater and marine sticklebacks (*Gasterosteus aculeatus*) of Eastern Europe. – Z. zool. Syst. Evolutionsforsch. **27:** 33–43.
RAJASILTA, M., 1980: Food consumption of the three-spined stickleback (*Gasterosteus aculeatus* L.). – Ann. Zool. Fenn. **17**(2): 123–126.
RANTA, E., and K. LINDSTROM, 1990: Assortative schooling in three-spined sticklebacks? – Ann. Zool. Fenn. **27**(2): 67–75.
REGAN, T., 1909: The species of three-spined sticklebacks (*Gasterosteus*). – Ann. Mag. Nat. Hist. **4:** 435–437.
REGAN, T., 1911: The Freshwater Fishes of the British Isles. Methuen, London. 287 pp.
REIMCHEN, T. E., 1980: Spine deficiency and polymorphism in a population of *Gasterosteus aculeatus*: an adaption to predators? – Can. J. Zool. **58:** 1232–1244.
REIMCHEN, T. E., 1984: Status of unarmoured and spine-deficient population (Charlotte unarmoured stickleback) of three-spined stickleback, *Gasterosteus sp.* on the Queen Charlotte Islands, British Columbia. – Can. Field Nat. **98**(1): 120–126.
REIMCHEN, T.-E., and J. S. NELSON, 1987: Habitat and morphological correlates to vertebral number as shown in a teleost, *Gasterosteus aculeatus*. – Copeia 1987(4): 868–874.
ROGERS, D. E., 1968: A comparison of the food of sockeye salmon fry and three-spined sticklebacks in the Wood River lakes. – Univ. Wash. Publ. Fish. **3:** 1–44.
ROSS, S. T., 1973: The systematics of *Gasterosteus aculeatus* (Pisces, Gasterosteidae) in Central and Southern California. – Nat. Hist. Mus. Los Angeles Cty. Contrib. Sci. **243:** 1–20.
ROTH, F., 1920: Über den Bau und die Entwicklung des Hautpanzers von *Gasterosteus aculeatus* L. – Anat. Anz. Jena **52:** 513–534.
SAAT, T. in prep.: Order Gasterosteiformes. In: E. TIHU (Ed.): Fishes of Estonia.

SALVANES, A. G. V., and P. J. B. HART, 1998: Individual variability in state-dependent feeding behaviour in three-spined sticklebacks. – Behaviour **55:** 1349–1359

SAUNDERS, J. T., 1914: A note on the food of freshwater fish. – Proc. Camb. Phil. Soc. **17:** 236–239.

SAUVAGE, H. E., 1874: Révision des espèces du groupe des épinoches. – Nov. Arch. Mus. Hist. Nat. Paris **10:** 5–38.

SCHAARSCHMIDT, T., E. MEYER, and K. JÜRSS, 1999: A comparison of transport-related gill enzyme activities and tissue-specific free amino acid concentrations of Baltic Sea (brackish water) and freshwater threespine sticklebacks, *Gasterosteus aculeatus*, after salinity and temperature acclimation. – Marine Biol. **135:** 689–697.

SCHNEIDER, G., and K. M. LEVANDER, 1900: Notizen über die an der Südküste Finnlands in den Skären des Kirchspieles Esbo vorkommenden Fische. – Acta Soc. Fauna Flora Fenn. **2**(1): 1–67.

SEVENSTER, P., 1960: The mechanism of a displacement activity. – Arch. Néerl. Zool. **13**(4): 576–579.

SEVENSTER, P., 1961: A causal analysis of a displacement activity (fanning in *Gasterosteus aculeatus* L.). – Behaviour Suppl. **9:** 1–170.

SIEBOLD, E., 1863: Die Süßwasserfische von Mitteleuropa. 430 pp., Leipzig.

SNYDER, R. J., 1990: Clutch size of anadromous and freshwater threespine sticklebacks: a reassessment. – Can. J. Zool. **68**(9): 2027–2030.

SNYDER, R. J., and H. DINGLE, 1990: Effects of freshwater and marine overwintering environments on life histories of threespine sticklebacks: evidence for adaptive variation between anadromous and resident freshwater populations. – Oecologia, Berlin **84**(3): 386–390.

STEPHANIDIS, A., 1971: Poissons d'eau douce du Péloponnèse. – Biologia Gallo-Hellen. **3**(2): 163–212.

STEVENS, M., 1987: Über eine Farbanomalie beim Dreistachligen Stichling. – DATZ **40**(5): 212–213.

"STT.", 1933: Oelgewinnung aus Stichlingen. – Die Umschau **37**(41): 797.

"STT.", 1934: Die Pillauer Stichlingsindustrie. – Mitt. Deutsch. Seefischereiver. **49**(4): 142.

SVETOVIDOV, A. N., 1964: (Fishes of the Black Sea.). Izd. Nauka, Moscow and Leningrad. 550 pp. (in Russian).

SWARUP, H., 1956: Production of heteroploidy in the three-spined stickleback [*Gasterosteus aculeatus* (L.)]. – Nature, London **178:** 1124–1125.

SWARUP, H., 1958: The reproductive cycle and development of the gonads in *Gasterosteus aculeatus* (L.). – Proc. Zool. Soc. Calcutta **11:** 47–55.

SWARUP, H., 1959a: Abnormal development in the temperature treated eggs of *Gasterosteus aculeatus* L. II. Gastrulation abnormalities. – J. Zool. Soc. India **11:** 1–6.

SWARUP, H., 1959b: Stages in the development of the stickleback *Gasterosteus aculeatus* (L.). – J. of Embryology and Experimental Morphology **6:** 373–383.

SWARUP, H., 1959c: Effect of triploidy on the body size, general organization and cellular structure in *Gasterosteus aculeatus* (L.). – J. Genetics **56:** 143–155.

TANIGUCHI, H., Y. HONMA, and K. KAWAMATA, 1990: Genetic differentiation of freshwater and anadromous threespine sticklebacks (*Gasterosteus aculeatus*) from northern Japan. – Jap. J. Ichthyol. **37**(3): 230–238.

THOMAS, G., 1977: The influence of eating and rejecting prey items upon feeding and food searching behaviour in *Gasterosteus aculeatus* L. – Anim. Behav. **25:** 52–66.

TINBERGEN, N., 1951: The study of instinct. Clarendon Press, Oxford. 228 pp.

TINBERGEN, N., 1957: The curious behaviour of the stickleback. In: Scientific American, Twentieth-century Bestiary, London **102:** 26–34.

TITSCHACK, E., 1922: Die sekundären Geschlechtsmerkmale von *Gasterosteus aculeatus*. – Zool. Jb. Abt. Allgem. Zool. Physiol. **39:** 83–148.

UKEGBU, A. A., and F. A. HUNTINGFORD, 1988: Diet consumption and stomach fullness in three-spined sticklebacks from three Scottish populations. – Scott. Nat. **100**(1): 5–16.

VAINU, E. M., 1961: Über den Stichling der Rigaer Bucht. – Loodus Selts. Tallinn **56:** 147–156.

VAN DEN ASSEM, J., 1967: Territory in the three-spined stickleback *Gasterosteus aculeatus* L. an experimental study in intraspecific competition. – Behaviour Suppl. **16:** 1–164.

VAN IERSEL, J. J. A., 1953: An analysis of the parental behaviour of the male three-spined stickleback, *Gasterosteus aculeatus* L. – Behaviour Suppl. **3:** 1–159.

VAN IERSEL, J. J. A., 1958: Some aspects of territorial behaviour of the male three-spined stickleback. – Arch. Néerl. Zool. **8,** Suppl. : 383–400.

VAN MULLEM, P. J., 1967: On synchronization in the reproduction of the stickleback (*Gasterosteus aculeatus* L. forma *leiura* Cuv.). – Arch. Néerl. Zool. **17:** 258–274.

VAN MULLEM, P. J., and J. C. VAN DER VLUGT, 1964: On the age, growth and migration of the anadromous stickleback *Gasterosteus aculeatus* L. investigated in mixed populations. – Arch. Néerl. Zool. **16:** 111–139.

VRAT, V., 1949: Reproductive behaviour and development of the eggs of the three-spined stickleback (*Gasterosteus aculeatus*) of California. – Copeia 1949: 252–260.

WASSMANN, R., 1987: Dreistachliger Stichling (*Gasterosteus aculeatus*) als Nahrung von Bleßrallen (*Fulica atra*). – Ornithol. Mitt. **39**(5): 124.

WHEELER, A., 1985: The Linnaean fish collection in the Linnaean Society of London. – Zool. J. Linn. Soc. **84**(1): 1–76.

WINNICKI, A., A. KORZELECKA, and A. TANSKI, 1998: New data on breeding and early ontogenesis of three-spined stickleback (*Gasterosteus aculeatus* L.). – Arch. Ryb. Pol. **6**(Fasc.1): 115–122.

WOHLGEMUTH, E., 1986: The occurrence of the three-spined stickleback in the territory of the town of Brno. – Ziva **34**(2): 60.

WOOTTON, R .J., 1972: The behaviour of the male three-spined stickleback in a natural situation: A quantitative description. – Behaviour **41:** 232–241.

WOOTTON, R. J., 1973: Fecundity of the three-spined stickleback, *Gasterosteus aculeatus* (L.). – J. Fish. Biol. **5:** 683–688.

WOOTTON, R. J., 1976: The biology of the sticklebacks. Academic Press London. pp. I–X + 1–387.

WOOTTON, R. J., 1984: A functional biology of sticklebacks. Croom Helm Ltd., London. 288 pp.

WOOTTON, R. J., 1986: Problems in the estimation of food consumption and fecundity in fish production studies. – Polskie Arch. Hydrobiol. **33**(3/4): 263–276.

WOOTTON, R. J., and G.W. EVANS, 1976: Cost of egg production in the three-spined stickleback (*Gasterosteus aculeatus* L.). – J. Fish. Biol. **8:** 85–395.

WOOTTON, R. J., D. A. FLETCHER, C. SMITH, and F. G. WHORISKEY, 1995: A review of the reproductive rates in sticklebacks in relation to parental expenditure and operational sex ratios. –Behaviour **132:** 915–933.

WUNDER, W., 1928: Experimentelle Untersuchungen an Stichlingen (Kämpfe, Nestbau, Laichen, Brutpflege). – Zool. Anz. Suppl. **3** (Verh. Deutsch. Zool. Ges. **32**): 115–127.

WUNDER, W., 1930: Experimentelle Untersuchungen am Dreistachligen Stichling (*Gasterosteus aculeatus* L.) während der Laichzeit (Kämpfe, Nestbau, Laichen, Brutpflege). – Z. Morphol. Ökol. Tiere **16**(3/4): 453–498.

ZANDER, C. D., J. MÖLLER-BUCHNER, and H.-D. TOTZKE, 1984: The role of sticklebacks in the food web of the Elbe and Eider Estuaries (Northern Federal Republic of Germany). – Zool. Anz. **212**(3/4): 209–222.

ZIUGANOV, V. V., 1978: Factors determining morphological differentiation in *Gasterosteus aculeatus* (Pisces, Gasterosteidae). – Zool. Zhurnal **57:** 1686–1694.

ZIUGANOV. V. V., 1979: Analysis of mechanisms which determine differences in the reaction of spermatozoa to water salinity in the threespined stickleback marine and freshwater forms. – Ontogenes **10:** 506–509.

ZIUGANOV, V. V., 1983: Genetics of osteal plate polymorphism and microevolution of threespine stickleback (*Gasterosteus aculeatus* L.). – Theor. Appl. Genet. **65:** 229–246.

ZIUGANOV, V. V., 1988: Studies on the mechanisms of ethological reproductive isolation of forms (*trachurus*, *leiurus*, *semiarmatus*) of threespine stickleback *Gasterosteus aculeatus* (Gasterosteiformes, Gasterosteidae) from two geographic regions. – Zool. Zhurnal **67**(5): 719–728.

ZIUGANOV, V. V., 1995: Reproductive isolation among lateral plate phenotypes (low, partial, complete) of the threespine stickleback, *Gasterosteus aculeatus*, from the White Sea basin and the Kamchatka Peninsula, Russia. – Behaviour **132:** 1173–1181.

Pungitius Coste, 1848

E = ninespined stickleback, F = épinochette, D = Neun- oder Zehnstachliger Stichling, R = dewjatyli koljuschki

HANS-JOACHIM PAEPKE

Gasterosteus Linnaeus, 1758: 296
Pygosteus Gill (ex Brevoort), 1861: 39
Gasterostea (subgenus) Sauvage, 1874: 729

The D_1 form a row of 7 to 12 short spines, which, except in *Pungitius hellenicus*, alternate left and right along the anterior part of the dorsal midline followed by the soft D_2 rays. The ventral fins are armed with one spiny and one soft ray. There is a shorter curved spine anterior to the anal fin. The caudal fin has a straight to somewhat convex posterior margin.

A lateral row of bony plates or scutes, which may form a keel on the caudal peduncle, is a fundamental structure that may not be fully developed by members of all populations. The pelvic girdle is visible externally and consists of a ventral shield and short lateral processes.

The species are Holarctic and distributed mainly in fresh water. However, they also occur in brackish water and less frequently penetrate into marine habitats. Frequently, a species of this genus is the only one or one of a few that live in habitats with particularly poor conditions, such as ditches through peat with a low pH and in water bodies of the Far North, which are covered by ice for long periods.

The genus encompasses a large number of different forms, about which there are many systematic and taxonomic interpretations. HAGLUND *et al.* (1992) interpreted an investigation of allozyme variations among specimens taken from Europe, North America, and East Asia and expressed the opinion that *Pungitius pungitius* (Linnaeus,1758), a species with a circumpolar distribution, may not form a single panmictic unit but rather three distinct evolutionary units, each of which is associated with one continental land mass and which he believed constitute three species: *Pungitius pungitius* (Linnaeus, 1758) in Europe, *Pungitius occidentalis* (Cuvier in Cuvier and Valenciennes, 1829) in North America, and *Pungitius sinensis* (Guichenot,

1869) in East Asia. However, this investigation was based on only three samples from Europe, five from East Asia, and one from North America. Therefore, additional investigations of a larger number of samples is desired in order to confirm the conclusions.

McPhail (1963), Münzing (1969), Wootton (1976), Bell (1984), Paepke (1996), and recently also Keivany and Nelson (2000) have assumed that the distribution of *Pungitius pungitius* is circumpolar. Keivany and Nelson (2000) distinguish five subspecies: *P. p. laevis, P. p. occidentalis, P. p. pungitius, P. p. sinensis,* and *P. p. tymensis.* In addition, the following nominal species inhabit Europe: *Pungitius platygaster* (Kessler, 1859) in the drainage areas of the Caspian Sea, Sea of Azov, and the Aral Sea, and *Pungitius hellenicus* Stephanidis, 1971, in the Sperchios River System in Greece.

As a fourth European species, Kottelat (1997) mentioned *Pungitius laevis* (Cuvier, 1829). He believes the species encompasses all ninespined sticklebacks inhabiting parts of England, Ireland, and western France that have no trace of lateral bony scutes and therefore no caudal keel. Under the concept set forth herein, the ninespined sticklebacks without scutes are treated simply as varieties of *Pungitius pungitius* (Linnaeus, 1758) without any taxonomic status, as discussed in the appropriate section.

Through breeding experiments in the laboratory, Ziuganov and Gomerlyuk (1985) were able to determine that there are apparently no isolation mechanisms between the allopatric species *P. pungitius* und *P. platygaster* other than zoogeography. The F1–F3 hybrids proved to be fertile. These results are important because routes for contact between the northern "*pungitius*" and southern "*platygaster*" populations have been produced by construction of artificial waterways. In this way, members of these two species, which are easily distinguishable phenotypically, have come together in the western Siberian Irtysh Basin, where it cannot be ruled out that the two have produced hybrid schools.

The special characteristics of *P. hellenicus* result exclusively from an extreme reduction in important structural features, such as spines, the pelvic girdle, and the lateral scutes, which also occur in other sticklebacks. In spite of this, Keivany *et al.* (1997) and Keivany and Nelson (1998, 2000) have proposed that *Pungitius hellenicus* be considered a valid, easily distinguished species.

Key to the European species

1 (2) The dorsal and ventral spines and the pelvic girdle are well developed 3

2 The dorsal spines and pelvic girdle are greatly reduced: the ventral spines and lateral scutes are completely absent. The species occurs only in Greece *Pungitius hellenicus* Stephanidis, 1971

3 (4) Lateral scutes are usually present on the caudal peduncle and are widened to a wedge shape. However, they are often lacking among Irish populations and sometimes also among populations in England, The Netherlands, and southwestern France. The range is Europe north of the Alps and Carpathian Mountains. .... *Pungitius pungitius* (LINNAEUS, 1758)

4 Lateral scutes are not widened to a wedge shape on the caudal peduncle; the ventral scutes are well developed. The range is limited to the watersheds of the Black Sea and Sea of Azov. ...... *Pungitius platygaster* (KESSLER, 1859)

Literature

See p. 266.

Pungitius hellenicus Stephanidis, 1971

E = Greek ninespined stickleback, G = Griechischer Neunstachliger Stichling

HANS-JOACHIM PAEPKE*

Synonyms

Pygosteus pungitius Stephanidis, 1950
Pungitius platygaster hellenicus Paepke, 1983

Holotype

No holotype has been designated. The description is based on one specimen collected in 1939 and 40 collected in 1940. These must be considered syntypes and have been deposited in the Zoological Museum of the University of Thessalonica.

Etymology

The name of the species refers its endemic occurrence in Greece.

Description

These data and subsequent information are based on the original description of the species as well as on the publications of KEIVANY *et al.* (1997) and KEIVANY and NELSON (1998) and the examination of six specimens from Askevi Spring bei Lamia, which the author is grateful to Prof. Dr. P. S. ECONOMIDIS for providing.

D I–VI/8–10, *A* I/8–10, P 9–11, (*V* 0/0), C 11–12., *Sp.br.* 7–9, the maximum *Tl* is 50 mm.

---

* The author thanks Prof. Dr. P. S. ECONOMIDIS, Thessaloniki, for providing valuable information and a series of specimens of *Pungitius hellenicus* (Table 1).

Morphology: Body slender, low, and laterally compressed. *H* equals about 19.7 % of *Sl** and falls in a range from 17.5 to 21.4 %. The head is larger than that of *Pungitius pungitius* from northern Europe. The *lc* is equal to about 31.2 % of *Sl** and falls in a range from 27.8 % to 36.0 %. *Oh* is relatively large, equalling about 31 % of *lc*. The mouth is directed diagonally upward. The body markings consist of a delicate brownish grey mottling on a light background. It is reminiscent of the pattern on *Pungitius platygaster*, but it lacks the coloured markings that suggest head or eye bands. The fin membranes are transparent, while the fin rays have a contrasting light pigmentation.

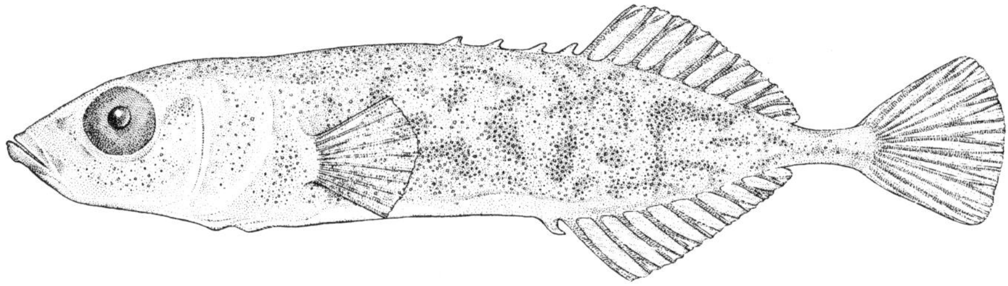

Fig. 51: *Pungitius hellenicus* from the river drainage area of Aggia Paraskevi Spring, Greece, (N°. ZMB 32342; drawn by H.-J. Behrendt).)

Karyotype: No investigations have been undertaken.

Protein specificity: Investigations on this subject have not yet been undertaken. They may provide important information for determining the relationship of this form to other ninespined sticklebacks, especially since osteological peculiarities provide only limited clues in this case.

Osteological characters: The stickleback is characterized by a more or less advanced reduction in its bony body structures, although such reductions have been observed among other sticklebacks, albeit not so extreme. The reductions affect mainly the dorsal spines (*Du*), their basal plates (*SD*), the bones of the pelvic girdle including the ventral fins, and the lateral bony scutes (*SL*). They convinced Stephanidis (1971) to describe this form as a new species, *Pungitius hellenicus*.

Of the 8 to 11 short *Du*, which are usually considered characteristic of the genus *Pungitius*, only the posteriormost 1 to 4 are present in *P. hellenicus*. The anterior spines are all rudimentary. It is remarkable that among the remaining tiny *Du*, the posteriormost immediately in front of *D* is the largest, while the anteriormost is

Table 38: Morphometric and meristic features of *Pungitius hellenicus* from the Sperchios River drainage area at Aggia Paraskevi Spring, Greece (ZMB N° 32342).

	n	average	s	ranges
*Sl** (mm)	6	28.77	1.56	26.00–34.60
% of *Sl**				
lc	6	29.70	1.62	27.10–31.70
h	6	19.85	1.17	18.50–21.40
lcp	6	11.30	0.37	11.00–11.90
pD	6	52.70	4.90	47.70–61.20
pA	6	61.80	0.72	61.00–62.80
lP	6	13.33	1.67	10.77–15.85
% of *lc*				
prO	6	22.37	1.31	20.90–24.40
Or	6	26.04	1.53	23.91–28.00
Io	6	17.74	1.82	16.2–21.05
% of *H*				
Du	6	14.20	2.20	11.70–16.70
Db	6	52.42	2.33	48.10–56.00
Au	6	16.38	3.13	12.96–20.00
Ab	6	50.20	3.53	44.20–54.70
Du	6	3.20	1.33	1–5
Db	6	9.50	0.55	9–10
Au	6	1.00	0.00	1
Ab	6	8.83	0.58	8–9
Cu	6	2.17	0.45	2– 3
Cb	6	10.00	0.00	10
Pb + Pu	6	10.00	0.00	10
Sp.br.	6	7.00	0.00	7

the smallest. Thus, the reduction tendency is highest in the anterior spines and lowest in the posterior ones. At all, *Pungitius hellenicus* has the fewest and smallest *Du* of all *Pungitius* species.

Those paired pelvic bones that are still present are very strongly reduced or distorted. They are concealed beneath the skin of the abdomen and not visible without special staining. They are usually separated from one another, and only in exceptional cases are they fused along the midline. However, even if fused, they do not form the ventral shield that is especially characteristic of *P. platygaster* but also present in most other sticklebacks. V is completely absent. Furthermore, the lateral body surfaces of *Pungitius hellenicus* are bare and free of any sort of bony scutes.

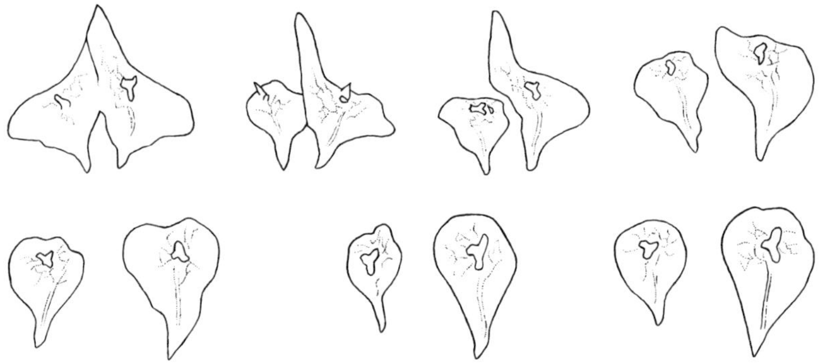

Fig. 52: Deformed pelvic bones of *Pungitius hellenicus* (after STEPHANIDIS 1971, drawn by H.-J. BEHRENDT).

According to KEIVANY *et al.* (1997) and KEIVANY and NELSON (1998), *Pungitius hellenicus* also has the fewest gill rakers (*Sp.br.*) and numbers of precaudal and total vertebrae of all *Pungitius* species.

Like the pelvic bones, the basals of the dorsal spines (SD) in *Pungitius hellenicus* are greatly reduced in size, distorted, and do not interlock with one another as they do in *Pungitius pungitius*. In addition, the ectocoracoid of *Pungitius hellenicus* is reduced in size to the extent that it is not joined to the cleithrum.

Similar but not such extreme reductions of the skeletal elements have been described for other sticklebacks, especially those in the Nearctic Region (BELL 1974, NELSON 1971, 1977, NELSON and ATTON 1971, ZIUGANOV 1989). The authors have noted that this is a case of parallel evolution. They have traced some aspects of this phenomenon back as far as the Upper Tertiary. The possible causes for this have not yet been fully explained.

During their detailed studies of *Pungitius hellenicus,* KEIVANY *et al.* (1997) and KEIVANY and NELSON (1998) came to believe that the fishes showing such a strong tendency toward reduction of many features belong to a valid species because they meet all criteria for this.

Distribution

Pungitius hellenicus is endemic to central Greece. It occurs only in the drainage area of the Sperchios River. At the *locus typicus*, which is the now dried up spring brooks of Kombotades and Mexiates, it has already been eliminated. However, the species has since been found in the karst water bodies at Aggia Paraskevi (ECONOMIDIS 1991, 1995).

Fig. 53: Geographical distribution of *Pungitius hellenicus*.

Ecology

Pungitius hellenicus prefers streams and small rivers through karst that have sandy or stony bottoms. The water is rich in vegetation, clear, rich in oxygen, and cool, with maximum summer temperatures of 12 °C. Like other ninespined sticklebacks, it seems to avoid larger water bodies. Because of its very limited range, in which anthropogenic influences have been increasing, the Greek ninespined stickleback is considered to be an endangered species. As such, it is protected by law (ECONOMIDIS 1991, 1995). Details concerning its diet, reproduction, population dynamics, and other features are unknown.

Literature

Bell, M. A., 1974: Reduction and loss of the pelvic girdle in *Gasterosteus* (Pisces): A case of parallel evolution. – Contribution in Science **257:** 1–36.

Economidis, P. S., 1991: Check List of Fresh Water Fishes of Greece, Recent Status of Threats and Protection. Hellenic Soc. Protection of Nature, Athens.

Economidis, P. S., 1995: Endangered freshwater fishes of Greece. – Biol. Conserv. **72:** 201–211.

Keivany, Y., and J. S. Nelson, 1998: Comparative osteology of the Greece nine-spine stickleback *Pungitius hellenicus* (Gasterosteidae). – Voprosy Ikhtiologii **38**(4): 458–468.

Keivany, Y., J. S. Nelson, and P. S. Economidis, 1997: Validity of *Pungitius hellenicus* Stephanidis, 1971 (Teleostei, Gasterosteidae), a stickleback fish from Greece. – Copeia 1997(3): 558–564.

Münzing, J., 1969: Variabilität, Verbreitung und Systematik der Arten und Unterarten der Gattung *Pungitius* Coste, 1848 (Pisces, Gasterosteidae). – Z. f. Systematik u. Evolutionsforsch. **7**(3): 208–233.

Nelson, J. S., 1971: Absence of the pelvic complex in ninespine sticklebacks, *Pungitius pungitius*, collected in Ireland and Wood Buffalo National Park Region, Canada, with notes on meristic variation. – Copeia **1971(4):** 707–717.

Nelson, J. S., 1977: Evidence of a genetic basis for absence of the pelvic skeleton in brook stickleback, Genus *Pungitius*, *Culaea inconstans*, and notes on the geographical distribution and origin of the loss. – Jour. Fish. Res. Board of Canada **34**(9): 1314–1320.

Nelson, J. S., and F. M. Atton, 1971: Geographic and morphological variation in the presence and absence of the pelvic skeleton in the brook stickleback, *Culaea inconstans* (Kirtland), in Alberta and Saskatchewan. – Canad. Journ. of Zool. **49**(3): 343–352.

Paepke, H.-J., 1983: Die Stichlinge. First ed. – Die Neue Brehm Bücherei, Vol. 10, Ziemsen Verlag, Wittenberg Lutherstadt. 144 pp.

Paepke, H.-J., 1996: Die Stichlinge. Second ed. – Die Neue Brehm Bücherei, Vol. 10, Westarp Wissenschaften, Magdeburg. 175 pp.

Stephanidis, A., 1943: Contribution à l'étude des poissons d'eau douce de la Grèce. – Praktika Akad. Athenon **18:** 200–210.

Stephanidis, A., 1971: On some fishes of the Greek fresh waters. – Biologia Gallo-Hellenica **3**(2): 213–240.

Ziuganov, V. V., 1989: The discovery of reproductive isolated parapatric forms of the nine spine stickleback, *Pungitius pungitius*, with and without a pelvic girdle. – J. Ichthyol. **29:** 134–142.

Pungitius platygaster (Kessler, 1859)

E = Ukrainian stickleback, F = –, D = Südlicher Neunstachliger Stichling, R = Malaya yushnaya kolyushka

HANS-JOACHIM PAEPKE

Synonyms (BERG 1949, MÜNZING 1969, and KOTTELAT 1997)

Gasterosteus pungitius Nordmann, 1840
Gasterosteus platygaster Kessler, 1859
Gasterosteus platygaster var.*kessleri* Yakovlev, 1870
Gasterosteus platygaster var.*niger* Yakovlev, 1870
Gasterosteus platygaster var. *caucasicus* Kessler, 1877
Gasterosteus platygaster var. *aralensis* Kessler, 1877
Gasterosteus platygaster var. *danubica* Steindachner, 1899
Pygosteus platygaster forma *trachua-brevis* Bertin, 1925
Pygosteus platygaster var. *nuda* Berg, 1905
Pygosteus platygaster var. *aralensis* Berg, 1905

Holotype

No holotype has been designated. The type series for *Gasterosteus platygaster* Kessler, 1859, which consists of seven specimens from Odessa and 14 from a backwater of the Dnieper at Aleschki, could be found neither in the Zoological Collection of the University of Moscow nor in the Zoological Institute of the Academy of Sciences at St. Petersburg (ZISP) and must therefore be regarded as lost (Y. SHCHERBACHEV and N. CHERNOVA, pers. comm.). According to KOTTELAT (1997), type specimens for the following taxa are available: 12 syntypes of *Gasterosteus platygaster* var. *caucasicus* Kessler, 1877 (LU), 6 syntypes of *Gasterosteus platygaster* var. *danubica* Steindachner, 1899 (NMW 6823), and syntypes of *Pygosteus platygaster* var. *nuda* Berg, 1905 (ZISP 10613).

Etymology

The name *platygaster* is a combination of the Greek words *platys* meaning wide or shallow and *gastér* meaning belly or stomach. It refers to the wide ventral shield, which is formed from the paired pelvic bones.

Description

D_1 VIII–XI (IX), D_2 7–10, *A* I 6–9, *C* 12–13, *V* I 0–1, *P* 9–11, *Vert.* 30–31, *Sp.br.* 10–13, *SL* 27–34. The *Sl** ranges from 40 mm to a maximum of 60 mm.

Morphology: The body is slender, low, and laterally compressed. *H* = 20.0 % to 22.0 % and averages 21.4 % of *Sl**. The *lc* = 28.3 % to 34.7 % and averages 30.7 % of *Sl**. The eyes are large, equalling about 24.0 % of *lc*. The mouth is small and directed diagonally upward. The body colouration varies from greyish green to dark brown with a dark mottling on a light ground colour. The dark pattern tends to suggest a band on the head that begins at the tip of the snout, runs through the eyes, crosses the operculum, and finally breaks up and blends into the pattern of mottling. A narrower light stripe runs through the dark band. During the spawning period, the nuptial colouration of the males is almost black. The membranes of the fins are transparent, and chromatophores are present only along the fin rays.

Karyotype: ZIUGANOV and GOMERLYUK (1985) mentioned 42 chromosomes without providing additional details.

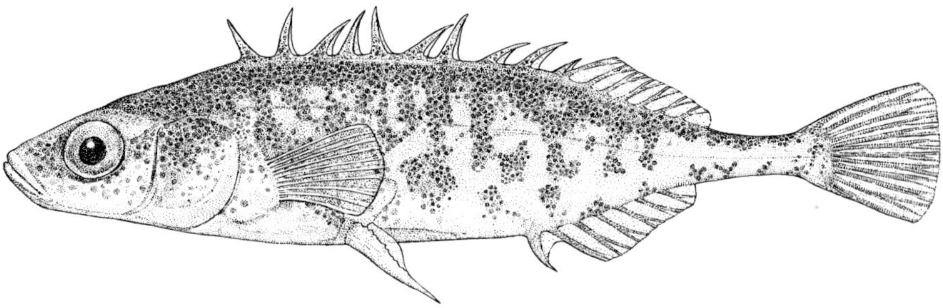

Fig. 54: *Pungitius platygaster* from Lake Caraorman, Romania, ZMB 22211 (drawn by H.-J. BEHRENDT).

Table 39: Morphometric and meristic features of *Pungitius platygaster* from Lake Caraorman, Romania, ZMB 22211.

	n	Average	s	Ranges
*Sl** (mm)	10	37.22	1.02	36.00–39.00
% of *Sl**				
lc	10	28.98	1.35	26.67–31.05
H	10	20.73	1.19	18.46–21.72
lpc	10	15.81	1.23	13.06–16.87
pD	10	31.14	1.45	29.44–33.89
pA	10	64.95	2.32	60.79–68.33
lV	10	14.63	0.87	13.68–16.11
lP	10	13.29	1.12	11.79–15.83
pelvic plate	10	22.75	1.38	19.47–25.00
% of *lc*				
prO	10	22.49	1.50	20.00–24.04
Or	10	27.36	1.40	25.62–30.00
io	10	19.50	0.81	17.27– 22.31
% of *H*				
Du	10	36.61	4.38	29.11–43.84
Db	10	42.29	5.12	35.00–48.61
Au	10	39.46	4.84	32.91– 47.56
Ab	10	50.26	3.47	41.77–53.85
Du	10	9.30	0.68	8–10
Db	10	7.60	0.70	7–9
Au	10	1.00	0.00	1
Ab	10	6.80	0.42	6–7
Cu	10	2.10	0.45	2–3
Cb	10	10.00	0.47	9– 11
Pb + Pu	10	10.00	0.00	10
Vu	10	1.00	0.00	1
Vb	10	1.00	0.00	1
SL	10	30.40	1.26	29– 33
Sp.br.	10	9.00	0.67	8– 10

Protein specificity: Unknown.

Osteological characteristics: A continuous row of 27 to 34 lateral bony scutes is characteristic of the populations in the Black Sea Watershed, including the rivers flowing into the sea. This row begins behind the head and runs to the base of the caudal fin. In comparison to those of *Pungitius pungitius sinensis,* which is also a

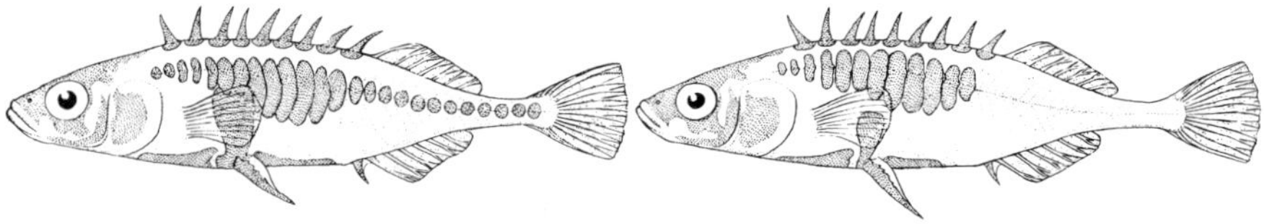

Fig. 55: Predominant arrangements of the lateral scutes (SL) of *Pungitius platygaster.* Left: completely plated morph from near the Black Sea; right: partially plated morph from the vicinity of the Aral Sea (drawn by H.-J. BEHRENDT).

ninespined stickleback with a continuous row of scutes, the individual scutes are relatively weak and arranged in less compact rows. The first four are very small and rounded, while the following seven to nine along the middle of the body are much larger, stronger, and vertically elongated. Toward the caudal region, the scutes become very small and resemble tubercles, while those on the caudal peduncle increase somewhat in size again. However, even these do not touch each other and remain flat, forming no ridge of keeled scutes like that which almost always characterizes *Pungitius pungitius.*

As the name *platygaster* suggests, the paired pelvic bones are especially strongly developed, firmly enmeshed with one another ventrally, and armed with similarly strong, wing-like lateral processes. In this way, a strong butress for the stable ventral spines is formed. Its lateral edges appear serrated among the populations in the watershed of the Black Sea.

Sexual dimorphism: Other than the dark spawning colouration of males and the body fullness of gravid females, no secondary sex characteristics are known.

Variability: In the Aral Sea Watershed, there are populations referred to as *P. platygaster aralensis,* which differ from the nominate form in the following ways. The body is often protected by six or more lateral scutes only on the anterior part. These usually form only a short row reaching about as far as the last dorsal spine. If scutes are absent, they are always missing from the caudal region, while those lacking on *Pungitius pungitius* are typically missing from the anterior part of the body. In addition, the edges of the ventral spines are smooth or very finely serrated in populations from the Aral Sea watershed, while they are obviously serrated in the nominate form. STEINDACHNER (1899) indicated that the variety from the Danube swamps near Negotin and at Save near Belgrade in Serbia, which he described as "variety *danubica*", has a very long ventral shield that is drawn out to a point and reaches the anus. In contrast, KESSLER (1859) stressed the fact that his type specimens from the harbour of Odessa and the drainage area of the Dnieper have very wide ventral shields that are rounded on the posterior margin. The specimens from

Romania that are available to the author resemble STEINDACHNER'S variety "*danubica*" in this respect. This variety, however, falls within the range of variability of *Pungitius platygaster*, just as do the varieties *kessleri* Yacovlev, 1870, from the mouths of the Volga and Emba; *niger* Yacovlev, 1870, also from the mouth of the Volga; and *caucasicus* Kessler, 1877, from Transcaucasia (MÜNZING 1969).

Subspecies: Following the proposal of BERG (1905), it has been customary to recognize two subspecies: (a) one with a complete row of lateral scutes called *P. platygaster platygaster* from the Black Sea Watershed, and (b) *P. platygaster aralensis* from the region of the Aral Sea, which often but not always possesses an incomplete row of scutes. Because of the poor definition of *P. platygaster* var. *aralensis* by KESSLER 1877, it is difficult to find a clear distinction between this form and the one with complete rows of scutes. For this reason, subspecies of *Pungitius platygaster* will not be recognized here. See also KEIVANY and NELSON (2000).

Hybrids: ZIUGANOV and GOMERLYUK (1985) have provided a detailed report on the hybrids between *Pungitius platygaster* and *Pungitius pungitius*. They were shown to be fertile and displayed a combination of the characteristics of both parent forms.

Distribution

The range of the completely plated form includes coastal regions of the Black, Caspian, and Aral Seas and the Sea of Azov as well as the middle and lower courses of the rivers flowing into those seas (ANTIPA 1909, BĂNĂRESCU 1964, BERG 1949, MARTINOV 1965, SVETOVIDOV 1964, MÜNZING 1969, PAEPKE 1996). The subspecies has moved westward as far as the Save at Belgrade and the Danube swamps near Negotin in Serbia (STEINDACHNER 1899). Two isolated populations inhabit the Axios and Aliákamón Rivers in Greece (ECONOMIDIS pers. comm.). The southeastern section of the range extends as far as the coastal plain of the Caspian Sea in Iran (ZIUGANOV and GOMERLYUK 1985, KEIVANY and NELSON 2000). The northern border of the range is located approximately along the lower courses of the Dniester, the Bug at Wosnessensk, the Dnieper at Dnieperpetrovsk, the Don and Kuban Delta, the Volga at Astrachen, and the Ural at Orenburg and Magnitogorsk. To the northeast, the range reaches at least as far as the Karaganda region, from which the author obtained specimens.

To the east, the range overlaps that of the partially plated populations, which are also known as *P. platygaster aralensis*.

KESSLER (1877) obtained his specimens of *Pungitius platygaster* var. *aralensis* near the shores of the Aral Sea and from the delta of the Amu-Darya. BERG (1949) mentioned additional sites at which the subspecies occurs: lakes along the lower course of the Syr-Darya, including Lakes Kamyshlybas and Karakul; Lake Saryssu;

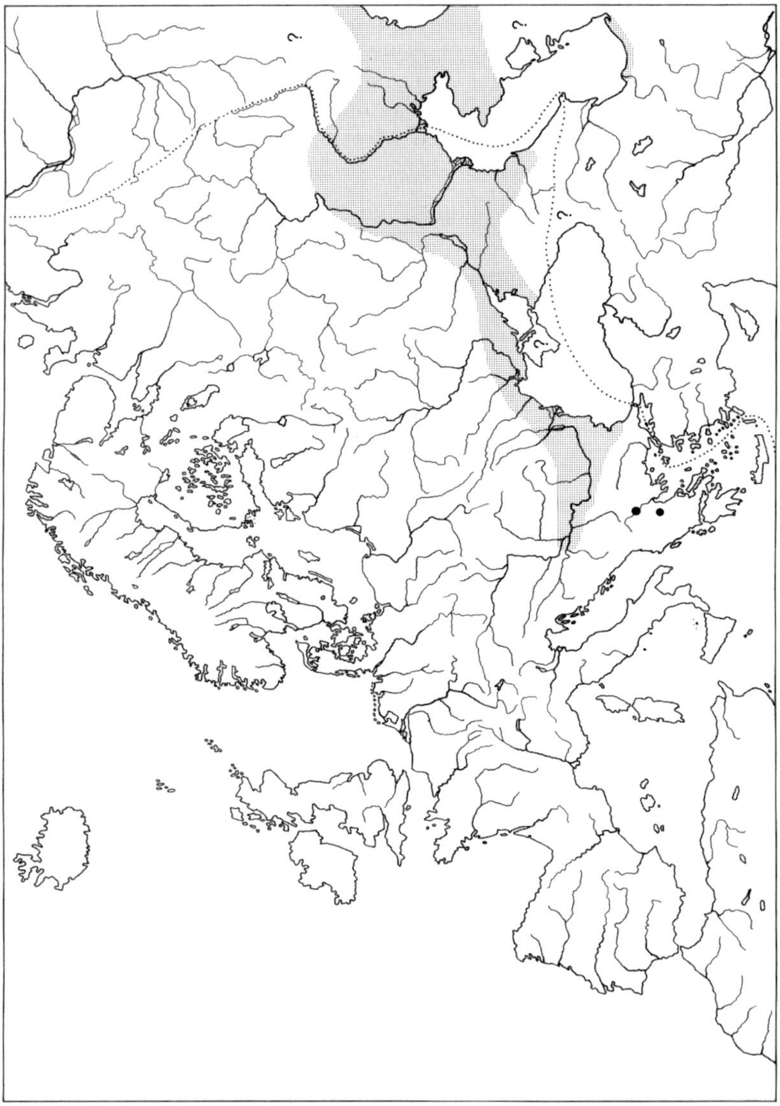

Fig. 56: Geographical distribution of *Pungitius platygaster*. Dots indicate the locations of isolated populations in Greece. Question marks refer to questionable occurrences.

and the Chu River upstream as far as Chumysh and at Frunse. It also occurs in the Issyk-kul and Lake Teniz in the Atbazar District, the Nura River, and Lake Kurgaldzhin. NIKOLSKY (1940) mentioned Lake Vostoschnyi-Karateren along the lower course of the Amu-Darya and Lake Tili-kul along the lower Sary-su. Additional details were provided by MÜNZING (1969).

Ecology

Pungitius platygaster settles predominantly highly structured freshwater habitats. However, it also lives in brackish water at such locations as the sea grass meadows in the harbour of Odessa in the Ukraine (KESSLER 1859) and in the coastal lagoons at Mamaia in Romania (H. D. KÜHLMANN, pers. comm.). With its more southerly range, *Pungitius platygaster* is adapted to higher water temperatures than such species as *Pungitius pungitius.* Individuals from the Black Sea region have been introduced to isolated freshwater ponds along the Karelian Coast of the White Sea but did not survive the cold seasons (ZIUGANOV and GOMERLYUK 1985). The fish prefers a diet of benthic organisms, which are captured in the habitats with many places of concealment.

Longevity: No detailed investigations of longevity have been undertaken. From findings of investigations on other small stickleback species, however, it is suspected that only a small percentage of the fishes reach an age between two and the maximum life span of three years.

Growth: Under laboratory conditions, *Pungitius platygaster* reaches an *Sl** of 20 mm in 75 days, thereby growing much slower than the hybrids between *Pungitius pungitius* and *Pungitius platygaster* (ZIUGANOV und GOMERLYUK 1985).

Population dynamics: Unknown

Sexual maturity: This species has been positively found to become capable of reproduction shortly before reaching the age of one year.

Spawning migrations: This form is largely a resident species that does not undertake any long spawning migrations.

Spawning sites: *Pungitius platygaster* prefers locations with dense vegetation for spawning.

Spawning season: The reproductive period begins when the water temperature reaches 17 °C (ZIUGANOV and GOMERLYUK 1985).

Reproductive behaviour

The reproductive behaviour is similar to that of other ninespine sticklebacks. During the spawning seasons, the males remain in one place and establish reproductive territories among dense submerged vegetation. From filamentous algae and submersed plants, they construct nests with an entrance below and an exit above. At that time, the males take on a black colouration, but they lack blue ventral spines. The spines are dark grey proximally and become lighter apically. The ripe females maintain their cryptic colouration and do not become "bright blinking mirrors" like female *Pungitius pungitius*. Females ready for spawning approach the nest-owning males with the heads held high. The males perform the zigzag dance less precisely than *Pungitius pungitius*. Leading the female to the nest or "lead swimming" occurs with fewer pauses than during comparable activity by other species. The contact with the partner also seems to be closer than in the case of other species during lead swimming, which is an advantage in a highly structured and not easily navigated habitat. The male indicates the location of the nest entrance. The female slips into the nest and is stimulated to spawn by the "snout vibrations" of the male. The male then fertilizes the clutch, which averages 90 eggs, and performs all of the parental care (GOMERLYUK 1979, ZIUGANOV and GOMERLYUK 1985).

Early ontogeny: Early developmental stages are described and illustrated by MAKKEVA *et al.* (1998).

Parasites

BĂNĂRESCU (1964) mentioned infestations with *Glugea anomala*, *Gyrodactylus arcuatus*, and *Gyrodactylus rarus*.

Economic importance

This is one of the few fish species that is able to settle the smallest water bodies and swamps. At such locations, *Pungitius platygaster* has a certain importance, which has apparently been underestimated, as a predator on mosquito larvae, thereby providing a biological control of malaria in the Black Sea region.

Literature

ANTIPA, G., 1909: Fauna ichthyologica a Romaniei, Academia Romana, Publ. Fondului Vasile Adamachi No. XVI: 44–52.

BĂNĂRESCU, P., 1964: Fauna Republicii Populare Romine (Pisces-Osteichthyes). Ed. Acad. R.P.R., Bucuresti, **13:** 579–582.

BERG, L. S., 1905: Ryby Turkestana (Fishes of Turkestan.). Izs. Turkest. Otd. Russk. Geogr.obšcestra Vol. 4, St. Petersburg. 216 pp. (in Russian).

BERG, L. S., 1932: Übersicht der Verbreitung der Süßwasserfische Europas. – Zoogeographica **1:** 107–208.

BERG, L. S., 1949: (Freshwater Fishes of the USSR and Neighbouring Countries.). Volume **1**, Part 3. Izd. Akad. Nauk SSSR, Moscow-Leningrad. pp. 969–971. (in Russian).

BERTIN, L., 1925: Recherches bionomiques, biometriques et systematiques sur les Épinochetes (Gastérostéides). Impressions Blondel la Rougery, Paris.

GOMERLYUK, V. E., 1979: Two *Pungitius* genus sticklebacks species mating behaviour divergence. Zoologijas Muz. Rak. P. Stuckas Latvij. Valsts. Univ., pp. 6–28.

KEIVANY, Y., and J. S. NELSON, 2000: Taxonomic review of the genus *Pungitius*. Ninespine sticklebacks (Gasterosteidae). – Cybium **24** (2): 107–122.

KESSLER, K. T., 1859: Auszüge aus dem Berichte über eine an die nordwestlichen Küsten des schwarzen Meeres und durch die westliche Krym unternommene Reise. – Bull. Soc. Impériale Nat. Moscow **32:** 202–208.

KESSLER, K. T., 1877: (Fishes inhabiting and encountered in the Aral-Caspian-Pontic ichthyological district). – Trudy Aralo-Kasp. Eksped. **4:** 1–360. Imper. Akad. Nauk, St. Petersburg. (in Russian).

KOTTELAT, M., 1997: European freshwater fishes. An heuristic checklist of the freshwater fishes of Europe (exclusive of former USSR), with an introduction for non-systematists and comments on nomenclature and conservation. – Biologia, Section Zoology **52**/Suppl. **5:** 1–271.

MAKEEVA, A. P., and D. S. PAVLOV, 1998: Ichthyo-plankton of the inland waters of Russia. Izd. Moskovskoyo Universiteta, Moscow, 215 pp.

MARINOV, B., 1965: *Pungitius platygaster* (KESSLER) – une nouvelle espèce de poisson pour notre bassin du Danube. – God. Sof. Univ. Biol. Fac. Sofia **58:** 125–127.

MÜNZING, J., 1969: Variabilität, Verbreitung und Systematik der Arten und Unterarten in der Gattung *Pungitius* COSTE, 1848 (Pisces, Gasterosteidae). – Z. Zool. Syst. Evolutionsforsch. 7: 208–233.

NIKOLSKY, G.,V., 1940: (Fishes of the Aral Sea.). – Contr. Counaiss. Fauna & Flora U.S.S.R. Moscow N.S. Sect. zool. **1:** 185. (in Russian).

NIKOLSKY, G. V., 1963: The Ecology of Fishes. Acad. Press, London and New York. 352 pp.

PAEPKE, H.-J., 1996: Die Stichlinge: Gasterosteidae. Second ed., Westarp Wissenschaften, Magdeburg. 175 pp.

STEINDACHNER, F., 1870: Ichthyologische Notizen (X). – Sitzungsber. Kaiserl. Akad. Wiss. Wien, Abt. 1, **61:** 623–642.

STEINDACHNER, F., 1899: Ueber das Vorkommen von *Gasterosteus platygaster* KESSL. im Stromgebiete der Donau. – Sitzungsber. Kaiserl. Akad. Wiss. Wien, Abt. 1, **108:** 539–542.

SVETOVIDOV, A. N., 1964: (Fishes of the Black Sea.). Izd. Nauka, Moskow and Leningrad. pp. 172–177. (in Russian).

YAKOVLEV, V. E., 1870: On the new and little known species of fish found in the mouth of the Volga. – Protokoly Zasedanii Obshchestva estestvoispytatelei pri Kazanskom Universitete, 1869–1870: 101–111 (not seen by the author).

ZIUGANOV, V. V., 1984: On penetration of the Aral stickleback *Pungitius platygaster aralensis* (KESSLER) (Gasterosteidae) into the Ob'River basin. – Vopr. Ikhtiol. **24**(4): 671–672.

ZIUGANOV, V. V., and V. E. GOMERLYUK, 1985: Hybridization of two forms of ninespine sticklebacks, *Pungitius pungitius* and. *P. platygaster*, under experimental conditions and an attempt to predict the consequences of their contact in nature. – Environmental Biol. Fishes **13**(4): 241–251.

Pungitius pungitius (Linnaeus, 1758)

E = ninespined stickleback, tenspined stickleback, pinfish, F = épinochette, petite épinoche, épinoche à neuf épines, G = Neunstachliger Stichling, Zehnstachliger Stichling, Zwergstichling, R = koljuschka severnaja dewjutilaja, dewjatiglaja koljuchka.

HANS-JOACHIM PAEPKE

Synonyms

(only for European populations)

Gasterosteus pungitius Linnaeus, 1758: 296
Gasterosteus laevis Cuvier, 1829: 170
Gasterosteus vulgaris Mauduyk, 1848
Gasterosteus burgundianus Blanchard, 1866: 240
Gasterosteus lotharingus Blanchard, 1866: 244
Gasterosteus breviceps Blanchard, 1866: 245
? *Gasterostea globiceps* Sauvage, 1874: 35
Gasterosteus (Gasterostea) pungitia Sauvage, 1874
Gasterosteus (Gasterostea) laevis Sauvage, 1874
Gasterosteus (Gasterostea) lotharinga Sauvage, 1874
Gasterosteus (Gasterostea) breviceps Sauvage, 1874
Gasterosteus (Gasterostea) pungitia var. *laevis* Moreau,1881
Gasterosteus (Gasterostea) pungitia var. *Lotharingus* Moreau, 1881
Gasterosteus (Gasterostea) pungitius var. *Breviceps* Moreau,1881
Gasterosteus (Pungitius)pungitius Fatio, 1882
Pygosteus pungitius Berg, 1907: 451
Pygosteus pungitius forma *carinata* Bertin, 1925: 122
Pygosteus pungitius forma *semiarmata* Bertin, 1925: 122
Pygosteus pungitius forma *hologymna* Bertin, 1925: 122
Pungitius pungitius Berg, 1932: 169

Holotype

A holotype was not designated. In the fish collection of the Linnaean Society of London, however, two skins from the left side of fishes are deposited, which had been in the personal possession of Linnaeus. These should be considered syntypes (WHEELER 1985).

Etymology

Pungitius means one that pricks. This name can be traced back as far as the 16th century (HERTEL 1978).

Description

Diagnosis: D_1 VIII–XI (X); D_2 9–12, (10.4); *A* I/8–11 (9.5); *C* 12; *P* 9–11 (10); V I/ 1; *Sp.br.* 10–14 (11.8); *SL* (3–11), (5.4); *Sl**36.0–64.0 mm (52.7 mm).

Morphology: The body is slender and laterally compressed. *H* 14.8 % – 22.4 % of *Sl**, averaging 19.6 %. Head length 23.4 % – 28.7 % of *Sl**, averaging 25.9 %. The eyes are relatively large. The mouth is small and directed diagonally upward. It usually does not reach as far posteriad as the anterior margin of the eye. Because the mouth opening is oriented diagonally, the lower jaw extends farther anteriad to form the apex of the head. D_1 consists of a row of relatively short spines, which can by raised to a vertical position. Their height amounts to 19.1 % to 28.7 % of *Sl** and averages 23.7 %. The posteriormost spine is somewhat larger than the one just anterior to it and is approximately as large as the strong spine anterior to the *A*. D_2 is located just dorsal to the *A*. Both are similar in form. The posterior margin of *C* is only slightly concave. Both of the outer *C* rays are unbranched, while all

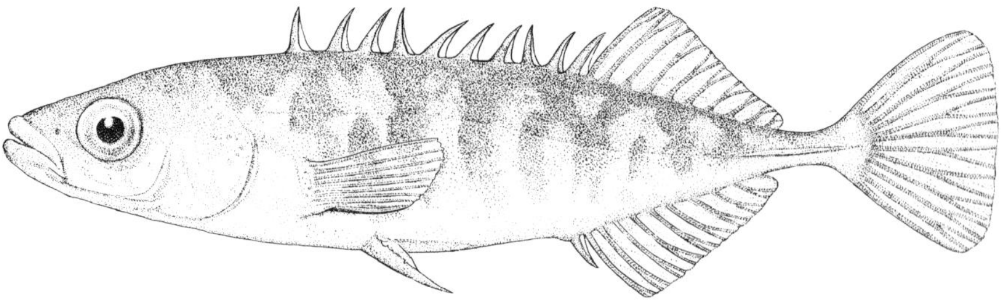

Fig. 57: *Pungitius pungitius* from Kläden, Germany, ZMB 24116 (drawn by H.-J. BEHRENDT).

others are forked. *P* is fan-shaped and has a slightly convex posterior margin. Its length equals 13.6 % to 19.4 % of *Sl** and averages 15.3 %. The *V* is supported by a strong spine and a tiny ray. Unlike *P. platygaster*, it possesses smooth spines. The spines reach a length of 6.8 % to 11.2 % of *Sl** and average 8.1 %. The predorsal distance (*pD*) varies according to the number and arrangement of the spines in D_1 and measures 26.5 % to 32.9 % of *Sl**; its average is 30.0 %. The preanal distance (*pA*) is 56.0 % to 65.9 % of *Sl** and averages 60.9 %. The caudal peduncle is short and very fragile. Its length (*lpc*) is 9.6 % to 18.3 % of *Sl** and averages 14.2 %.

The colouration varies according to the environment and season. Generally, it is darker dorsally than ventrally, especially in winter. Usually, yellow green or olive is the dominant ground colour. Individuals from water bodies rich in ferric hydroxide or from sea grass meadows in the Baltic Sea can have a brassy shine. Laterally, brownish grey flecks arranged irregularly or forming a regular pattern can be evident to some degree.

Karyotype: The number of chromosomes is 2n = 42 (Makino 1934, Post 1965, Muramoto *et al.* 1969, Chen and Reisman 1970, Klinkhardt and Buuk 1990). Klinkhardt and Buuk (1990) emphasized the "great degree of inhomogeneity and variability" of the chromosome structure. Hypodiploid values are encountered more frequently than among such species as *Gasterosteus aculeatus*. This makes it more difficult to prepare an ideogram characteristic of the species, which possesses about seven metacentric, six submetacentric, four subtelocentric, and four acrocentric chromosome pairs (Klinkhardt *et al.* 1995).

Protein specificity: By 1961, Penczak had determined the serological distances between *Pungitius pungitius* und *Gasterosteus aculeatus* in Poland and found them to be considerable, with an average value of 53.3 %. The same two species were also studied in Poland by Zietara (1989) to determine peculiarities in their lactate dehydrogenase (LDH). These are controlled in both species by three independent loci, Ldh-A, Ldh-B, and Ldh-C. Polymorphism was not found at any of the loci. Both stickleback species have the same LDH isozyme in the heart and skeletal muscles: Ldh-A4, and in the eye and brain: Ldh-B4.

Haglund *et al.* (1992) determined the enzyme variation among nine populations of *Pungitius pungitius* in Europe, East Asia, and North America and found polymorphism at 19 of the 21 loci. As already suggested by the generic definition of *Pungitius*, these findings provide evidence against the postulated conspecificity of those fishes which have been hitherto been grouped in the circumpolar species, *Pungitius pungitius*. Whether the far-reaching taxonomic conclusions from these investigations are justified must be determined by future studies using a wider range of specimens.

A series of additional biochemical studies have been conducted mainly on *Pungitius* populations outside of Europe, predominantly in East Asia. These can only be briefly cited: MURAMATO and IGARASHI (1969), MURAMATO *et al.* (1969), HUDON and GUDERLEY (1984), NIWA (1987), and TAKATA *et al.* (1987a, b).

Osteological characteristics: The number of vertebrae ranges from 31 to 35 and usually includes 14 precaudal and 18 or 19 caudal vertebrae. The ratio of dorsal spines to pterygophores is about 1:1. Usually, the first two pterygophores bear no spines, and there are fewer pterygophores than vertebrae (NELSON 1971a). As in the case of all other sticklebacks except for *Gasterosteus aculeatus,* the urostyle is fused to form a single plate. In addition, the postzygapophysis and basapophysis are larger and longer than those of *G. aculeatus*, making it possible to distinguish them among the food remains of predators (MEHNER 1990). The pelvic bones form a relatively small ventral shield with a median suture and lateral wings on both sides. The length of the ventral shield equals only 11.4 % to 18.2 % and averages 14.8 % of Sl*, making it shorter than that of *Pungitius platygaster.* The length of the lateral wings corresponds to that of a ventral spine. In Ireland, Russia, on Hokaido Island in Japan, and in the region of Alberta, Canada, there are populations with more or less strongly reduced pelvic girdles, some of which lack ventral fins (NELSON 1971 a, b; TAKATA *et al.* 1985; ZIUGANOV 1998, ZIUGANOV and ZOTIN 1995). Such reductions in the skeletons occur among many sticklebacks, including *Pungitius pungitius tymensis* in Japan (TAKATA *et al.* 1985). In Europe, it is especially evident in *Pungitius hellenicus.* According to MÜNZING (1969), who completed a fundamental study of the lateral scutes on the ninespined stickleback, members of the Eastern and Central European populations have remains of the lateral scute rows confined to the caudal peduncle. The last three to five bony scutes, located just anterior to the base of the caudal fin, are broadened to a wedge shape. In addition, two or three small, round bony plates less than 1 mm in diameter may be present along the anterior part of the lateral line, either dorsal or anterior to the pectoral fins. MÜNZING (1969) first called attention to these structures, which can only be made visible by staining.

The basals of the dorsal spines, the caudal skeleton, the keel scutes, the pelvic girdle, and the ectocoracoids were recently studied by KEIVANY and NELSON (2000) in order to find differenences between the species and subspecies of the genus *Pungitius*.

Sexual dimorphism: During the spawning season, the males are completely black or at least display a black throat and section of the abdomen. The ventral spines are brilliant turquoise. The females can be distinguished from the more slender males by their simple colour pattern during the spawning season and also by their well rounded bodies. An analysis of the population at Kläden near Stendal, Germany, showed that there are no significant quantitative or qualitative differences between the sexes. In contrast to the situation observed in the Birket River, England (JONES and HYNES 1950), the average for the standard lengths of both sexes are equal (Table 40).

Table 40: Morphometric and meristic features of male and female *Pungitius pungitius* from Kläden, near Stendal, Germany, collected on April 26, 1978, and maintained under the number ZMB 24116.

	Males				Females			
	n	average	s	ranges	n	average	s	ranges
*Sl** (mm)	63	52.55	6.29	35.00–64.00	54	52.74	6.22	39.00–63.00
% of *Sl**								
lc	63	26.22	0.93	24.14–28.70	54	24.23	1.04	22.50–27.84
H	63	18.84	1.12	14.80–22.55	54	19.28	1.68	17.10–25.00
lpc	63	13.47	1.42	9.60–17.78	54	13.89	1.48	11.65–18.31
pD	63	29.68	1.19	27.36–32.86	54	28.72	1.40	26.54–34.48
pA	63	60.44	1.69	56.00–65.51	54	60.52	1.70	57.54–67.11
lV	63	7.76	1.00	6.70–11.60	54	7.98	0.84	6.50–10.10
lP	63	15.10	1.31	12.50–19.35	54	15.36	1.22	11.63–18.46
pelvic plate	63	14.48	1.42	11.41–18.20	54	14.68	1.09	12.75–17.45
% of *lc*								
prO	63	24.06	2.00	20.40–28.90	54	23.43	1.44	20.70–27.80
Or	63	25.45	1.38	21.00–29.70	54	25.62	1.20	22.80–30.20
io	63	21.56	1.35	18.20–26.10	54	22.07	1.69	18.80–26.50
% of *H*								
Du	63	22.30	3.19	13.30–30.60	54	22.66	2.65	17.40–28.30
Db	63	51.81	4.72	37.90–64.20	54	51.08	5.15	39.10–59.40
Au	63	24.08	2.02	20.00–29.50	54	23.36	2.49	18.00–29.90
Ab	63	54.60	2.74	41.00–68.50	54	54.07	4.33	40.80–63.00
Du	63	9.97	0.17	9–11	54	10.02	0.65	8–11
Db	63	10.43	0.59	9–12	54	10.45	0.57	9–12
Au	63	1.00	0.00	1	54	1.00	0.00	1
Ab	63	9.57	0.63	8–11	54	9.46	0.64	8–11
Cu	63	2.03	0.25	2–4	54	2.00	0.00	2
Cb	63	9.98	0.38	8–11	54	10.02	0.02	10–11
Pb + *Pu*	63	9.57	1.24	5–10	54	10.02	0.41	9–11
Vu	63	1.00	0.00	1	54	1.00	0.00	1
Vb	63	1.00	0.00	1	54	1.00	0.00	1
SL	(males + females)				117	5.43	1.38	3–11
Sp.br.	(" ")				25	11.08	0.75	10–13

Variations: While a certain amount of geographical variation in the numbers of vertebrae, dorsal and anal pterygophores, and dorsal spines can be detected, the body height, number of gill rakers, and the length of the ventral spines seem to depend instead upon adaptations to specific peculiarities of the habitat, local food supplies, and predator pressure (GROSS 1979). In Japan, the relative lengths of the posteriormost dorsal and anal spines as well as the ventral spines decrease with the distance of the population inland from the seacoast (TANAKA 1982). Among North American populations, the benthos-feeding inhabitants of flowing water have fewer gill rakers than planktivorous lake-dwellers (COAD and POWER 1973). On this continent, a "Mississippi form" can also be distinguished from a "Bering form" according to geographic variations in the number of dorsal spines and gill rakers (MCPHAIL 1963). LINDSEY (1962) determined experimentally that members of an English population developed more vertebrae as well as more dorsal and anal pterygophores at higher temperatures than at lower ones. In Europe, the number of lateral bony scutes decreases along a cline from northeast to southwest. MÜNZING (1969) provided a considerable number of data on this subject. These show that the number of lateral bony scutes on the fishes near Pojakonda on the White Sea is about 14. Personal examination by the author of 41 specimens collected from the White Sea by WINKLER and DEBUS in 1995 and catalogued under the numbers ZMB 32658 and 32659 revealed a range from 13 to 20 scutes and an average of 15.1. Near Helsinki, Finland, the range reported by MÜNZING (1969) is only 6 to 13, and the average is 8.6. Near Potsdam in eastern Germany, the range is 3 to 7, and the average is 4.8, numbers that are similar to those for populations in The Netherlands, where individuals completely lacking scutes also occur. Such "naked" sticklebacks, which are not found east of the Weser River, are predominant in France and parts of the British Isles, especially Ireland.

Age and size variability: Although certain proportions of the body apparently change during the course of development, few reliable data on this subject are available. An examination of data on the fishes from Kläden, Germany, showed that smaller individuals from 37 to 48 mm in *Sl** do not differ from larger ones between 61 and 64 mm in *Sl** in the relative head length and body height as well as in the numbers of dorsal spines and soft rays. However, the largest ones had relatively large ventral shields, their length avaraging 15.1 % of *Sl** compared with 13.4 % for smaller individuals.

Subspecies

The "naked" ninespine sticklebacks without any lateral scutes, which occur in Western and Southwestern Europe, were first described as a species distinct from *Pungitius pungitius* and named *Gasterosteus laevis* by CUVIER (1829). Later, they were only tentatively classified as a subspecies (Münzing 1969), but recently again confirmed (Keivany and Nelson 2000). In view of the clinal variability of this feature, Gross (1979) and Haglund *et al.* (1992) proposed not even recognizing this form as a distinct subspecies. The author has accepted this proposal recently (Paepke 1996) and also in this account. The opposite view was held by Kottelat (1997), who considers *Pungitius laevis* (Cuvier, 1829) a valid taxon according to the phylogenetic species concept (PSC). For discussions of the East Asian subspecies of *Pungitius pungitius,* see Tanaka (1982), Niwa (1987), Takata *et al.* (1987a, b), Haglung *et al.* (1992), and Paepke (1996).

Distribution

Pungitius pungitius, as defined in its traditional taxonomic sense, is the only stickleback with an almost fully contiguous Holarctic range. In the Palearctic Region, the range covers a large part of Europe north of the Pyrenees, Alps, and Balkans, including Ireland and the southern parts of Great Britain and Skandinavia. It is notably absent from Iceland and Greenland. However, it ranges from Northern Europe across the northern and northeastern regions of Siberia and from Kamchatka southward through the Kuriles, Japanese Islands, the Amur region and Korea, and Northeast China. In the Nearctic Region, the species is distributed from Alaska across a large part of Canada, and into the Northeastern United States (Berg 1932, Münzing 1969, Wootton 1976, Paepke 1996).

European distribution: The species is found along the European Coast of the Arctic Ocean, the White Sea (Andrijashev 1964, Münzing 1969, Sawwaitowa and Winkler pers. comm.), and the Baltic Sea. It inhabits brackish water near the coasts at such locations as Rügen and Hiddensee (personal observations), the Darß-Zingster Boddenkette (Behnke *et al.* 1986, Winkler pers. comm.), the Kiel Bight (Möbius and Heincke 1883), Riga Bay, the vicinity of Tallinn (Schneider 1926), and around the rocky islands off the Swedish coast at Stockholm (Wittmack 1875). In the North Sea, it has only been found sporadically at such locations as the Sylt Waddensea, along the coast of West Jutland, and at Cuxhaven (Kourist 1963, Duncker and Ladiges 1960, Münzing 1969).

It also occurs more or less regularly in water bodies of the European lowlands from the Ural River westward to the French Atlantic Coast at Cherbourg and Bordeaux, north of the Pyranees, Alps, and Balkans (Münzing 1969). Its spread southward has been limited by the Mittelgebirge. In Westfalia, Germany, almost all of its habitats are located at elevations lower than 100 m above sea level, and its highest habitat is 155 m above sea level. Therefore, the 200 m contour line is not exceeded anywhere in the region (Feldmann 1980). In the Thuringian Basin of Germany, there is a vertical limitation between the 150 m and 200 m contour lines, as well

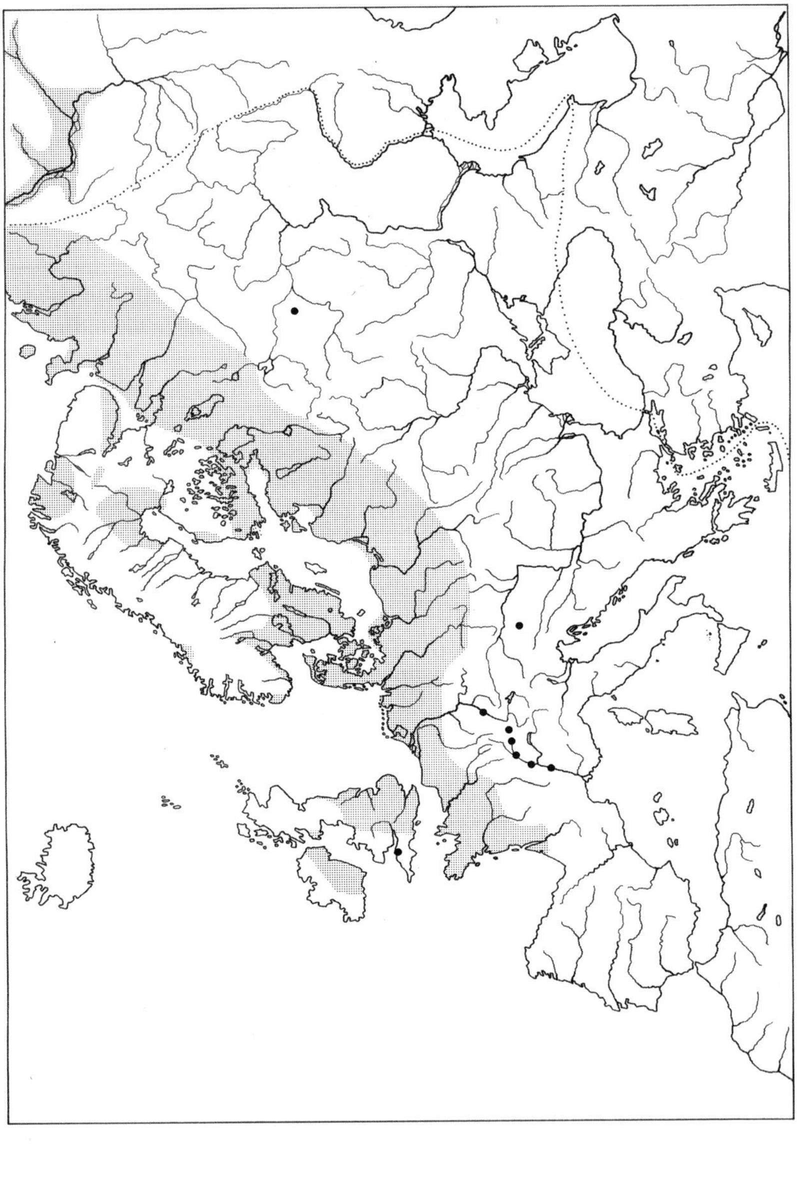

Fig. 58: Geographical distribution of *Pungitius pungitius* in Europe. Dots indicate the locations of isolated populations, some of which were introduced, at Salzburg, Austria, and the Skhodnya River near Moscow, Russia.

(BELLSTEDT pers. comm.). In Russia, the species also inhabits the middle and lower watersheds of the Pechora and Dvina Rivers, including the vicinty of Archangelsk, as well as Lakes Ladoga and Onega and the rivers flowing into them (BERG 1907, 1932, 1949). In Poland, it is found in the drainage area of the Vistula and Oder Rivers (JÀSKOWSKI 1962, KULAMOWICZ 1964, PLISZKA 1953, REMBISZEWSKI 1964, WOLSKI 1927). KOTUSZ *et al.* (1997) also found the species in Lower Silesia. Its range in Germany includes the watersheds of the Elbe, Weser, and Rhine as far upstream as Speyer; the Benelux countries and France including the watersheds of the Meuse, Schelde, Seine, Loire, and Dordogne. It has also been found at several sites in the Rhone watershed as far south as the vicinity of Lyon (SIEBOLD 1863, WITTMACK 1875, BERTIN 1925, HEUTS 1943, MÜNZING 1969). In Scandinavia, it is encountered on the Kola Peninsula (BERG 1949), in parts of Finland (MÜNZING 1969, RANTA and NUUTINEN 1984), in southern and northeastern Sweden (FRIES *et al.* 1895), and on Öland (WAHLGREN 1915). It is widespread in Denmark, where it is common in streams near Skagen (KRØYER 1838). In Norway, it is known from only three isolated locations: the Glommen River System (HUITFELDT-KAAS 1923), the vicinity of Trondheim, and the Finnmark (COLLETT 1902). It inhabits Great Britain except for the Scottish Highlands and western regions, and western Ireland (MÜNZING 1969, WHEELER 1969). Introduced populations have been confirmed near Salzburg, Austria (AHNELT and PATZNER 1992, personal observations of the author) and in the Skhodnya River, located in the Moscow River basin (ZIUGANOV and GOMERLYUK 1985; TSEPKIN and SOKOLOV 1998).

Ecology

Habitat: *Pungitius pungitius* is undoubtedly distributed primarily in fresh water. However, HEUTS (1943) and NELSON (1968) have shown that its salinity tolerance is considerably greater than that of primarily freshwater fishes. It is therefore not surprising that ninespine sticklebacks occasionally occur in brackish water and even pure seawater in the colder parts of their range. Along the coasts of the White Sea, they spawn in fresh water. Their hatchlings migrate into pure seawater (WINKLER, pers. comm.). Their reproduction in brackish water pools has been confirmed on several occasions (PAEPKE 1996). In an experiment using members of a Canadian population from a small river under tidal influence near Quebec, AUDET *et al.* (1986) detected a slight preference for weak brackish water. While *Gasterosteus aculeatus* and *Apeltes quadracus* react to increases in light intensity and temperature with an increasing preference for fresh water, an influence of these factors on salinity preference could not be detected for *Pungitius pungitius*.

Pungitius pungitius is apparently better adapted than *Gasterosteus aculeatus* to poor environmental conditions, including low water temperature, low oxygen concentrations, and the low pH in such habitats as moors. There is no other explanation for its circumpolar distribution, especially its occurrence in Siberia and Alaska. Even its relatively short ontogenetic development is suitable for survival in habitats of the Far North with short frost-free periods. LACHANCE *et al.* (1987) determined that ninespined sticklebacks from tidal pools at the mouth of the St. Lawrence River in Canada show a bimodal temperature preference in May with peaks at 9 °C to 10 °C and 15 °C to 16 °C. He attributed this to competition for the site by two separate syntopic stickleback species, *G. aculeatus* and *G. wheatlandi.*

In inland water bodies of Europe, *Pungitius pungitius* shows a clear preference for small, narrow streams, ditches through meadows, and along shores of water bodies, where the vegetation is dense, often fully overgrowing the habitat. Due to its preference for such locations, where the presence of fishes might not have been suspected, the species was often overlooked in faunal surveys. However, there are many exceptions to this habitat preference. For example, *Pungitius pungitius* occurs in spacious habitats of the Barth-Zingster Boddenkette in the southwestern Baltic Sea (BEHNKE *et al.* 1986). In Crooked Lake, Indiana, ninespined sticklebacks occur in water as deep as 30 m (NELSON (1968).

Migrations: Ninespined sticklebacks do not undertake migrations as spectacular as those of the migratory form of *Gasterosteus aculeatus.* Where it is possible, however, the reproductive community commutes between the deeper winter quarters in rivers and lakes and the shallower spawning grounds. Thus, BEHNKE *et al.* (1986) determined that during the spawning season, *P. pungitius* leaves the open water zone in the central part of the Barther Bodden, Germany, in order to spawn in its littoral zone or in the freshwater streams flowing into it. Due to the terracing and construction along many streams, the migration from the main rivers into small tributaries is no longer possible in many places (CAMERON *et al.* 1973, KOBAYASHI 1933).

Hardiness: In the water bodies of Thuringia, Germany, *P. pungitius* has proven more resistant to increases in the nitrate concentration and other kinds of chemical contamination of the water than *Gasterosteus aculeatus* (BELLSTEDT, pers. comm.). Its presence in water bodies of moors, which are otherwise poor in species, has already been mentioned.

Feeding habits: *Pungitius pungitius* feeds on all invertebrates which it can recognize as prey and catch. Thus, small crustaceans as well as aquatic insects and their larvae form a large proportion of the diet. Similar conclusions were reached by ZANDER *et al.* (1984) and BEHNKE *et al.* (1986) that syntopic *P. pungitius* and *G. aculeatus* occupy generally the same dietary niches in the habitats they share. In the central part of the Barther Bodden, both species feed upon the copepod, *Eurytemora affinis*, which occurs there in enormous numbers, as well as on harpacticid copepods. They also prey to a lesser extent upon *Neomysis integer*, the pupae of chironomids that come to the surface to emerge, and other water insects. The maximal stomach fullness indices have been calculated for fishes taken between 3 and 11 p.m., while the minimal indices have been found for the fishes between 1 a.m. and 3 p.m. This clearly speaks for a diurnal feeding pattern. Ninespined sticklebacks that sometimes visit pelagic zones clearly differ in their nutritional ecology from those inhabiting shallow water bodies along the Baltic Sea Coast in that they feed almost exclusively on plankton rather than on benthic organisms, such as chironomid larvae and ostracods. In the small Birket River in England, *P. pungitius* feeds upon tubificids, as well as insects in the Coleoptera, Lepidoptera, and Chironomidae, and to a lesser extent upon species of Collembola, Homoptera, Thysanoptera, and spiders (JONES and HYNES 1950). As BLEGVAD (1917) was able to demonstrate, the eggs and larvae of sticklebacks temporarily form a remarkably large proportion of the diet. They are mainly captured by plundering nests. Amphibian larvae are also occasionally consumed, mainly when no alternative prey is available, such as during experiments and in captivity (GLANDT 1983).

Longevity: In the small Birket River in England, *Pungitius pungitius* can reach an age of 3.5 years, during which it experiences several spawning periods (JONES and HYNES 1950). In the small Kamokawa River in Japan, *Pungitius pungitius sinensis* usually dies while it is only one year old (TANAKA and MASAYOSHI 1979). The life span of the ninespined stickleback varies between these two extremes.

Growth: Like longevity, the growth rate can vary greatly according to the environmental conditions. Usually, ninespined sticklebacks grow very rapidy during the first year of life. After sexual maturity is attained, growth stagnates, and the increase in length becomes slow. In the small Birket River, members of the species reach a *Tl* of 55 mm, corresponding to an *Sl** of about 48.4 mm (JONES and HYNES 1950). In the Kläden Reservoir in Germany, average *Sl** values of 52.55 mm for the males and 52.74 for the females were calculated. The maximal *Sl** was 64.0 mm. The largest specimen measured by LEINER (1934) was 76 mm in *Tl*, which corresponds to about 66.8 mm in Sl*. In contrast, in the nutrient-poor peat holes near Potsdam, Germany, the very small *P. pungitius* reach at most a *Tl* of 37 mm and an *Sl** of 32 mm (PAEPKE 1996). Because of the extended spawning period, it is scarcely possible to separate the members of the individual year and age classes.

Population dynamics: From a sample of fishes collected at Kläden, Germany, on April 26, 1978, the author calculated a nearly equal sex ratio of 63 males to 54 females.

Reproduction biology

Sexual maturity: LEINER (1934), PAEPKE (1996), and KUMMER (1995) came to the same conclusion that males in captivity manifest their maturity through the capability to construct nests 90 to 100 days after hatching. However, in northern regions, where the climate is not favourable, an entire year is required for the onset of sexual maturity (CAMERON *et al.* 1973). In Quebec, Canada, the populations inhabiting the Matamek River System spawn at an age of one year, while those in lakes do not spawn until their second year of life (COAD and POWER 1973).

Gonads: KUMMER (1995) provided the following description of the oogenesis of *Pungitius pungitius*. Stage 1: the juvenile contains gonads that are simply weak strands containing only young oocytes with nuclei located in the middle. Stage 2: the maturing juvenile resembles mature individuals after spawning. The gonads are still very small. In the ovaries, there are already somewhat larger eggs with large nuclei and cortical vacuoles among the less developed oocytes. Stage 3: larger eggs in the second stage of maturity become differentiated from the others in the ovaries. Vitellogenesis begins, and oocytes containing clumps of yolk appear. The testes enlarge but do not release sperm when removed by dissection. Stage 4: the gonads are almost fully mature. The ovaries contain many oocytes with peripheral nuclei, from which the ring vacuoles have disappeared. When the testes are removed by dissection, motile sperm cells are released. Stage 5: ovulation occurs, and ripe eggs in which the nuclei have already dispersed lie free in the body cavities of the females and can be stripped. Sperm can also be stripped from the males. Stage 6: the fishes have already spawned. Females may still contain a few ovulated eggs. At the end of the spawning season, the gonads are limp, and the fishes revert to the Stage 2 condition. Development stages are described and depicted also by MAKEEVA and PAVLOV (1998).

Personal investigations have revealed that the number of eggs in a female is dependent upon size. Young females produce fewer eggs than larger females several years old. For example, 26 eggs were found in individuals 32 mm in *Tl*, while those 71 mm in *Tl* contained 248 eggs. The number of eggs in seven one-year-old females captured at Potsdam, Germany, on April 20, 1981, varied from 95 to 156 and averaged 122. The weight of the ovaries ranged from 11.1 % to 16.7 % of the total weight of the females and averaged 13.3 %. The diameter of the eggs before swelling was 1.0 to 1.3 mm.

Spawning season: The dwarf sticklebacks in Central Europe have an extended spawning season which can begin by the end of March under ideal conditions, but which usually begins in earnest during April and can last until some time in September. During this long period, the females spawn several times, and the males repeatedly build nests. In captivity, a female spawned 11 times between December 30, 1960, and September 17, 1961, before dying of exhaustion (PAEPKE 1996). In the Berlin Zoological Museum, the collection of specimens taken from Lake Madü in western Poland on September 17, 1904, and maintained under No. ZMB 20104 includes some young *P. pungitius*. The smallest of the juveniles is 6 mm in *Sl** and must have been at least one week old. TANAKA and MASAYOSHI (1979) was even able to confirm the presence of individuals in the subspecies *P. pungitius sinensis* that were ripe for spawning during the period from February to September. However, most were encountered during the warm summer months. KOBAYASHI (1933) found individuals ready to spawn during the last ten days of January at Kyoto. In Alaska, however, the short spawning season is confined to the two frost-free months, June and July (CAMERON *et al.* 1973).

Spawning sites: Because of their special habit of anchoring their nests to plants and for protection against predators, the ninespined stickleback clearly prefers to spawn in sections of very small streams that are overgrown with water plants, reducing the flow velocity. Nevertheless, when sufficient submerged vegetation is not available, they are capable of anchoring their nests to the sediment. They are able to wedge them in cracks between the stones (MCKENZIE and KEENSLYSIDE 1970) or embed them in soft leafy detritus (GRISWOLD *et al.* 1972, 1973; STALLKNECHT 1975).

Reproductive behaviour: Since the work of COSTE (1846) was published, numerous researchers have provided very detailed papers on the reproductive behaviour of *Pungitius pungitius*, such as those of LANDOIS (1871), LEINER (1931), and MORRIS (1958). These were summarized by WOOTTON (1976) and PAEPKE (1996).

At the beginning of the spawning season, the agressiveness of the males increases. They assume the well-known black nuptial colouration and defend their spawning territories against other members of their own species and, as far as they are able, against members of other species. The size of the territories depends upon the structure of the habitat and the density of the spawning population. For construction of the nests, much filamentous green algae and some soft plant sprouts are gathered together and anchored in the branches of water plants or, if necessary, to the sediment. In this way, a coarsely meshed tube is formed, which is directed diagonally upward with an entrance below and exit above. It is surrounded and supported by string-like kidney secretions. In addition to such glue strands, the secretion from the kidney is occasionally released as clumps, which are then snapped up by the fish while swimming in a spiral pattern (PAEPKE 1996) and used to construct a special support for the nest entrance. The nest material is not only gathered within the fish's own territory but also carried from very distant locations or even stolen from the nests of its neighbors. After the contours of the nest become visible, movement of the pectoral fins increases in intensity. This provides the male with information concerning the stability of the structure.

Ripe females patrol through the territories of the males and are welcomed with the well-known zig-zag dance when the male is ready or are chased away if the male is not yet prepared for spawning. The male performs the zig-zag dance with the head directed steeply downward, thereby making the bright colours of the widely extended ventral spines as well as the black colouration of the ventral side plainly visible. The female is recognized by her upwardly directed anterior end and swollen body. She has a notably pale colouration. The female is attracted by the lead swimming of the male to the nest, which she approaches by swimming along the bottom with her snout held between the ventral spines of the male in order not to lose contact with her partner among the tangled plants, which are often dense.

After arriving at the nest, the male assumes a position with the head facing downward in front of the entrance, fanning with the fins. The female then enters, usually after making several unsuccessful attempts. Spawning usually occurs rapidly, while the male repeatedly prods the protruding tail of the female with his mouth. After the female has left the nest through the exit and is chased away, the male enters to fertilize the eggs. Frequently, the male takes the eggs in his mouth and moves them deeper into the nest in preparation for spawning by another female. If any clumps of eggs fall out of the nest entrance, the male carries them back into the nest. In addition to repairing the nest and defending the territory, one of the most important activities of the male at this time is fanning with the fins. The intensity of this activity reaches a maximum shortly before the eggs hatch and then decreases again. After the young leave the eggs, the male builds a "kindergarten" out of loose masses of algae directly above the nest, sometimes including parts of the nest in this structure. The hatchlings are either carried into this structure or they arrive there themselves during their first attempts at swimming, which are usually directed vertically. They remain there for three to four days prior to swarming out into the water. After learning to swim, the young first attempt to reach the surface and fill their swim bladders with air. When this is accomplished, they are ready to begin independent life. While caring for the young, the intraspecific aggressiveness of the male goes through a series of variations. After spawning, it first decreases, permitting neighbors to expand their own territories. Shortly before the eggs hatch, the aggressiveness suddenly increases again.

The "normal" reproductive behaviour can vary in many ways. Occasionally, the nest and its contents are fully torn down to be reconstructed at another site. This frequently involves the loss of zygotes or freshly hatched larvae. Spontaneous spawning without a nest is also possible. In such cases, the nest, which is vital for survival, is constructed above the egg mass by the male (Paepke 1996).

Early ontogeny: The time required for the development of the eggs depends upon the temperature. At 15 °C to 16 °C, six to seven days are required (Paepke 1996). The freshly hatched larvae are about 5.65 mm in *Tl*. After 6 days, they have reached 6.75 mm, and after 11 days, they are 7.56 mm. The length then increases to 8.64 mm after 13 days, 10.77 mm at 18 days, and 12.91 mm at 22 days (Griswold *et al.* 1972). In any case, the initial size of the larvae and their growth rate are dependent upon the size of the mother and the eggs she produces as well as upon the availability of the first food of the young.

At an *Sl** of 6.30 mm and an age of about one week, the larvae still have a complete embryonic fin seam, and their yolk sacs are already empty. Their swim bladders and pectoral fins are well developed, and the anus has already opened to the outside. The eyes are remarkably large, and the length of the caudal region is shorter than that of the adults. The end of the spinal column in the region from which the caudal fin will arise is evident as an elongate, upwardly directed, acutely tipped urostyle. Ventral to this, the broadened, fused, posteriormost haemal spines form the hypural plate. At an *Sl** of 7.00 mm, which is attained about 13 days after hatching, the anal spine becomes the first spine to appear. By the time an Sl* of 7.50 mm is reached, the posteriormost dorsal spine just anterior to D2 is already present. At an *Sl** of 10.50 mm, which is attained about 19 days after hatching, the caudal portion of the skeleton is fully developed, and the fin rays of the *C*, *A*, and D_2 are evident. The dorsal spines develop in a craniad direction, meaning that the posterior ones have a head start in growth over those farther anterior. At this length, seven dorsal spines are already present. At an *Sl** of 15.50 mm, about 30 days after hatching, the young fishes are fully developed, and the remains of the embryonic fin seam have disappeared. Only the caudal keel is still lacking. As determined by personal observations, this structure is first evident as a raised band on each side of the tail of individuals 21.50 mm in *Sl**. AYVAZIAN and KRUEGER (1992) demonstrated that the raised scutes in the caudal region of *P. pungitius* in North America, now frequently referred to as *P. occidentalis,* appear earlier during ontogeny than the tiny abdominal scutes anterior to them.

Important parasites and diseases

According to BYKOVSKY *et al.* (1964), SCHÄPERCLAUS (1979), and AMLACHER (1986), 41 diseases caused by parasites have been identified in *Pungitius pungitius.* Of these, 13 are caused by species of Protozoa, 16 by Plathyhelminthes, 1 by Aschelminthes, 8 by Acanthocephala, 1 by Annelida, and 2 by Arthropoda (Table 41). Among the Protozoa, a sporozoan, *Henneguya pungitii,* and a ciliate, *Tripartiella pungitii,* are species specific for the host. Similarly, among the Plathyhelminthes, the tape worm, *Schistocephalus pungitii,* is apparently also specific for the host. According to SMITH and KRAMER (1987), the pleurocercoid of *Schistocephalus* induces the parasitized ninespine stickleback to remain close to the surface more frequently during periods of oxygen shortages than members of the species free of the parasites. This makes the capture of individuals bearing the parasites easier for the piscivorous birds, which are the final hosts of the tape worm. This phenomenon was first observed among *Gasterosteus aculeatus,* which is parasitized by a different species of *Schistocephalus.*

Table 41: Parasites of *Pungitius pungitius, P. platygaster,* and *Gasterosteus aculeatus* (Bykovskaya Bykohvsky, B. E. (ed.) 1962, Amlacher 1986, Schäperclaus *et al.* 1990)

	Pungitius pungitius	*Pungitius platygaster*	*Gasterosteus aculeatus*
PROTOZOA			
Flagellata			
Cryptobia branchialis	+	–	+
Sporozoa			
Eimeria gasterostei	–	–	+
Glugea anomala	+	–	+
Henneguya pungitii	+	–	–
Myxobilatus gasterostei	+	–	+
Myxobilatus medius	+	+	+
Pleistophora typicalis	+	–	–
Sphaerospora elegans	+	–	+
Ciliata			
Apiosoma amebae	+	–	+
Apiosoma conica	+	–	–
Hemiophrys branchiarum	+	–	+
Trichodina domerguai	+	–	+
T. d. f. *latispina*	+	–	+
Trichodina gracilis	–	–	+
Trichophrya intermedia	–	–	+
Tripartiella pungitii	+	–	–
PLATHYHELMINTHES			
Monogenia			
Gyrodactylus arcuatus	–	–	+
Gyrodactylus armatus	+	–	–
Gyrodactylus bychovskyi	+	–	+
Gyrodactylus rarus	+	+	+
Cestoidea			
Bothriocephalus scorpii	+	–	+
Diphyllobothrium dendriticum	+	–	+
Proteocephalus cernuae	–	+	+
Proteocephalus filicollis	+	–	+
Schistocephalus pungitii	+	–	–
Schistocephalus solidus	–	–	+
Triaenophorus nodulosus	+	–	+
Trematoda			
Azygia lucii	+	–	–
Brachyphallus crenatus	–	+	+
Bucephalus polymorphus	+	–	–
Bunodera luciopercae	+	–	–
Diplostomum spathaceum	+	–	+

Table 41: (cont.)

	Pungitius pungitius	*Pungitius platygaster*	*Gasterosteus aculeatus*
Trematoda cont.			
Ichthyocotylurus pilaetus	+	–	+
Lecithaster gibbosus	–	–	+
Phyllodistomum sp.	+	–	–
Podocotyle atomon	–	–	+
Podocotyle reflexa	–	–	+
Sphaerostoma bramae	+	–	–
ASCHELMINTHES			
Nematoda			
Cystidicola farionis	–	–	+
Raphidascaris acus	+	–	+
ACANTHOCEPHALA			
Acanthocephalus lucii	+	–	+
Camallanus lacustris	+	–	+
Camallanus truncatus	–	–	+
Corynosoma semerme	+	–	+
Corynosoma strumosum	+	–	+
Metechinorhynchus salmonis	–	–	+
Neochinorhynchus cristatus	+	–	–
Neochinorhynchus rutili	+	–	+
Paracanthocephalus curtis	+	–	–
Pomphorhynchus laevis	–	–	+
Pseudochinorhynchus clavula	+	–	+
ANNELIDA			
Hirudinea			
Piscicola geometra	+	–	+
MOLLUSCA			
Bivalva			
Unionidae sp.	–	–	+
ARTHROPODA			
Crustacea			
Argulus foliaceus	+	–	+
Caligus lacustris	–	–	+
Ergasilus auritus	–	–	+
Lernaea esocina	–	–	+
Thersitina gasterostei	+	+	+

Economic importance

Because of its small size, low population density, and preference for small habitats, *Pungitius pungitius* does not become a food competitor of useful fish species. It has no economic importance for man in Europe, but, like all other sticklebacks, is it a frequent subject of behavioural and evolutionary biology studies. It is less important as an aquarium fish. In the Far East, the Yakuts use this fish for food, while members of other ethnic groups feed them to their dogs (Nikolski 1957).

The local importance of the ninespined stickleback should not be underestimated, because it is capable of colonizing the narrowest streams and smallest moor pools. Frequently, this is the only fish species that can survive in such habitats, where it assumes the function of a link between the arthropod fauna in such water bodies and piscivorous vertebrates, such as herons, kingfishers, and water shrews.

In spite of large-scale projects that have led to the destruction or ecological degradation of many small streams, the populations of *Pungitius pungitius* in most lowlands of Central Europe are not yet endangered. Nevertheless, this species has been included in a research project at the Institut für Gewässerökologie und Binnenfischerei Berlin (Berlin Institute for the Ecology of Water Bodies and Inland Fishery) in Germany designed to develop artificial reproduction of endangered small fish species.

Literature

Ahnelt, H. and R. A. Patzner, 1992: Über das Vorkommen des Neunstachligen Stichlings (*Pungitius pungitius*, Teleostei: Gasterosteidae) in Österreich. – Österreichs Fischerei **45:** 48–50.

Amlacher, E., 1986: Taschenbuch der Fischkrankheiten. 5. ed. VEB G. Fischer, Jena.

Andryiashev, A. P., 1954: (Fishes of the northern seas of the U.S.S.R.). Vol. **53,** Isd. Akademii NAUK U.S.S.R., Moscow und Leningrad. (in Russian).

Ayvazian, S., and W. H. Krueger, 1992: Lateral plate ontogeny in the North American ninespine stickleback, *Pungitius occidentalis*. – Copeia 1992(1): 209–214.

Behnke, H., H. M. Winkler, and L. Debus, 1986: Untersuchungen zum Bestand und zur Nahrungsökologie der Stichlinge (*Gasterosteus aculeatus* und *Pungitius pungitius*) im Freiwasser des Barther Boddens. – Wiss. Z. Wilh.-Pieck-Univ. Rostock **35:** 54–60.

Berg, L. S., 1907: A review of the species of the ten-spined sticklebacks or *Pygosteus* from East Asia. – Proc. U. S. Nat. Mus. **32:** 451–454.

Berg, L. S., 1912: Über die Zusammensetzung und Herkunft der Fischfauna des Amurflusses mit Bezug auf die Frage von den zoogeographischen Regionen für die Süßwasserfische. – Zool. Jahrb. Abt. Syst. **32:** 475–521.

Berg, L. S., 1932: Übersicht der Verbreitung der Süßwasserfische Europas. – Zoogeographica **1:** 107–208.

Berg, L. S., 1949: (Freshwater fishes of the U.S.S.R. and neighboring countries). – Isd. Akademii NAUK U.S.S.R. **3:** 965–971. (in Russian).

BERTIN, L., 1925: Recherches bionomiques, biométriques et systématiques sur les épinoches (Gasterosteidae). – Ann. Inst. Océanogr. Monaco **2:** 1–204.

BLANCHARD, E., 1866: Poissons des eaux douces de France. Paris. pp. 174–246.

BLEGVAD, H., 1917: On the food of fish in the Danish waters within the Skaw. – Rep. Danish Biol. Stat. **24:** 19–72.

BYKOHVSKY, B. E. (Ed.), 1962: Key to the parasites of the USSR (in Russian). – Isd. Akad. Nauk. SSR, Moscow, Leningrad [in cooperation with B. BYKOVSKAYA, A. V. SHEKIN, S. S. SHULMAN, and V. M. EPISHTEJN], 866pp.

CAMERON, J. N., J. KOSTORIS, and P. A. PENHALL, 1973: Preliminary energy budget of the nine-spine stickleback (*Pungitius pungitius*) in an Arctic lake. – J. Fish Res. Board Canada **30:** 1179–1189.

CHEN, T. R., and H. M. REISMAN, 1970: A comparative chromosome study of the North American species of sticklebacks (Teleostei: Gasterosteidae). – Cytogenetics **9:** 321–332.

COAD, B.-W., and G. POWER, 1973: Observations on the ecology and meristic variation of the ninespine stickleback, *Pungitius pungitius* (L. 1758) of the Matamek River System, Quebec. – American Midl. Naturalist **90:** 498–503.

COLLETT, R., 1902: Meddelelser om Norges Fiske i Aarene 1884–1901. – Chra. Vid.-Selsk. Forh. for 1902, No.1, Part 1.

COSTE, P., 1846: Nidification des épinoches et des épinochettes. – Mém. Savants Etrang. Paris **10:** 574–588.

CUVIER, G., 1829: La règne animal distribué d'après son organisation. Nouv. éd. Vol. 2: 170, Paris. 406 pp.

DUNCKER, G, 1926: Die Verbreitung der Süßwasserfische der Nordmark. – Arch. Hydrobiol. **17**(4): 736–750.

DUNCKER, G., and W. LADIGES, 1960: Die Fische der Nordmark. – Abh. Verh. Naturwiss. Verein. Hamburg N.F. **3.** Suppl.

FATIO, V. 1882: Faune des verterbrés de la Suisse. Vol. 4: Histoire naturelle des poissons. Genevè.

FELDMANN, R., 1980: Zur Verbreitung und Ökologie des Dreistachligen Stichlings und des Zwergstichlings in Westfalen. – Natur u. Heimat **40:** 99–109.

FRIES, B. F., C. U. EKSTRÖM, and C. J. SUNDEVALL, 1895: A history of Scandinavian fishes. P.a. Norstedt & Söna, Stockholm. pp. 644–660. (Reprint by Friedländer & Sohn, Berlin).

GLAND, D., 1983: Experimentelle Untersuchungen zum Räuber-Beute-Verhältnis zwischen Stichlingen, *Gasterosteus aculeatus* L. und *Pungitius pungitius* (L.) (Teleostei) und Grasfroschlarven, *Rana temporaria* L.(Amphibia). – Zool. Anz. **211**(5–6): 277–284.

GRISWOLD, B., J. SMITH, and L. LLOYD, 1972: Early survival and growth of the ninespine stickleback, *Pungitius pungitius* L. – Trans. Amer. Fish. Soc. **101**(2): 350–352.

GROSS, H. P., 1979: Geographic variation in European ninespine sticklebacks, *Pungitius pungitius*. – Copeia 1979(3): 405–412.

GUICHENOT, M., 1869: Notice sur quelques poissons inédits de Madagaskar et de la Chine. – Arch. Mus. Hist. Nat. Paris **5:** 204–206.

HAGLUND, T. R., D. G. BUTH, and R. LAWSON, 1992: Allozyme variation and phylogenetic relationship of Asian, North-American, and European populations of the ninespine stickleback, *Pungitius pungitius*. In: R. L. MAYDEN (Ed.), Systematics and Historical Ecology of North American Freshwater Fishes. Stanford Univ. Press, pp. 438–452.

HEUTS, M. J., 1943: La regulation osmotique chez l'Epinochette (*Pygosteus pungitius* L.). – Ann. Soc. Zool. Belg. **74:** 99–105.

HERTEL, R., 1978: Über die "Ichthyographie der Elbe" des JOHANNES KENTMANN. – Zool. Abh. Mus. Tierkd. Dresden **35:** 75–100.

HUDON, J., and H. GUDERLEY, 1984: An electrophoretic study of the phylogenetic relationships among four species of sticklebacks (Pisces: Gasterosteidae). – Can. J. Zool. **62**(11): 2313–2316.

HUITFELDT-KAAS, H., 1923: Einwanderung und Verbreitung der Süßwasserfische in Norwegen, mit einem Anhang über den Krebs. – Arch. Hydrobiol. **14:** 223–314.

JASKOWSKI, J., 1962: Materialy do Zuajomosei ichthyofauny Warty i jej doplywów. – Fragm. Faunistika **9**(28): 449–499.

JONES, J. W., and H. B. N. HYNES, 1950: The age and growth of *Gasterosteus aculeatus*, *Pygosteus pungitius* and *Spinachia vulgaris*, as shown by their otoliths. – J. Ann. Ecol. **19**(1): 59–73.

KEIVANY, Y., and J. S. NELSON, 1998: Comparative osteology of the Greece nine-spine stickleback *Pungitius hellenicus* (Gasterosteidae). – Voprosy Ikhtiologii **38**(4): 458–468.

KEIVANY, Y., and J. S. NELSON, 2000: Taxonomic review of the genus *Pungitius*, ninespined sticklebacks (Gesterosteidae). – Cybium **24**(2): 107–122.

KEIVANY, Y., J. S.NELSON, and P. S. ECONOMIDIS, 1997: Validity of *Pungitius hellenicus* STEPHANIDIS, 1971 (Teleostei, Gasterosteidae), a stickleback fish from Greece. – Copeia 1997(3): 558–564.

KLINKHARDT, M. B., and B. BUUK, 1990: Karyologische Studien an verschiedenen Süßwasserfischen aus brackigen Küstengewässern der südwestlichen Ostsee. 3. Der Dreistachlige Stichling (*Gasterosteus aculeatus* Linnaeus, 1758) und der Neunstachlige Stichling [*Pungitius pungitius* (Linnaeus, 1758)]. – Zool. Anz. **225**(5–6): 341–352.

KLINKHARDT, M., M. TESCHE, and H. GREVEN, 1995: Database of Fish Chromosomes, p. 132. Westarp Wiss., Magdeburg, 237 pp.

KOBAYASHI, J., 1933: Ecology of a stickleback *Pungitius pungitius* var. *kaibarae* (Tanaka). – J. Sci. Hiroshima Univ. B. Div. **1**(2): 71–90.

KOTTELAT, M., 1997: European freshwater fishes. An heuristic checklist of the freshwater fishes of Europe (exclusive of former USSR), with an introduction for non-systematists and comments on nomenclature and conservation. – Biologia, Section Zool. **52**/Suppl. **5:** 1–271.

KOTUSZ, J., J. KUSZNIERZ, and J. BLACHUTA, 1997: Occurrence of *Pungitius pungitius* (L.) in Silesia (SW Poland). – Preceglad Zoologiczny **16**(3–4): 247–250.

KOURIST, W., 1963: Zwergstichlinge – *Pungitius pungitius* (Linné, 1758) – im Cuxhavener Wattenmeer. – Mitt. d. FAG, H. **1:** 14–16.

KRØYER, H., 1838–40: Danmarks Fiske. Vol. 1: 31 Copenhagen, 616 pp.

KULANOWICZ, A., 1964: Cierniczek, *Pungitius pungitius* L. (Gasterosteidae) w dorzeczu Pilicy. – Przeglad Zoologiczny **3**(3): 265–267.

KUMMER, T., 1995: Der Einfluß von Licht und Temperatur auf die Fortpflanzung des Zwergstichlings, *Pungitius pungitius* (Linnaeus, 1758). Diplom-Arbeit, Humboldt-Univ., Berlin. 118 pp.

LACHANCE, S., P. MAGMAN, and G. J. FITZGERALD 1987: Temperature preference of three sympatric sticklebacks (Gasterosteidae). – Can. J. Zool. **65**(6): 1573–1576.

LEINER, M., 1931: Der Laich- und Brutpflegeinstinkt des Zwergstichlings, *Gasterosteus* (*Pygosteus*) *pungitius* L. – Z. Morphol. Ökol. Tiere **21:** 765–788.

LEINER, M., 1934: Die drei europäischen Stichlinge (*Gasterosteus aculeatus* L., *Gasterosteus pungitius* L. und *Gasterosteus spinachia* L.) und ihre Kreuzungsprodukte. – Z. Morphol. Ökol. Tiere **28:** 107–134.

LINDSEY, C. C., 1962: Observations on meristic variation in ninespine sticklebacks, *Pungitius pungitius*, reared at different temperatures. - Can. J. Zool. **4:** 1237–1247.

LINNAEUS, C., 1758: Systema naturae per regne tria naturae...Holmiae Laurentii Salvii.

MAKEEVA, A. P., and D. S. PAVLOV, 1998: Ichthyoplancton of the inland waters of Russia. Izd. Moskovskogo Universiteta, Moscow, 215 pp.

MAKINO, S., 1934: The chromosomes of the sticklebacks *Pungitius tymensis* (NIKOLSKY) and *P. pungitius* (LINNAEUS). - Cytologia **5:** 155–168.

MAUDUYK, M., 1849: Ichthyologie de la Vienne. - Bull. Soc. Acad. Agri. Belle Lettres Sci. Arts Poitiers, Nouv. Sér. **3:** 8–49 (not seen).

MCKENZIE, J.A., and M. H. A. KEENLEYSIDE, 1970: Reproductive behaviour of ninespine sticklebacks [*Pungitius pungitius* (L.)] in South Bay, Manitoulin Island, Ontario. - Can. J. Zool. **48:** 55–61.

MCPHAIL, J. D., 1963: Geographic variation in North American ninespine stickleback, *Pungitius pungitius*. - J. Fish. Res. Bd. Canada **20:** 27–44.

MEHNER, T., 1990: Zur Bestimmung der Beutefische aus Fragmenten der Wirbelsäule bei der Nahrungsanalyse (Osteichthyes, Teleostei). - Zool. Anz. **225**(3–4): 210–222.

MÖBIUS, K., and F. HEINCKE, 1883: Die Fische der Ostsee. P. Parey, Berlin, 208 pp.

MOREAU, E., 1881: Histoire naturelle des poissons de la France Vol. 1. Masson, Paris. 480 pp.

MORRIS, D., 1958: The reproductive behaviour of the ten-spined stickleback (*Pygosteus pungitius* L.). - Behaviour Suppl. **6:** 1–154.

MÜNZING, J., 1966: Variabilität, Verbreitung und Ökologie von *Pungitius pungitius* (L.), (Pisces) in Norddeutschland. -Faun.-Ökol. Mitt. **3**(1–2): 36–50.

MÜNZING, J., 1969: Variabilität, Verbreitung und Systematik der Arten und Unterarten der Gattung *Pungitius* Coste, 1848 (Pisces, Gasterosteidae). - Z. Zool. Syst. u. Evolutionsforsch. **7**(3): 208–233.

MURAMATO, J., and K. IGARASHI, 1969: A preliminary note on the chromosomes and enzymatic patterns of three forms of sticklebacks. - Jap. Fac. Sci. Hokkaido-Univ. (Ser. Zool.) **17:** 266–270.

MURAMATO, J., K. IGARASHI, M. ITOH, and S. MAKINO, 1969: A study of the chromosomes and enzymatic patterns of sticklebacks in Japan. - Proc. Japan. Acad. **45:** 803–807.

NELSON, J. N., 1968: Ecology of the southernmost sympatric population of the brook stickleback, *Culaea inconstans* and the ninespine stickleback, *Pungitius pungitius*, in Crooked Lake, Indiana. - Proc. Indiana Acad. Sci. *77:* 185–192.

NELSON, J. N., 1971a: Comparison of the pectoral and pelvic skeleton and of some other bones and their phylogenetic implication in the Aulorhynchidae and Gasterosteidae (Pisces). - J. Fish. Res. Bd. Canada **28**(3): 427–442.

NELSON, J. N., 1971b: Absence of pelvic complex in ninespine stickleback, *Pungitius pungitius* collected in Ireland and Wood Buffalo National Park Region, Canada, with notes on meristic variation. - Copeia 1971(4): 707–717.

NIKOLSKY, G. W., 1957: Spezielle Fischkunde. VEB Deutsch. Verl. Wiss., Berlin. pp. 505–506.

NIKOLSKY, N., 1889: (Sachalin Island and its fauna of vertebrate animals.). - Zapinski Imperatorskoi Akademii NAUK **60:** 293. (in Russian).

NIWA, T., 1987: Comparison of the gene frequency between sympatric populations of ninespine sticklebacks, genus *Pungitius*, in Hokkaido, Japan. - Japan. J. Ichthyol. **34**(2): 184–190.

PAEPKE, H.-J., 1996: Die Stichlinge. Die Neue Brehm-Bücherei No. 10. 2nd ed. Westarp Wiss., Magdeburg, 175 pp.

PALLAS, P.S., 1811: Zoographia Rosso-Asiatica. Part 3: 228–231. Petersburg.

Penczak, T., 1961: Serological relationship between the nine-spine and the three-spined stickleback. – Nature (London) **192:** 673–674.

Pliszka, F., 1953: Dynamika stosunków pokarmowych ryb jeziora Harsz. – Polskie Arch. Hydrobiol. **1:** 271–300.

Post, A., 1965: Vergleichende Untersuchungen der Chromosomenzahlen bei Süßwasser-Teleosteern. – Z. Zool. Syst. Evolutionsforsch. **3:** 47–93.

Ranta, E., and V. Nuutinen, 1984: Zooplankton predation by rock-pool fish (*Tinca tinca* L. and *Pungitius pungitius* L.): an experimental study. – Ann. Zool. Fenn. **21**(4): 431–449.

Rembiszewski, J.M., 1964: Ryby (Pisces) rzeki Jeziorski i Czarney Strugi koto Warszawy. – Fragmenta Faunistica **11**(7): 83–102.

Sauvage, H.-E., 1874: Revision des espèces du groupe des épinoches. – Nouv. Arch. Mus. Paris **10:** 5–38.

Schäperclaus, W. (Ed.), 1990: Fischkrankheiten. 5th ed., Akademie-Verlag. Berlin.

Schneider, G., 1926: Die Süßwasserfische des Ostbaltikums und ihre Verbreitung innerhalb des Gebietes. – Arch. Hydrobiol. **16:** 133–155.

Siebold, T. von, 1863: Die Süßwasserfische von Mitteleuropa. Engelmann Verlag, Leipzig, 431 pp.

Smith, R. S., and D. L. Kramer, 1986: Effects of a cestode (*Schistocephalus* sp.) on the response of ninespine. – Canad. Journ. Zool. **65**(7): 1862–1865.

Takata, K., A. Goto, and K. Hamada, 1984: Geographic distribution and variation of three species of ninespine stickleback (*Pungitius tymensis, P. pungitius* and *P. sinensis*) in Hokkaido. – Jap. J. Ichthyol. **31**(3): 312–326.

Takata, K., A. Goto, and K. Hamada, 1985: Pelvic spine deficiency of *Pungitius tymensis* in Hokkaido. – Jap. J. Ichtyol. **32**(1): 100–183.

Takata, K., A. Goto, and F. Yamazaki, 1987a: Biochemical identification of a brackish water type of *Pungitius pungitius,* and its morphological and ecological features in Hokkaido Japan. – Jap. J. Ichthyol. **34**(2): 176–183.

Takata, K., A. Goto, and F. Yamazaki, 1987b: Genetical differences of *Pungitius pungitius* and *P. sinensis* in a small pond of the Omono River system Japan. – Jap. J. Ichthyol. **34**(2): 384–386.

Tanaka, S., 1982: Variation in ninespine sticklebacks *Pungitius pungitius* and *P. sinensis,* in Honshu, Japan. – Jap. J. Ichthyol. **29**(2): 203–212.

Tanaka, S., and H. Masayoshi, 1979: Growth and maturity of nine-spine stickleback, *Pungitius sinensis* Guichenot in the Kamokawa, a streamlet in Toyama Prefecture. – Bull. Toyama Sci. Mus. **1:** 19–29.

Tsepkin, E. A., and L.I . Sokolov, 1998: About the quick spreading nine-spine stickleback *Pungitius pungitius* (L.) in the Moscow River basin. – Vestnik Moscowskogo Universitete, Seriya XVI Biol. 3. July–Sept. 1998: 37–40.

Wahlgren, E., 1915: Det Öländska Alvarets Djurvärld. I. – Arkiv Zool. **9:** 22.

Wheeler, A., 1969: The Fishes of the British Isles and North-West Europe. Macmillan, London, Melbourne, Toronto, 613 pp.

Wheeler, A., 1985: The Linnean fish collection in the Linnean Society of London. – Zool. J. Linnean Soc. **84:** 1–76.

Wittmack, L., 1875: Beiträge zur Fischerei-Statistik des Deutschen Reichs sowie eines Theiles von Österreich-Ungarn und der Schweiz im Auftrage des Deutschen Fischerei-Vereins. Hofdruckerei W. Moeser, Berlin. 18 pp.

WOLSKI, T., 1927: Przyczynek do znajomosci cierniczka *Pygosteus pungitus* w Polsce. – Arch. Hydrobiol. i Rybactwa **2:** 119–141.

WOOTTON, R. J., 1976: The biology of the sticklebacks. Academic Press, London, New York, San Francisco, 387 pp.

ZANDER, C. D., J. MÖLLER-BUCHNER, and H.-D. TOTZKE, 1984: The role of sticklebacks in the food web of the Elbe and Eider Estuaries (Northern Federal Republic of Germany). – Zool. Anz. **212**(3–4): 209–222.

ZIETARA, M. S., 1989: Lactate and malate dehydrogenase polymorphism in populations of the threespine (*Gasterosteus aculeatus*) and the ninespine (*Pungitius pungitius*) sticklebacks from the Vistula River and the Gdansk Gulf. – Biochem. Syst. Ecol. **17**(3): 253–259.

ZIUGANOV, V. V. 1989: Discovery of a reproductively isolated parapatric form of nine spine stickleback, *Pungitius pungitius* in Northern Karelia with loss of the pelvic girdle. – Vopr. Ikthiol. **29**(3): 448–455 (in Russian).

ZIUGANOV, V. V., and GOMERLYUK, G. J. 1985: Hybridization of two forms of ninespine stickleback, *Pungitius pungitius* and *Pungitius platygaster*, under experimental conditions and an attempt to predict the consequences of their contact in nature. – Environmental Biol. Fishes **13:** 241–251.

ZIUGANOV, V. V., and A. A. ZOTIN, 1995: Pelvic girdle polymorphism and reproductive barriers in the ninespine stickleback *Pungitius pungitius* (L.) from northwest Russia. – Behaviour **132:** 1095–1105.

Index

This index contains all scientific and vernacular names of fishes, mentioned in this volume. Generic, subgeneric, specific and subspecific names – including also units as natio (n.), forma (f.), morpha (m.) and varietas (var.) – are set in *italics*. Page numbers set in **bold faces** indicate the main sections, those set in *italics* refer to figures and maps. Synonyms and partly hybrids (x) are also included.

Contributors to vol. 5/III

Dr. Petru Bănărescu, Laboratorul Biologie Evolution, Spl. Independentei 296, RO-79651 Bucuresti, Romania

Dcs. Vlastimil Baruš, Academy of Sciences of the Czech Republic, Institute of Vertebrate Biology, Kvetna 8, CZ-603 65 Brno, Czech Republic

Dr. Nina G. Bogutskaya, Zoological Institute, Russian Academy of Sciences, Universitetskaya emb. 1, RUS-199034 St. Petersburg, Russia

Dr. Klaus Kohlmann, Institut für Gewässerökologie und Binnenfischerei, Müggelseedamm 310, D-12587 Berlin, Germany

Dr. Alexander M. Naseka, Zoological Institute, Russian Academy of Sciences, Universitetskaya emb. 1, RUS-199034 St. Petersburg, Russia

Dr. Hans-Joachim Paepke, Lennéstr. 14b, D-14469 Potsdam, Germany

Dcs. Milan Peňáz, Academy of Sciences of the Czech Republic, Institute of Vertebrate Biology, Kvetna 8, CZ-603 65 Brno, Czech Republic

Vladimir G. Poznjak, Russian Academy of Sciences, Zoological Institute, Universitetskaya emb. 1, RUS-199034 St. Petersburg, Russia

Dr. Andrzej J. Szczerbowski, The Stanislaw Sakowicz Inland Fisheries Institute, Ul. M. Oczapowskiego 10, PL-10-719 Olsten 5, Poland

Prof. Dr. Jan A. Szczerbowski, The Stanislaw Sakowicz Inland Fisheries Institute, Ul. M. Oczapowskiego 10, PL-10-719 Olsten 5, Poland